Making Starships and Stargates
The Science of Interstellar Transport and Absurdly Benign Wormholes

James F. Woodward

Making Starships and Stargates

The Science of Interstellar Transport and Absurdly Benign Wormholes

With illustrations by Nembo Buldrini
and
Foreword by John G. Cramer

 Springer

James F. Woodward
Anaheim, California
USA

SPRINGER-PRAXIS BOOKS IN SPACE EXPLORATION

ISBN 978-1-4614-5622-3   ISBN 978-1-4614-5623-0 (eBook)
DOI 10.1007/978-1-4614-5623-0
Springer New York Heidelberg Dordrecht London

Library of Congress Control Number: 2012948981

Cover illustration: Going Home, designed by Nembo Buldrini

Springer is part of Springer Science+Business Media (www.springer.com)

This book is dedicated to all those whose lives have been affected by my seemingly quixotic quest to find a way around spacetime quickly, especially Carole and the cats.

Foreword

When I arrived in Seattle in the mid-1960s to assume my new faculty position at the University of Washington, I remarked to one of my new Physics Department colleagues that the magnificent views of Mt. Rainier, available from many parts of the city, gave Seattle a special and unique flavor and ambiance.

"You know," he said, "Rainier is dormant at the moment, but it's still an active volcano. It erupts every 5,000 years. The geological record shows regular thick ash falls from Rainier and giant mud flows down the glacier-fed river valleys, where a lot of people live now."

"Really," I said. "When was the last eruption?"

"Five thousand years ago," he said with a quirky smile.

That anecdote provides a good analogy to our present situation, as residents of this planet. All of our "eggs," our cities, our people, our art and culture, our accumulated knowledge and understanding, are presently contained in one pretty blue "basket" called Planet Earth, which orbits in a Solar System that is configured to hurl very large rocks in our direction at random intervals.

Between the orbits of Mars and Jupiter lies the Asteroid Belt, a collection of large rocky leftovers from the formation of the Solar System. Within the Asteroid Belt there are empty bands known as the Kirkwood Zones, broad regions in which no asteroids are observed to orbit. The Kirkwood Zones are empty because when any asteroid is accidentally kicked into one of them, orbital resonances with Jupiter will propel it chaotically into a new orbit, sometimes one that dives into the inner Solar System and has the potential of impacting Planet Earth. Thus, our planetary system has a built-in shotgun that, on occasion, sends large rocks in our direction.

Meteors come in all sizes, and it is estimated that a total of about 14 million tons of meteoritic material falls upon Planet Earth each year, much of it from the debris of asteroids and comets. About 65 million years ago a large rock, probably an asteroid, impacted Earth in the vicinity of the Yucatan Peninsula, creating the Chicxulub crater, 180 km wide and 900 m deep. Some three quarters of Earth's species became extinct during the aftermath of the Chicxulub impact, including the dinosaurs. This is one of the more recent examples of the approximately 60 giant meteorites 5 or more kilometers in diameter that have impacted Earth in the past 600 million years. Even the smallest of these would have carved a crater some 95 km across and produced an extinction event.

The geological fossil records show evidence of "punctuated equilibrium," periods in which life-forms expand and fit themselves into all the available ecological niches, punctuated by extinction events in which many species disappear and the survivors scramble to adapt to the new conditions.

Life on Planet Earth may have been "pumped" to its present state by this cycle of extinction and regeneration, but we do not want to get caught in the next cycle of the pump. We, as a species, need to diversify, to place our eggs in many baskets instead of just one, before the forces of nature produce another extinction event.

The basic problem with such "basket diversification" is that we reside at the bottom of a deep gravity well, from which the laws of physics make it very difficult for us to escape. The only escape method presently in use involves giant chemical rockets that burn and eject vast volumes of expensive and toxic fuel in order to lift tiny payloads out of the gravity well of Earth.

And even if we can escape most of Earth's gravity well, things are not much better in near-Earth orbit. The Solar System, outside Earth's protective atmosphere and shielding magnetic field, is a fairly hostile place, with hard vacuum and the Sun's flares and storms sending out waves of sterilizing radiation. And human biology seems to require the pull of gravity for a healthy existence. A micro-gee space station is an unhealthy place for long-term habitation, and astronauts return from extended stays there as near-invalids.

The other planets and moons of the Solar System, potential sources of the needed pull of gravity, are not promising sites for human habitation. Mars is too cold, too remote from the Sun, and has a thin atmosphere, mostly carbon dioxide with a pressure of 1/100 of an Earth atmosphere. Venus is much too hot, with a surface temperature around 870 °F and an atmosphere of mostly carbon dioxide, with pressures of about 90 times that of Earth. Moons, asteroids, and artificial space habitats may be better sites for human colonies, but they all have low gravity and other problems. To find a true Earth-like habitat, we need to leave the Solar System for the Earth-like planets of other stars.

But if escaping Earth's gravity well is difficult, travel to the stars is many orders of magnitude more difficult. Fairly optimistic studies presented at the recent 100 Year Starship Symposium in Orlando, Florida, sponsored by DARPA and NASA, showed conclusively that there was little hope of reaching the nearby stars in a human lifetime using any conventional propulsion techniques, even with propulsion systems involving nuclear energy. The universe is simply too big, and the stars are too far away.

What is needed is either trans-spatial shortcuts such as wormholes to avoid the need to traverse the enormous distances or a propulsion technique that somehow circumvents Newton's third law and does not require the storage, transport, and expulsion of large volumes of reaction mass. In short, the pathway to the stars requires "exotic" solutions.

This brings us to the work of James Woodward. Jim, armed with a thorough theoretical grounding in general relativity and a considerable talent as an experimental physicist, has pioneered "outside the box" thinking aimed at solving the propulsion problem and perhaps even the problem of creating exotic matter for wormholes and warp drives. His work has included investigations in general relativity as applied to Mach's principle and experimental investigations of propulsion systems that propose to circumvent the third law need to eject reaction mass in order to achieve forward thrust. This line of inquiry has also led to a plausible way of satisfying the need for "exotic matter" to produce wormholes and warp drives. The book you hold may offer the key to human travel to the stars.

Let's consider the problem of reactionless propulsion first. Woodward extended the work of Sciama in investigating the origins of inertia in the framework of general relativity by consideration of time-dependent effects that occur when energy is in flow while an object is being accelerated. The result is surprising. It predicts large time-dependent variations in inertia, the tendency of matter to resist acceleration. Most gravitational effects predicted in general relativity are exceedingly small and difficult to observe, because the algebraic expressions describing them always have a numerator that includes Newton's gravitational constant G, a physical constant that has a very small value due to the weakness of gravity as a force. The inertial transient effects predicted by the Woodward–Sciama calculations are unusual and different, in that they have G in the *denominator*, and dividing by a small number produces a large result.

Woodward has devoted many years to demonstrating that these inertial transient effects exist, and recently he has been able to demonstrate tens of micronewton-level thrusts delivered to a precision torsion balance, effects that show the correct behavior in the presence of variations and reversals of the experimental parameters. The latest results of this work have not yet been published in peer-reviewed physics journals or reproduced in other laboratories, but nevertheless they represent convincing evidence that Woodward–Sciama inertial transients are a real physical phenomenon and that the underlying calculations behind them should be taken seriously.

The propulsion effects observed so far are quite small, but not so small as to be useless. Some of the ion thrusters employed by NASA for long-duration space missions produce comparable thrusts. Further, because of the G-in-denominator and their strong frequency dependence, the inertial transients can in principle produce very large propulsion forces. Developing Woodward–Sciama units that operate large volumes of inertia-varying mass at high frequencies could, in principle, produce macroscopic thrust and lift. Personal flying cars and reactionless heavy-lift Earth-to-orbit space vehicles cannot be ruled out if such technology could be realized.

But perhaps the most interesting Woodward–Sciama inertial transient is the "second term," which is always negative and can in principle be used to drive the inertial mass to zero or negative values. As carefully outlined in the book, this could, with a bit of luck and clever engineering design, be used to provide the "exotic mass" needed to stabilize wormholes and produce superluminal warp drives.

When I was studying physics in graduate school, the calculations of general relativity were done by hypothesizing a configuration of mass and then calculating the "metric" or distortion of spacetime that it produced. This approach has led to many interesting results, but none that could be considered "exotic" or "unphysical." But there is another way to do such calculations in general relativity, an approach that has been labeled "metric engineering." One specifies a spacetime metric that will produce some desired result, for example, a wormhole or warp drive, and then calculates the distribution of masses that would be required to produce such a metric, with all its consequences. General relativity, used in this way, had suggested the possibility of wormholes, time machines, and warp drives that could transport a local mass at speeds far faster than the speed of light.

Many of the theoretical physicists who work with general relativity have fundamental objections to the very idea of wormholes and warp drives, which they consider to be unphysical. Some of them have decided that one should erect a "picket fence" around those solutions of Einstein's equations that are considered to be physically reasonable, and to place exotica such as stable traversable wormholes, faster-than-light warp drives, and time machines in the forbidden area outside the fence, excluded because it is presumed that nature does not allow such disreputable objects to exist. They are, in effect, attempting to discover new laws of physics that would place restrictions on GR solutions.

Their first attempt at building such a fence was called the Weak Energy Condition (WEC). In essence, the WEC assumes that negative energy is the source of "problems" with GR and requires that for all observers, the local energy in all spacetime locations must be greater than or equal to zero. In other words, if any possible observer would see a negative energy, that solution of Einstein's equations is excluded by the WEC. A less restrictive variant of the WEC is the Average Weak Energy Condition (AWEC), which requires that when time-averaged along some arbitrary world line through all time, the net energy must be greater than or equal to zero, so that any time period when the energy is negative must be compensated by a period of positive energy.

The WEC, AWEC, and the other similar energy rules are "made-up" laws of nature and are not derivable from general relativity. They appear to be obeyed for observations of all known forms of matter and energy that do not fall within the domain of quantum mechanics. However, even for simple situations involving quantum phenomena (examples: the Casimir effect, squeezed vacuum, and the Hawking evaporation of black holes), the WEC and AWEC are both violated.

More recently, certain quantum inequalities (QI) to be applied to wormholes and warp drives have been derived from quantum field theory. Basically, one chooses a "sampling function," some bell-shaped curve with unit area and width T that specifies a particular restricted region of time. This function is then used with quantum field theory methods to average the energy per unit volume of a field within the time-sampling envelope and to place limits on how much negative energy can exist for how long.

The QI are bad news for would-be practitioners of metric engineering. Taken at face value, the QI say that stable wormholes may be impossible and that a warp drive might, at best, exist for too short a time to go anywhere. While a wormhole might wink into existence during the short time that the negative energy is present, it would wink out of existence again before any matter could pass through it. It appears that within the QI conditions, when negative energy is created, it is either too small in magnitude or too brief in duration to do anything interesting.

However, it is not clear whether the techniques outlined in this book for employing inertia transients are subject to the QI limitations. Further, there are reasons to think that quantum field theory cannot be trusted in its application to the field-energy situations envisioned by the QI calculations. Quantum field theory must be wrong, in some fundamental way, because it attributes far too much positive energy to spacetime itself. The density of "dark energy" deduced from the observations of astrophysicists investigating Type Ia supernovae and the space-frequency structure of the cosmic microwave background is about 6.7×10^{-10} J per cubic meter. The same quantity, as calculated by quantum field theory, is about 10^{40} J per cubic meter. Thus, quantum field theory missed the mark in this very fundamental calculation by about 50 orders of magnitude. Therefore, until quantum field theory can accurately predict the energy content of the vacuum, the restrictions that it places on metric engineering cannot be taken seriously.

In conclusion, James Woodward has been able to pry a previously closed doorway to the stars open by a small crack. It is up to the rest of us to open it further and, perhaps, walk in and use this opportunity to remove some of our "eggs" from the pretty blue basket in which they presently reside and move them to safer locations among the stars.

Westport, NY   John G. Cramer

Preface

Until the summer of 2011, I didn't think that I'd write the book now in your possession. I'd thought of writing a book about "advanced and exotic propulsion" in the mid-1990s. But Tom Mahood showed up in then the new Master's program in physics at Cal State Fullerton where I taught and did research,[1] and thoughts of book writing faded away with increased activity in my lab.

The 1990s were the heyday of speculations about advanced and exotic propulsion. But not long after the turn of the millennium, with the "war on terrorism," a looming energy crisis, financial shenanigans, climate change, and assorted political developments, advanced and exotic propulsion faded into the background. Ironically, it was during the 1990s and the first decade of this century that the real motivations for the exploration of advanced and exotic propulsion came to be appreciated: the inevitability of an extinction-level asteroid impact and, if clever critters elsewhere in the cosmos have mastered exotic propulsion, the likely eventual arrival of aliens interested in exploiting the resources of our planet. These threats may sound remote and romantic, the stuff of science fiction, and grade B screen epics with lots of special effects. However, they are quite real and, literally, deadly serious.

In the first decade of this century, chemical rocketeers and their supporters in positions of power in government and industry set about stripping out anything with even a whiff of exotic propulsion from programs with serious funding.[2] This was especially true when NASA was headed by Michael Griffin. "Advanced" propulsion didn't fare quite so badly, for it was widely defined as "electric" propulsion of various sorts, and that had long been understood not to be a threat to the dominance of the chemical propulsion community. After all, electric propulsion only held out any promise for deep space missions if launched from orbital craft with very modest masses. There is no chance that electric propulsion is practicable for Earth to orbit launchers and deep space manned spacecraft. But times have changed. Notwithstanding the resistance of the bureaucracies that deal with spaceflight, the realization that exotic propulsion is the only realistic method for reaching out to the

---

[1] Though I no longer teach, I still do research.

[2] MS Word's auto speller kept trying to change "rocketeer" into "racketeer" when I wrote this. I was tempted.

stars, and getting significant numbers of people off the planet and out of the Solar System should that prove desirable, has sparked a revival of interest in exotic technologies.

It seems that the revival of interest in advanced propulsion is serious. Why do we say that? Well, because chemical propulsion types are attacking it. They likely wouldn't waste their time doing that if it weren't perceived as a serious issue. An example: When I recently returned from an advanced and exotic propulsion conference (with about 15 attendees), I was greeted by the latest issue of the American Institute of Aeronautics and Astronautics (AIAA) publication *Aerospace America* (the March 2012 issue). On page 24 appears a "viewpoint" piece by Editor-at-Large Jerry Grey entitled, "The ephemeral 'advanced propulsion.'" The sidebar reads, "New technologies with the promise of more affordable, more efficient, and safer propulsion for space launch currently seem to be out of reach. That, however, does not mean that we should stop searching." Wanna bet? Only three paragraphs into the piece Grey allows that, "Unfortunately, advanced propulsion with sufficient thrust for Earth-based launchers requires concepts involving esoteric materials (often described as 'unobtainium') or other new (or as yet unknown) principles of physics such as antigravity, modifying the structure of space-time, employing electro-magnetic zero-point energy, faster-than-light drive, or 'wormholes.' None of these is likely to be operational in the foreseeable future." The unspoken inference is that it is a waste of resources to invest in any of these technologies.

Grey's impressive credentials are presented in another sidebar. The piece is quite long, almost entirely devoted to explaining why chemical rocketeering is the only reasonable way to proceed at this time. He wraps up his piece mentioning the recent 100 Year Starship project and the resuscitation of NASA's National Institute for Advanced Concepts (NIAC), closing with, "But don't expect anything approaching *Star Trek's* faster-than-light 'warp drive' for many years to come." Not if you are counting on funding by the government, anyway.

The 100 Year Starship project was a kiss-off of government funding for starship investigations. NASA put up 100 kilobucks and Defense Advanced Research Projects Agency (DARPA) 1 megabuck. They spent 600 kilobucks cranking people up and, then, gave the remaining half megabuck to a consortium of people almost completely unknown to most of the people who had actually been working on "advanced" propulsion. They allowed that there would be no more money from the government to support these activities. As for NIAC, it has never funded anything more challenging than solar sails and space elevators. You may think those pretty challenging. But by comparison with wormhole tech, they aren't. All of this would seem to suggest that Grey's assessment is correct.

Grey's assessment of the state of "advanced propulsion" appears to be justified by what is arguably one of the very best books on time machines and warp drives, by Allen Everett and Thomas Roman and recently published (2011) by the University of Chicago Press. Everett and Roman's book is, in a word, outstanding. If you are looking for a book that covers the theory of wormhole physics developed in the last several decades, Everett and Roman's book is the one you'd want to read. Their take on wormhole physics is strongly influenced by arguments developed by Roman and his colleague Larry Ford and others, loosely called "quantum inequalities" and "energy conditions." Quantum inequalities – which lead to the appearance of the negative energy needed to make wormholes – lead

Everett and Roman to the conclusion that the discovery of the laws of quantum gravity will be required for wormhole physics to advance farther than its present state. You might think this unimportant, but as Everett and Roman note in their epilogue:

> An efficient method of space travel could be an important issue for the survival of the human race. For example, we know that asteroid impacts have occurred numerous times in the history of our planet. One such impact sixty-five million years ago quite probably ended the reign of the dinosaurs. We know that if we remain on this planet long enough, eventually another such catastrophic impact will happen and possibly herald the end of our species.... So it would seem that it should be a fundamental goal for us to develop the capability to get off the planet (and out of the solar system).

They go on, several pages later, to remark:

> A theory of quantum gravity could, and many believe would, be as scientifically revolutionary as quantum mechanics, but will it affect humanity to the same extent? The energy scale of quantum gravity [the "Planck scale"] is so enormous [really, really enormous] that we may not be able to manipulate its effects in the near future, if ever.

Some speculative comments follow based on the supposition that mastery of the Planck scale might eventually prove possible.

This book is not a competitor to Everett and Roman's excellent contribution to wormhole physics. It is predicated on very different circumstances from those that they imagine. Where they, and Grey, assume that overall our present understanding of physics is pretty thorough and well worked out, and that "new" physics in the form of quantum gravity or something equivalent will be required to make wormhole tech a reality, this book is predicated on the supposition that our understanding of present theory is not so thorough and complete that we can assume that it precludes the development of wormhole tech. As you will find in the following pages, this view was not expected. Many of the key insights were not actively sought. In a very real sense, much of what is described in what follows was little more than a sequence of accidents, such as blundering onto a paper that happened to have just the right argument presented in an easily accessible way and stumbling onto a flaw in an apparatus that made the system perform in some unexpected but desirable way. Having tolerant friends and colleagues willing to listen to sometimes inchoate remarks and ask good questions helped. The metaphor that comes to mind is the well-known joke about the drunk looking for his or her keys under a streetlamp.

Kip Thorne, prodded by Carl Sagan, transformed advanced and exotic propulsion in 1988 with the publication (with his then grad student Michael Morris) of the foundational paper on traversable wormholes (in the *American Journal of Physics*). That work made plain that if you wanted to get around the galaxy quickly, you were going to have to find a way to assemble a Jupiter mass of negative rest-mass matter in a structure at most a few tens of meters in size. And to be practical, the method would have to depend only on the sort of energy resources now available that could be put onto a small craft. That prospect was so daunting that those of us working on advanced and exotic propulsion just ignored wormholes – and kept on working under our personal streetlamps as we had before. The path traversed by most of us to our streetlamps was a search of the professional literature for anomalous observations on gravity and electromagnetism and for speculative theories that coupled gravity and electromagnetism in ways not encompassed by general relativity and standard electrodynamics (classical or quantum).

When Thorne published his wormhole work, nothing anyone working on advanced propulsion was doing looked even remotely like it might produce the needed technology. The options were either to give up or to keep on looking for less ambitious propulsion schemes, illuminated by the streetlamps we had found, that would nonetheless improve our ability to explore space. After all, even a drunk knows that it's pretty stupid to look for your keys in the dark, no matter where they may actually be.

In the fall of 1989, after finding a flaw in a calculation done a decade earlier, I abandoned the streetlamp I had been working under for many years for another. That was not a pleasant experience. Abandoning a research program done for more than a decade is like divorce. Even in the best of circumstances, it's no fun at all. Though I didn't appreciate it at the time, several keys were in the gravel at the base of the new streetlamp I had chosen. No penetrating insight was required to see them. Just fabulous good luck. It is said that the Great Spirit looks out for drunks and fools.

If you are plugged into the popular space science scene at all, from time to time you hear commentators remark that given the mind-boggling number of Sun-like stars in the galaxy, and the number of galaxies in the observable universe, the likelihood that we are the only intelligent life-forms in the galaxy, much less the universe, is essentially zero. If there really are other intelligent life-forms present, and the physics of reality enables the construction of starships and stargates, the obvious question is: Why haven't we been visited by more advanced life-forms or life-forms of roughly our level of intelligence or greater that mastered high tech long before us? This is known in the trade as the Fermi paradox, for Enrico Fermi posed the question on a lunch break at Los Alamos in the early 1950s. His words were, "Where are they?"

A non-negligible number of people today would answer Fermi's question with, "They're already here, and they are abducting people and doing other sorts of strange things." Most serious scientists, of course, don't take such assertions seriously. Neither do they take seriously claims of crashed alien technology secreted by various governments and reverse engineered by shadowy scientists working on deep black projects.

Good reasons exist for scientists not taking popular fads and conspiracy theories seriously. Even if there are a few people who have really been abducted by aliens, it is obvious that the vast majority of such claims are false, regardless of how convinced those making the claims may be that their experience is genuine. In the matter of alleged conspiracies, it is always a good idea to keep in mind that we, as human beings, are wired to look for such plots in our experiences. Finding patterns in events that might pose a threat to us is something that has doubtless been selected for eons. When such a threat actually exists, this trait has survival value. When no threat is present, thinking one to be so is unlikely to have a negative survival impact. Others will just think you a bit odd or paranoid. Maybe. But you are still alive.

A more fundamental reason exists, though, that discredits the conspiracy schemes. It is predicated on the assumption that even if crashed alien tech exists, and our smartest scientists have had access to it, they would be able to figure out how it works. Is this reasonable? You can only figure out how something works if you understand the physical principles on which it is based. The fact of the matter is that until Thorne did his work on wormholes and Alcubierre found the warp drive metric, no one really understood the physical principles involved in starships and stargates. And even then, no one had a clue as to how you might go about inducing Jupiter masses of exotic matter to do the requisite spacetime warping. Though you might be the brightest physicist in the world, you could pore over the wreckage of an alien craft and still not have a clue about how it worked. Imagine giving the brightest physicists of the early nineteenth century a modern solid-state electronic device and asking them to reverse engineer it. How long do you think that would take?

Actually, there is an important point to be made in all of this talk of understanding and being able to master a technology. Although most of us might be willing to admit that dealing with the unknown might be challenging, indeed, perhaps very challenging, we would likely not be willing to admit that dealing with the unknown might prove completely insuperable. After all, we deal with unknowns all the time in our everyday lives. Our experiences and prior education, however, equip us to deal with the sorts of unknown situations we routinely encounter. As Thomas Kuhn pointed out in his *Structure of Scientific Revolutions* more than half a century ago, the sciences function in much the same way by creating "paradigms," collections of theories, principles, and methods of practice that guide practitioners in the field in handling the problems they address. Actually, paradigms even guide practitioners in the selection of problems sanctioned by their peers as worthy of investigation.

This may sound like the practitioners of a discipline collude to circumscribe things so that they only have to work on tractable problems that assure them of the approbation of their colleagues when they successfully solve one. But, of course, that's not the case. The practice of what Kuhn calls "normal" science can be exceedingly challenging, and there is no guarantee that you will be able to solve whatever problem you choose to tackle.

That said, there is another order entirely of unknowns and problems. In the quirky turn of phrase of a past Secretary of Defense, there are "unknown unknowns" in contrast to the "known unknowns" of paradigms and everyday experience. They are essentially never tackled by those practicing normal science. And when they are tackled by those with sufficient courage or foolhardiness, they usually try to employ the techniques of the normal science of the day. An example would be "alternative" theories of gravity in the age of Einstein.

As the importance of Special Relativity Theory (SRT) became evident in the period of roughly 1905–1915, a number of people realized that Newtonian gravity would have to be changed to comport with the conceptualization of space and time as relative. Perhaps the earliest to recognize this was Henri Poincaré. In a lengthy paper on relativity and gravity written in 1905, but published more than a year later, he did precisely this. His theory was not the precursor of General Relativity Theory (GRT). It was constructed using standard techniques in the flat pseudo-Euclidean spacetime of SRT. Not long after, others, notably Gustav Mie and Gunnar Nordstrom, also tackled gravity in the context of what would be called today unified field theory. They, too, used standard techniques and flat spacetime.

When Einstein told Planck of his intent to mount a major attack on gravity early in the decade of the teens, Planck warned him off the project. Planck told Einstein that the problem was too difficult, perhaps insoluble, and even if he succeeded, no one would much care because gravity was so inconsequential in the world of everyday phenomena. Einstein, of course, ignored Planck's advice. Guided by his version of the Equivalence principle and what he later called Mach's principle, he also ignored the standard techniques of field theory of his day. Rather than construct his field theory of gravity as a force field in a flat background spacetime, he opted for the distortion of spacetime itself and the non-Euclidean geometry that entails as his representation of the field.

It is easy now to look back and recognize his signal achievement: GRT. But even now, most do not appreciate the fundamentally radical nature of Einstein's approach. If you look at the history of gravitation in the ensuing century, much of it is a story of people trying to recast GRT into the formalism of standard field theory where the field is something that exists in a flat spacetime background and is communicated by gravitons. That's what it is, for example, in string theory. String theory is just the most well known of these efforts. GRT, however, is "background independent"; it cannot meaningfully be cast in a flat background spacetime. This property of GRT is pivotal in the matter of wormhole tech. It is the property that makes wormholes real physical structures worth trying to build.

The point of this is that if Einstein had not lived and been the iconoclast he was, the odds are that we today would *not* be talking about black holes and wormholes as real geometric structures of spacetime. Instead, we would be talking about the usual sorts of schemes advanced in discussions of deep space transport: electric propulsion, nuclear propulsion, and so on. Radical speculation would likely center on hypothetical methods to reduce the inertia of massive objects, the goal being to render them with no inertia, so they could be accelerated to the speed of light with little or no energy. That is, the discussion would be like that before Kip Thorne did his classic work on wormholes.

You sometimes hear people say that it may take thousands, if not millions, of years of development for us to figure out how to do wormhole tech. Perhaps, but probably not. The key enabling ideas are those of Einstein and Thorne. Clever aliens, if they did not have an Einstein and a Thorne, may well have taken far longer to figure out wormhole tech than, hopefully, we will. We have been fabulously lucky to have had Einstein, who recognized gravity as fundamentally different from the other forces of nature, and Thorne, who had the courage to address the issue of traversable wormholes, putting his career at serious risk.

If you've not been a professional academic, it is easy to seriously underestimate the courage required to do what Thorne did. As a leading figure in the world of gravitational physics, to stick your neck out to talk about traversable wormholes and time machines is just asking for it. Professionally speaking, there just isn't any upside to doing this sort of a thing. It can easily turn out to be a career ender. Those of lesser stature than Thorne were routinely shunned by the mainstream community for much less and often still are. It is likely, though, that in the future Thorne will chiefly be known for his work on wormholes. And both his work and his courage will be highly regarded.

The plan of this book is simple. The material is divided into three sections. The first section deals with the physics that underlie the effects that make the reality of stargates possible. The principles of relativity and equivalence are discussed first, as the customary treatments of these principles do not bring out their features that are important to the issue of the origin of inertia. Next, Mach's principle and the gravitational origin of both inertial reaction forces and mass itself are dealt with. Derivation of "Mach effects" – transient mass fluctuations that can be induced in some objects in special circumstances – complete the first section.

In the second section, after an overview of past experimental work, recent experimental results are presented and examined in some detail. Those results suggest that whether or not stargates can be made, at least a means of propellant-free propulsion can be created using Mach effects.

The first two sections are not speculative. The physics involved is straightforward, though the emphasis differs from the customary treatments of this material. Experimental results can be questioned in a number of ways. But in the last analysis, they are the touchstones and final arbiters of reality.

The third section is different. The central theme of this section is the creation of an effective Jupiter mass of exotic matter in a structure with typical dimensions of meters. This discussion is impossible unless you have a theory of matter that includes gravity. The Standard Model of relativistic quantum field theory – that is, the widely accepted, phenomenally successful theory of matter that has dominated physics for the past half century – does not include gravity. Indeed, this is widely regarded as its chief defect. For the purpose of making stargates, that defect is fatal. Fortuitously, a theory of the simplest constituents of matter, electrons, that includes general relativity has been lying around for roughly 50 years. It was created by Richard Arnowitt, Stanley Deser, and Charles Misner (commonly referred to as ADM) in 1960. It has some problems (which is why it didn't catch on either when proposed or since). But the problems can be fixed.

When fixed, the ADM electron model allows you to calculate how much exotic matter is available in everyday matter, normally screened by the gravitational interaction with chiefly distant matter in the universe, if a way to expose it can be found. Such exposure can be achieved by canceling the gravitational effect of the chiefly distant matter with nearby exotic, negative rest-mass matter. The amount of exotic matter needed to trigger this process is minuscule by comparison with the Jupiter mass of exotic matter that results from exposure. Mach effects provide a means to produce the exotic matter required to produce exposure. All of this is spelled out in some detail in the third section. And we finish up with some comments on how you would actually configure things to make a real starship or stargate.

There may be times, as you wend your way through the following chapters, when you ask yourself, "Why in God's name did this stuff get included in a book on stargates?" Some of the material included is a bit confusing, and some of it is a bit arcane. But all of the material in the main body of the text is there because it bears directly on the physics of starships and stargates. So please bear with us in the more difficult parts.

So who exactly is this book written for? Strictly speaking, it is for professional engineers. You might ask: Why not physicists? Well, physicists don't build starships and stargates. They build apparatus to do experiments to see if what they think about the world is right. You'll find some of this sort of activity reported in the second section. But moving beyond scientific experiments requires the skills of engineers; so they are the target audience. That target audience justifies the inclusion of some formal mathematics needed to make the discussion exact. But grasping the arguments made usually does not depend critically on mathematical details. So if you find the mathematics inaccessible, just read on.

You will find, as you read along, in the main part of the book, that it is not written like any engineering (or physics) text that you may have read. Indeed, much of the main part of this book is written for an educated audience who has an interest in science and technology. This is not an accident. Having read some truly stultifying texts, we hope here not to perpetrate such stuffiness on anyone. And the fact of the matter is that some, perhaps much, of the scientific material belongs to arcane subspecialties of physics, and even professional engineers and physicists in different subspecialties are not much better prepared to come to grips with this material than members of the general public.

If you are an engineer or a physicist, though, you should not get the idea that this book is written for nonprofessionals. Mathematics where it is needed is included for clear communication and to get something exactly right. Nonetheless, we hope that general readers will be able to enjoy much, if not most, of the content of this book. For if the material in this book is essentially correct, though some of us won't see starships and stargates in our lifetime, perhaps you will in yours.

About the Author

Jim Woodward earned Bachelor's and Master's degrees in physics at Middlebury College and New York University, respectively, in the 1960s. From his undergraduate days, his chief interest was in gravitation, a field then not very popular. So, for his Ph.D., he changed to the history of science, writing a dissertation on the history of attempts to deal with the problem of "action-at-a-distance" in gravity theory from the seventeenth to the early twentieth century (Ph.D., University of Denver, 1972).

On completion of his graduate studies, Jim took a teaching job in the history of science at California State University Fullerton (CSUF), where he has been ever since. Shortly after his arrival at CSUF, he established friendships with colleagues in the Physics Department who helped him set up a small-scale, tabletop experimental research program doing offbeat experiments related to gravitation – experiments that continue to this day. In 1980, the faculty of the Physics Department elected Jim to an adjunct professorship in the department in recognition of his ongoing research.

In 1989, the detection of an algebraic error in a calculation done a decade earlier led Jim to realize that an effect he had been exploring proceeded from standard gravity theory (general relativity), as long as one were willing to admit the correctness of something called "Mach's principle" – the proposition enunciated by Mach and Einstein that the inertial properties of matter should proceed from the gravitational interaction of local bodies with the (chiefly distant) bulk of the matter in the universe. Since that time, Jim's research efforts have been devoted to exploring "Mach effects," trying to manipulate them so that practical effects can be produced. He has secured several patents on the methods involved.

Jim retired from teaching in 2005. Shortly thereafter, he experienced some health problems that have necessitated ongoing care. But, notwithstanding these, he remains in pretty good health and continues to be active in his chosen area of research.

Acknowledgments

No work is done alone. I have enjoyed the support of many people (and a few cats) and more than one academic institution – too many people to individually name them all here. My family and friends are part of my private life and, for their privacy, won't be mentioned by name. They know who they are, and they know how deeply indebted I am to them. In a semiprivate sort of way, I am indebted to several members of the staff of USC Norris Cancer Hospital. They have kept me alive since shortly after my retirement in the summer of 2005 – treating first metastatic lung cancer, then heart failure, then metastatic Hodgkins lymphoma, then metastatic non-Hodgkins lymphoma, then relapsed metastatic Hodgkins lymphoma, and so on. They are Barbara Gitlitz, Ann Mohrbacher, Jerold Shinbane, Ching Fei Chang, and my oncology nurse Elisa Sousa. They have been helped by many of the outstanding staff of USC's Keck School of Medicine. They have repeatedly and literally saved my life, making me an outlier among outliers. I do not have the words to thank them properly.

The Mach effects project has attracted the attention of a number of people who have provided support in one form or another over the years. Many of them are mentioned in the following pages. They are an unusual group, for a project like this doesn't attract the sort of people you normally associate with "exotic" propulsion schemes. The reason for this lack of interest on the part of the amateur community seems to have been the mathematics and difficult physics involved. Actually, this has been a blessing, since the sorts of distractions that kind of attention brings have been avoided. Those that found their way to this project long ago are Nembo Buldrini (who is also the illustrator of this book), Paul March, Sonny White, and Andrew Palfreyman. Thomas Mahood, who did a Master's thesis on Mach effects in the late 1990s, made substantial contributions to the advancement of the project for many years. Marc Millis, Paul Murad, and Tony Robertson have understood that the project was one involving serious science being pursued as serious science. Those with management experience, looking to transform the project from one of tabletop science into a serious engineering enterprise, are Jim Peoples, Gary Hudson, and David Mathes. They've not yet succeeded. But this book is the first step in that direction, insisted upon by Jim for the past several years. Management types are usually deeply risk averse. These three are very remarkable for their willingness to take risks that others shun.

The manuscript of this book went to a small group of readers before being turned in to Maury Solomon of Springer for professional scrubbing. They include Terry and Jane Hipolito, Nembo Buldrini, Ron Stahl, Greg Meholic, David Jenkins, and David Mathes. They all provided useful suggestions and wise counsel. The arguments relating to Mach's principle, chiefly found in Chap. 2, were sharpened by email exchanges with Paul Zielinski and Jack Sarfatti. Those exchanges, often contentious, eventually produced a little enlightenment. Chiefly, however, they contributed significantly in identifying those things that needed careful explanation.

A small group of people merit special mention, as their support was truly pivotal. First among them is Wolfgang Yourgrau, mentor and friend, who saw in me promise not evident to others at the time. And those who helped me through graduate study at the University of Denver are Bob Aime, Eric Arnold, Jim Barcus, Allen Breck, Larry Horwitz, and Alwyn van der Merwe. Jim gave me space in his lab and made machine shop access possible so I could pursue some tabletop experiments in my spare time while doing a Ph.D. in the history of science. It was there that I learned the skills that led me to spending as much time in front of lathes as in libraries in subsequent years.

On my first day of work at California State University Fullerton (CSUF) in the fall of 1972, I met Ron Crowley, the unlikely general relativist at the university. (State universities in those days did not hire general relativists as a general rule.) The next year, Dorothy Woolum was hired from a postdoc position at Caltech. They became lifelong friends and supporters – and honest, skeptical critics who kept me grounded. Though my home department was History, my colleagues and friends in Physics tolerated my sideline experimental activities for years (and still do). The forbearance of my History colleagues has been truly remarkable. Alan Sweedler, Keith Wanser, Heidi Fearn, Morty Khakoo, Stephen Goode, Louis Shen, and Mark Shapiro, skeptics all, were there when I needed someone to talk to while grappling with the theoretical and practical problems of gravitational physics. Heidi's friend, Peter Milonni, made a crucial contribution, calling my attention to the fact that the bare masses of electrons are infinite and *negative* in the Standard Model of relativistic quantum field theory. And he had seen that the putative quantum vacuum had more than one representation.

Unbeknownst to me, John Cramer was keeping an eye on those doing advanced and exotic propulsion physics for many years. I came across his work on the interpretation of quantum mechanics while reading a biography of Schrödinger in the early 1990s. He had applied the ideas of John Wheeler and Richard Feynman – so-called "action-at-a-distance" or "absorber" electrodynamics – to quantum theory to reconcile seemingly instantaneous actions with the principle of relativity that demands that nothing travel faster than the speed of light. Instantaneous action is a feature of inertial reaction effects; so I included mention of his work in the first paper on Mach effects where almost everything was pulled together. Squaring instantaneous action with relativity is required if your physics is to be plausible. John returned the favor by mentioning Mach effects in one of his Alternate View columns for *Analog*. That was the event that lifted Mach effects from obscurity and triggered much of what you will find in the following pages.

Contents

Foreword   vii

Preface   xiii

About the Author   xxi

Acknowledgments   xxiii

Part I

1   The Principle of Relativity and the Origin of Inertia   3

2   Mach's Principle   29

3   Mach Effects   65

Part II

4   Getting in Touch with Reality   89

5   In Reality's Grip   133

Part III

6   Advanced Propulsion in the Era of Wormhole Physics   183

7   Where Do We Find Exotic Matter?   207

8   Making the ADM Electron Plausible 225

9   Making Stargates 235

10   The Road Ahead 259

Bibliography 269

Author Index 273

Subject Index 275

Part I

1

The Principle of Relativity and the Origin of Inertia

GETTING AROUND QUICKLY

When you think of traveling around the Solar System, especially to the inner planets, a number of propulsion options arguably make sense. When the destination involves interstellar distances or larger, the list of widely accepted, plausible propulsion schemes involving proven physical principles drops to zero. If a way could be found to produce steady acceleration on the order of a "gee" or two for long periods without the need to carry along vast amounts of propellant, interstellar trips within a human lifetime would be possible. But they would not be quick trips by any stretch of the imagination. If a way to reduce the inertia of one's ship could be found, such trips could be speeded up, as larger accelerations than otherwise feasible would become available. But such trips would still be sub-light speed, and the time dilation effects of Special Relativity Theory (SRT) would still apply. So when you returned from your journeys, all of your friends and acquaintances would have long since passed on.

As is now well-known, wormholes and warp drives would make traversing such distances in reasonable times plausible. And returning before your friends age and die is possible. Indeed, if you choose, you could return before you left. But you couldn't kill yourself before you leave. A wide range of "traversable" wormholes with a wide range of necessary conditions are possible. The only ones that are manifestly practical are, in the words of Michael Morris and Kip Thorne, "absurdly benign." Absurdly benign wormholes are those that restrict the distortion of spacetime that forms their throats to modest dimensions – a few tens of meters or less typically – leaving the surrounding spacetime flat. And their throats are very short. Again, a few tens of meters or less typically. Such structures are called "stargates" in science fiction. The downside of such things is that their implementation not only requires Jupiter masses of "exotic" matter, they must be *assembled* in a structure of very modest dimensions. Imagine an object with the mass of Jupiter (about 600 times the mass of Earth) sitting in your living room or on your patio.

J.F. Woodward, *Making Starships and Stargates: The Science of Interstellar Transport and Absurdly Benign Wormholes*, Springer Praxis Books, DOI 10.1007/978-1-4614-5623-0_1,
© James F. Woodward 2013

Even the less daunting methods of either finding a way to accelerate a ship for long intervals without having to lug along a stupendous amount of propellant or reduce its inertia significantly do not seem feasible. Sad to say, solutions to none of these problems – vast amounts of propellant, or inertia reduction, or Jupiter masses of exotic matter to make wormholes and warp drives – are presently to be found in mainstream physics. But Mach effects – predicted fluctuations in the masses of things that change their internal energies as they are accelerated by external forces – hold out the promise of solutions to these problems.

To understand how Mach effects work, you first have to grasp "Mach's principle" and what it says about how the inertial properties of massive objects are produced. You can't manipulate something that you don't understand, and inertia is the thing that needs to be manipulated if the goal of rapid spacetime transport is to be achieved.

Mach's principle can only be understood in terms of the principle of relativity and Einstein's two theories thereof. While the theories of relativity, widely appreciated and understood, do not need a great deal of formal elaboration, the same cannot be said of Mach's principle. Mach's principle has been, from time-to-time, a topic of considerable contention and debate in the gravitational physics community, though at present it is not. The principle, however, has not made it into the mainstream canon of theoretical physics. This means that a certain amount of formal elaboration (that is, mathematics) is required to insure that this material is done justice. The part of the text that does not involve such formal elaboration will be presented in a casual fashion without much detailed supporting mathematics. The formal material, of interest chiefly to experts and professionals, will usually be set off from the rest of the narrative or placed in appendixes. Most of the appendixes, however, are excerpts from the original literature on the subject. Reading the original literature, generally, is to be preferred to reading a more or less accurate paraphrasing thereof.

THE RELATIVITY CONTEXT OF MACH'S PRINCIPLE

Ernst Mach, an Austrian physicist of the late nineteenth and early twentieth centuries, is now chiefly known for Mach "numbers" (think Mustang Mach One, or the Mach 3, SR71 Blackbird). But during his lifetime, Mach was best known for penetrating critiques of the foundations of physics. In the 1880s he published a book – *The Science of Mechanics* – where he took Newton to task for a number of things that had come to be casually accepted about the foundations of mechanics – in particular, Newton's notions of absolute space and time, and the nature of inertia, that property of real objects that causes them to resist changes in their states of motion.

Einstein, as a youngster, had read Mach's works, and it is widely believed that Mach's critiques of "classical," that is, pre-quantum mechanical, physics deeply influenced him in his construction of his theories of relativity. Indeed, Einstein, before he became famous, had visited Mach in Vienna, intent on trying to convince Mach that atoms were real. (The work Einstein had done on Brownian motion, a random microscopic motion of very small particles, to get his doctoral degree had demonstrated the fact that matter was atomic). Mach had been cordial, but the young Einstein had not changed Mach's mind.

Nonetheless, it was Mach's critiques of space, time, and matter that had the most profound effect on Einstein. And shortly after the publication of his earliest papers on General Relativity Theory (GRT) in late 1915 and early 1916, Einstein argued that, in his words, Mach's principle should be an explicit property of GRT. Einstein defined Mach's principle as the "relativity of inertia," that is, the inertial properties of material objects should depend on the presence and action of other material objects in the surrounding spacetime, and ultimately, the entire universe. Framing the principle this way, Einstein found it impossible to show that Mach's principle was a fundamental feature of GRT. But Einstein's insight started arguments about the "origin of inertia" that continue to this day. Those arguments can only be understood in the context of Einstein's theories of relativity, as inertia is an implicit feature of those theories (and indeed of any theory of mechanics). Since the issue of the origin of inertia is not the customary focus of examinations of the theories of relativity, we now turn briefly to those theories with the origin of inertia as our chief concern.

Einstein had two key insights that led to his theories of relativity. The first was that if there really is no preferred reference frame – as is suggested by electrodynamics[1] – it must be the case that when you measure the speed of light in vacuum, you always get the same number, no matter how you are moving with respect to the source of the light. When the implications of this fact for our understanding of time are appreciated, this leads to Special Relativity Theory (SRT), in turn, leads to a connection between energy and inertia that was hitherto unappreciated. The curious behavior of light in SRT is normally referred to as the speed of light being a "constant." That is, whenever anyone measures the speed of light, no matter who, where, or when they are, they always get the same number – in centimeter-gram-second (cgs) units, 3×10^{10} cm/s. Although this works for SRT, when we get to General Relativity Theory (GRT) we will find this isn't quite right. But first we should explore some of the elementary features of SRT, as we will need them later. We leave detailed consideration of Einstein's second key insight – the Equivalence Principle – to the following section, where we examine some of the features of general relativity theory.

---

[1] The simple case analyzed by Einstein in his first paper on special relativity theory – titled "On the Electrodynamics of Moving Bodies" – is the motion of a magnet with respect to a loop of wire. If the relative motion of the magnet and wire causes the "flux" of the magnetic field through the loop of wire to change, a current flows in the loop while the flux of the magnetic field passing through the loop is changing. It makes no difference to the current in the loop whether you take the loop as at rest with the magnet moving, or vice versa. The rest of the paper consists of Einstein's demonstration that the mathematical machinery that gets you from the frame of reference where the magnet is at rest to the frame where the loop is at rest requires that the speed of light measured in both frames is the same, or "constant." This is only possible if space and time are inextricably interlinked, destroying Newton's absolute notions of space and time as physically distinct, independent entities. The concept underlying the full equivalence of the two frames of reference is the principle of relativity: that all inertial frames of reference are equally fundamental and no one of them can be singled out as more fundamental by any experiment that can be conducted locally.

THE PRINCIPLE OF RELATIVITY

Mention relativity, and the name that immediately jumps to mind is Einstein. And in your mental timescape, the turn of the twentieth century suffuses the imagery of your mind's eye. The *principle* of relativity, however, is much older than Einstein. In fact, it was first articulated and argued for by Galileo Galilei in the early seventeenth century.

A dedicated advocate of Copernican heliocentric astronomy, Galileo was determined to replace Aristotelian physics, which undergirded the prevailing Ptolemaic geocentric astronomy of his day, with new notions about mechanics. Galileo hoped, by showing that Aristotelian ideas on mechanics were wrong, to undercut the substructure of geocentric astronomy. Did Galileo change any of his contemporaries' minds? Probably not. Once people think they've got something figured out, it's almost impossible to get them to change their minds.[2] As Max Planck remarked when asked if his contemporaries had adopted his ideas on quantum theory (of which Planck was the founder), people don't change their minds – they die. But Galileo did succeed in influencing the younger generation of his day.

Galileo's observations on mechanics are so obvious that it is, for us, almost inconceivable that any sensible person could fail to appreciate their correctness. But the same could have been said of Aristotle in Galileo's day. Arguing from commonplace experience, Aristotle had asserted that a force had to be applied to keep an object in motion. If you are pushing a cart along on a level road and stop pushing, not long after the cart will stop moving. However, even to a casual observer, it is obvious that how quickly the cart stops depends on how smooth and level the road is and how good the wheels, wheel bearings, and axle are. Galileo saw that it is easy to imagine that were the road perfectly smooth and level, and the wheels, wheel bearings, and axle perfect, the cart would continue to roll along indefinitely.

Galileo, in his *Science of Mechanics* (published in 1638, a few years before he died), didn't put this argument in terms of carts. He used the example of a ball rolling down an incline, then along a smooth level plane, eventually ending rolling up an incline. From this he extracted that objects set into motion remain in that state of motion until influenced by external agents. That is, Newton's first law of mechanics. Newton got the credit because he asserted it as a universal law, where Galileo only claimed that it worked below the sphere of the Moon. After all, he was a Copernican, and so assumed that the motions of heavenly bodies were circular.

Galileo figured out most of his mechanics in the 1590s, so when he wrote the *Dialog on the Two Chief World Systems* in the 1620s (that got him condemned by the Inquisition a few years later for insulting the Pope in one of the dialogs), he had his mechanics to draw upon. One of the arguments he used involved dropping a cannonball from the crow's nest on the mast of ship moving at steady speed across a smooth harbor. Galileo claimed that the cannonball would fall with the motion of the ship, and thus land at the base of the mast, whereas Aristotle would have the cannonball stop moving with the ship

---

[2] Galileo himself was guilty of this failing. When Kepler sent him his work on astronomy (the first two laws of planetary motion anyway), work that was incompatible with the compounded circular motions used by Copernicus, Galileo, a convinced Copernican, ignored it.

when it was released. As a result, according to Aristotle, if the ship is moving at a good clip, the cannonball should land far from the base of the mast as the ship would keep moving horizontally and the cannonball would not. Anyone who has ever dropped something in a moving vehicle (and a lot who haven't) knows that Galileo was right. Galileo was describing, and Newton codifying, "inertial" motion. Once Galileo's take on things is understood, Aristotelian ideas on mechanics become features of the intellectual landscape chiefly of interest to historians.

Galileo did more than just identify inertial motion. He used it to articulate the principle of relativity. Once you get the hang of inertial motion, it's pretty obvious that there is, as we would say today, no preferred frame of reference. That is, on the basis of mechanics with inertial motion, there is no obvious way to single out one system as preferred and at rest, with respect to which all other systems either move or are at rest. Galileo's way of making this point was to consider people shooting billiards in the captain's cabin of the ship where the cannonball got dropped from the crow's nest. He posed the question: if all of the portholes were covered up so you couldn't see what's going on outside the cabin, can you tell if the ship is moving across the harbor at constant speed and direction, or tied up at the dock, by examining the behavior of the balls on the billiards table? No, of course not. Any inertial frame of reference is as good as any other, and you can't tell if you are moving with respect to some specified inertial frame by local measurements. You have to go look out the porthole to see if the ship is moving with respect to the harbor or not. This is the principle of relativity.

Galileo's attack on Aristotelian mechanics didn't stop at identifying inertial motion. Aristotle, again on the basis of casual observations, had asserted that heavier objects fall faster than light objects. It had been known for centuries that this was wrong. But Aristotelians had either ignored the obvious, or concocted stories to explain away "anomalous" observations. Galileo brought a cannonball and a musket ball to the top of the leaning Tower of Pisa and dropped them together. (But not in front of the assembled faculty of the local university.) He noted that the musket ball arrived at the ground within a few fingers' breadth of the cannon ball. The cannonball, being more than ten times more massive than the musket ball, should have hit the ground far in advance of the musket ball. It didn't. Galileo surmised that the small difference in the arrival times of the two balls was likely due to air resistance, and inferred that in a vacuum the arrivals would have been simultaneous. Moreover, he inferred that the time of fall would have been independent of the compositions, as well as the masses, of the two balls. This is the physical content of, as Einstein later named it, the Equivalence Principle.

Isaac Newton, one of the best physicists of all time,[3] took on the insights of Galileo, asserted them as universal principles, and codified them into a formal system of mechanics.

---

[3] Most historians of science would probably name Newton the greatest physicist of all time. Most physicists would likely pick Einstein for this honor (as did Lev Landau, a brilliant Russian theoretical physicist in the mid-twentieth century). Getting this right is complicated by the fact that Newton spent most of his life doing alchemy, biblical studies, pursuing a "patent" of nobility, and running the government's mint after the mid-1690s. Physics and mathematics were sidelines for him. Einstein, on the other hand, aside from some womanizing, spent most of his life doing physics, albeit out of the mainstream after the late 1920s. It's complicated.

He worked out the law of universal gravitation, and saw that his third law – the requirement of an equal and opposite reaction force for all "external" applied forces – was needed to complete the system of mechanics. He did experiments using pendula to check up on Galileo's claim that all objects fall with the same acceleration in Earth's gravity field.[4] His synthesis of mechanics and gravity, published in 1687 as the *Principia Mathmatica Philosophia Naturalis,* ranks as one of the greatest achievements of the human intellect.

However, if Newton incorporated the principle of relativity and the Equivalence Principle into his work, one might ask, why didn't he figure out the *theory* of relativity? Absolute space, and absolute time. Newton was nothing if not thorough. So he provided definitions of space and time, which he took to be completely separate physical entities (as indeed they appear to us today on the basis of our everyday experience of reality). Alas, it turns out that this is wrong. And if you make this assumption, as Newton did, you can't discover the theory of relativity.

Before turning to relativity theory, a small digression on the nature and manifestation of inertia as understood in Newtonian mechanics seems advisable. The notion has gotten abroad since the advent of general relativity that inertia – the property of massive objects that makes them resist accelerations by external forces – does not involve force. Common sense tells you that if some agent exerts a force on you, the way to resist it is to, in turn, exert a force back on the thing pushing you. But in general relativity, inertial "forces" are deemed "fictitious," and this led, in the twentieth century, to a systematic effort to claim that inertia does not involve "real" forces.[5] In the seventeenth century, such a claim would not have been taken seriously.

The commonplace language of that era was to talk about "vis viva" and "vis inertia" – that is, "living force" and "dead force." Living forces were those that acted all the time: electrical forces, magnetic forces, and gravity (in the Newtonian worldview). Vis inertia, dead, or inert force (vis is Latin for force), in contradistinction, was normally not in evidence. That is, it normally did not act. Indeed, the only time vis inertia did act was when a body was acted upon by an external force to accelerate the body. Then the dead force would spring to life to resist the live force by exerting an equal and opposite reaction force *on the accelerating agent.*

It is important to note that the inertial reaction force, resident in the body acted upon, does not act on the body itself; rather it acts on the accelerating agent. Were it to act on the body itself, in Newtonian mechanics the total force on the body would then be zero, and the body would not accelerate.[6] But the inertial reaction force – as a force – is an

---

[4] The period of a pendulum depends only on its length if the Equivalence Principle is true, so you can put masses of all different weights and compositions on a pendulum of some fixed length, and its period should remain the same. Newton did this using a standard comparison pendulum and found that Galileo was right, at least to about a part in a thousand. Very much fancier experiments that test this principle have been (and continue to be) done to exquisitely high accuracy.

[5] The technical definition of a "fictitious" force is one that produces the same acceleration irrespective of the mass of the object on which it acts. It has nothing to do with whether the force is "real" or not.

[6] This is a common problem encountered in teaching Newtonian mechanics, and evidently part of the reason for the program of rejecting the idea that inertia involves forces in physics pedagogy. That program is misguided at best.

essential part of Newtonian mechanics. It is the force on the accelerating agent that ensures that Newton's third law of mechanics is obeyed, and in consequence that momentum conservation is not violated in isolated systems. If an unconstrained body acted upon by an external force did not exert an inertial reaction force on an agent trying to accelerate it, the body would accelerate, acquiring momentum, but the accelerating agent would not be forced to accelerate in the opposite direction, acquiring an equal measure of momentum in the opposite direction.

You may think that the reaction force in this case can be ascribed to the electromagnetic contact forces that operate at the junction of the agent and the body, but this is a mistake. Those electromagnetic contact forces communicate the forces present to and from the agent and body. But they are not themselves either the accelerating force or the inertial reaction force.

Consider a simple example often used in discussions of inertia: centrifugal force. We take a rock, tie a string to it, and swing it around our head in steady circular motion. We ignore things such as air resistance and the action of gravity. Where the string attaches to the rock there is an action-reaction pair of electrical forces in the string. One of those electrical forces communicates a "centripetal" (toward the center) force on the rock, causing it to deviate from inertial motion in a straight line. That force arises in and is caused by our muscles. It is communicated to the rock by electrical force in the string.

The other electrical force in the string, the other part of the action-reaction pair, gets communicated through the string to our arm. It is a real force. Where does it arise? What causes it? The inertia of the rock causes the force. It is the inertial reaction *force* that springs into existence to resist the acceleration of the rock by the action of your arm through the string. Note that while it originates in the rock, it acts through the string on you, not the rock. The reason why it is called an "inert" or dead force is that it only manifests itself when an external force acts on the rock to force it out of inertial motion.

For Newton and most of his contemporaries and successors, inertia was a "primary" property of matter. That is, it was regarded as fundamental and did not need further explanation. But this view of inertia and inertial forces was, even in Newton's day, not universal. George Berkeley, a younger contemporary of Newton, criticized Newton's notion of inertia by posing the question: If a body is alone in an empty universe, can you tell if it is rotating? Newton's view on this situation was contained in his "bucket" experiment. You fill a bucket with water and suspend it with a twisted cord. When the bucket is released, the twisted cord causes it to rotate. At first, the water in the bucket does not rotate with the bucket, though eventually it will because of friction at the bucket walls. Newton explained this by asserting that the water was inertially at rest with respect to absolute space, and were there no friction at the bucket walls, the water would remain at rest by virtue of its inertia while the bucket rotated. Whether the water in the bucket was rotating, Newton noted, could always be ascertained by a local measurement, namely, whether the surface of the water was flat or concave. There matters stood until Ernst Mach and Albert Einstein came along nearly 300 years later.

SPECIAL RELATIVITY THEORY

Nowadays, everyone knows that SRT takes the physically independent, absolute Newtonian notions of space and time and inextricably mixes them up together to get "spacetime." That is, in the Newtonian world-view, all observers, no matter where they are or how they are moving with respect to each other (or any other specified frame of reference), see the same space and measure the same time.

Einstein's profound insight was to see that if all observers measure the same value for the speed of light (in a vacuum), this can't be true, for if one observer measures a particular value in Newtonian space and time, and another observer is moving with respect to him, that other observer *must* measure a different value for the speed of light, c. But if this is so, then we can pick out some frame of reference, for whatever reason, and call it the fundamental frame of reference (say, the frame of reference in which nearby galaxies are, on average, at rest, or the frame in which the speed of light has some preferred value in a particular direction), and we can then refer all phenomena to this fundamental frame. The *principle* of relativity, however, requires that such a frame with preferred physical properties that can be discovered with purely local measurements not exist, and the only way this can be true is if the measured speeds of light in all frames have the same value, making it impossible on the basis of local experiments to single out a preferred frame of reference.

So, what we need is some mathematical machinery that will get us from one frame of reference to another, moving with respect to the first, in such a way that the speed of light is measured to have the same value in both frames of reference. The "transformation" equations that do this are called the "Lorentz transformations" because they were first worked out by Hendrick Antoon Lorentz a few years before Einstein created SRT. (Lorentz, like Einstein, understood that the "invariance" of the speed of light that follows from electrodynamics required the redefinition of the notions of space and time. But unlike Einstein, he continued to believe, to his death roughly 20 years after Einstein published his work on SRT, that there were underlying absolute space and time to which the "local" values could be referred).

Many, many books and articles have been written about SRT. Some of them are very good. As an example, see Taylor and Wheeler's *Spacetime Physics*. We're not going to repeat the customary treatments here. For example, we're not going to get involved in a discussion of how time slows when something is moving close to the speed of light and the so-called "twins paradox." Rather, we're going to focus on the features of SRT that we'll need for our discussion of Mach's principle and Mach effects. Chief among these is what happens to the physical quantities involved in Newtonian mechanics such as energy, momentum, and force. The way in which SRT mixes up space and time can be seen by choosing some spacetime frame of reference, placing some physical quantity at some location, and examining how it looks in two different frames of reference.

Mathematically speaking, physical quantities come in one of three types: scalars, vectors, or tensors. Scalars are those things that have only magnitude, like temperature or energy, and thus can be specified by one number at every event in spacetime. Vectors are things that have both magnitude and point in some direction, like momentum and force.

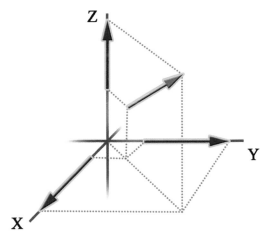

Fig. 1.1 A three-dimensional vector (*red*) and its projections on three Cartesian coordinate axes, X, Y, and Z. The "norm" or length of the vector is the square root of the sum of the squares of the lengths of the component vectors

They are customarily represented by arrows that point in the spatial direction of the quantity. Their length represents the magnitude of the quantity at the point where their back end is located (see Fig. 1.1). In Newtonian physics, with its absolute space and time, that means that they point in some direction in space, and thus require three numbers to be fully specified – the projected lengths of the vector on three suitably chosen coordinate axes, one for each of the dimensions of space.

Since time is treated on the same footing as space in spacetime, four-vectors in spacetime need four numbers to specify their projections on the four axes of spacetime. Tensors are used to specify the magnitudes that characterize more complicated things such as elasticity, which depend on the direction in which they are measured. We'll not be concerned with them at this point.

The things we will be most interested in are those represented by vectors. So we start with a vector in a simple three-dimensional space, as in Fig. 1.1. We include some Cartesian coordinate axes of an arbitrarily chosen "frame of reference."[7] The projections of the vector on the coordinate axes are the component vectors shown in Fig. 1.1.

The distinctive feature of the vector is that its length (magnitude) shouldn't depend on how we choose our coordinates. That is, the length of the vector must be "invariant" with respect to our choice of coordinates. This will be the case if we take the square of the length of the vector to be the sum of the squares of the component vectors, because the vector and its components form a right triangle, and the sum of the squares of the

---

[7] Named for their inventor, Rene Descartes, these axes are chosen so that they are (all) mutually perpendicular to each other. He got the idea lying in bed contemplating the location of objects in his room and noting that their places could be specified by measuring how far from a corner of the room they were along the intersections of the floor and walls.

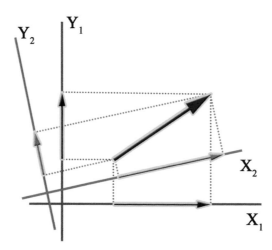

Fig. 1.2 The lengths of the components of a vector differ from one set of coordinates to another. But their norm does not, as vectors are "invariant" under coordinate transformations, as shown here

shorter sides of the triangle is equal to the square of the longest side. Note, as the Pythagorean theorem informs us, that this is true even if we choose some other coordinates, say those in red in Fig. 1.2, a two dimensional simplification of Fig. 1.1. The component vectors are different from those in the black coordinates, but the sum of their squares is the same. This is true when space is "flat," or "Euclidean." It is not in general true when space is "curved."

To make the transition to spacetime we need to be able to treat space and time on an equal footing. That is, if we are to replace one of our two space axes in Fig. 1.2 with a time coordinate, it must be specified in the same units as that of the remaining space coordinate. This is accomplished by multiplying time measurements by the speed of light. This works because the speed of light is an invariant – the same for all observers – and when you multiply a time by a velocity, you get a distance. So, with this conversion, we end up measuring time in, say, centimeters instead of seconds. Should you want to measure time in its customary units – seconds – to get everything right you'd have to divide all spatial distances by the speed of light. Spatial distances would then be measured in light-seconds. It doesn't matter which choice you make, but we'll use the customary one where times are multiplied by c.

We now consider a vector in our simple two-dimensional spacetime in Fig. 1.3. Were spacetime like space, we would be able to specify the length of our vector as the sum of the squares of its projections on the space and time axes. But spacetime isn't like space. The requirement that the speed of light be measured to have the same value in all spacetime frames of reference forces us to accept that spacetime is "pseudo-Euclidean." Pseudo-Euclidean? What's that?

Well, as in a Euclidean (or flat) space, we still use the squares of the projections of vectors on their coordinate axes to compute their lengths. And when there is more than one

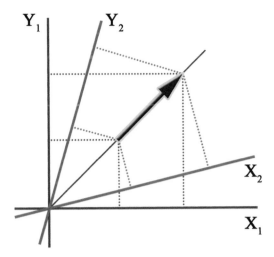

Fig. 1.3 If we take X_1 as a spatial coordinate and Y_1 as the time coordinate, and we note that the time registered by a clock moving at the speed of light is zero, then the squares of the components of the *blue* vector lying along a light path must sum to zero for an observer in this reference frame, and one must be *subtracted* from the other, making spacetime pseudo-Euclidean. This must also be true for another observer moving with respect to the first (who happens here to have coordinates with the same origin as those of the first observer). For this to be true, the moving observers' coordinates must look like those designated with the subscript 2 here. The mathematics that gets you from one frame of reference to the other is called the "Lorentz transformations"

space dimension, the "spacelike" part of the vector is just the sum of the squares of the projections on the space axes. But the square of the projection on the time axis (the "timelike" part of the vector) is *subtracted* from the spacelike part of the vector to get its total length in spacetime.

Why do you subtract the square of the timelike component of the vector from the square of the spacelike part? Because time stops for things traveling at the speed of light (as Einstein discerned by imagining looking at a clock while riding on a beam of light, since moving away from the clock on the light beam carrying the clock time, time stops). Look at the two-dimensional spacetime and ask, how can we construct a path for light so that its length in spacetime is zero but its distances in space and time separately aren't zero?

Well, it's impossible to add two non-zero positive numbers to get zero. And the squares of the component vectors are always positive. So it must be that we have to subtract them. And to get zero, the two numbers must be the same. This means that the path of light rays in our two-dimensional spacetime is along the line at a $45°$ angle to the two coordinate axes so that the distance in space along the path is the same as the distance in time. Since a clock taken along this path registers no passage of time, it is called the path of zero "proper" time. A "proper" measurement is one that is made moving with the thing measured – that is, the measurement is made in the "rest" frame of the thing measured.

In Fig. 1.2 we saw how vectors in different sets of coordinates preserved their lengths. You may be wondering, what happens when you rotate (and "translate" if you choose) the

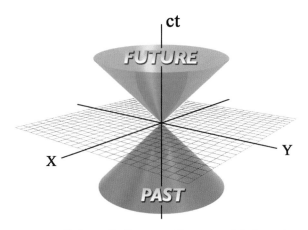

Fig. 1.4 The customary "lightcone" diagram with two spatial dimensions and one time dimension. Light rays emitted at the origin of coordinates propagate along the surface of the future lightcone; and light rays arriving at the origin propagate in the surface of the past lightcone. In the three-dimensional generalization of this diagram the lightcones become spherical surfaces converging on the origin of coordinates (*past*), or emanating away from the origin (*future*) at the speed of light

coordinates of spacetime in Fig. 1.3? The answer is that you can't do the simple sort of rotation done in Fig. 1.2, because as soon as you rotate the space and time axes as if they were Euclidean, the lengths of the spacelike and timelike component vectors for a vector lying along a light path have unequal lengths, so the difference of their squares is no longer zero – as it must be in spacetime.

Consider an observer who is moving with respect to an observer at rest at the origin of the black coordinates in Fig. 1.3 at the instant when the moving observer passes the one at rest. They both see the same path for a light ray – the path at a 45° angle to the black coordinates of the observer at rest. The question is: How do the space and time coordinates for the moving observer (at this instant) change while at the same time preserve the condition that the difference of the squares of the projections of any part of the path of the light ray on his/her coordinates is zero? The only possibility is that the moving observers coordinates look like the red coordinates in Fig. 1.3. That is the effect of the pseudo-Euclidean nature of relativistic spacetime.

Oh, and a word about "lightcones." If we imagine our two-dimensional spacetime to now have another spacelike dimension, we can rotate our 45° light path around the timelike axis, creating a conical surface in which light that passes through the origin of coordinates propagates. That surface is the future lightcone of the event at the origin of coordinates if it lies in the direction of positive time. The lightcone that lies in the direction of negative time is the past lightcone of the event at the origin of coordinates. Events that lie within the past and future lightcones of the event at the origin of coordinates can communicate with the event at the origin at sublight speeds. Those that lie outside the lightcones cannot. These are said to be "spacelike" separated from the origin of coordinates (Fig. 1.4).

The principle of relativity forces us to accept that the speed of light is measured by all observers to have the same value, irrespective of their motion. And the invariance of the speed of light in turn forces us to accept that space and time are interconnected, and that the geometry of spacetime is pseudo-Euclidean. The question then is: What does this do to Newtonian mechanics? Well, not too much. The first and third laws of mechanics aren't affected at all. Bodies in motion at constant velocity, or at rest, in inertial frames of reference not acted on by external forces keep doing the same thing (first law). And when forces act on objects, they still produce equal and opposite inertial reaction forces (third law).

Customarily, it is said that the second law is only affected in that the mass must be taken to be the "relativistic" mass (as the mass of an object, as measured by any particular observer, depends on the velocity of the object with respect to the observer). This is all well and good, but we want to take a bit closer look at the second law.

The most famous equation in all of physics that $E = mc^2$ replaced 20 or 30 years ago was $\mathbf{F} = m\mathbf{a}$, or force equals mass times acceleration – the simple version of Newton's second law. ($E = mc^2$, or energy equals mass times the square of the speed of light, is identified by Frank Wilczek as Einstein's first law, terminology we adopt.) Boldface letters, by the way, denote vectors, and normal Latin letters denote scalars. The correct, complete statement of Newton's second law is that the application of a force to a body produces changing momentum of the body in the direction of the applied force, and the rate of change of momentum depends on the magnitude of the force, or $\mathbf{F} = d\mathbf{p}/dt$.[8] (Momentum is customarily designated by the letter "p." The "operator" d/dt just means take the time rate of change of, in this case, $\mathbf{p}$).

Now, there is a very important property of physical systems implicit in Newton's second law. If there are no "external" forces, the momentum of an object (or collection of objects) doesn't change. That is, momentum is "conserved." And this is true for all observers, since the lengths of vectors in space are measured to be the same by all observers in Newtonian physics. Moreover, you can move vectors around from place to place and time to time, preserving their direction, and they don't change. (Technospeak: vectors are invariant under infinitesimal space and time translations.[9])

The question, then, is: How do we generalize this when we make the transition to spacetime required by relativity? Evidently, the three-vector momentum in absolute space

---

[8] Since $d\mathbf{p}/dt = d(m\mathbf{v})/dt = m\mathbf{a} + \mathbf{v}\,dm/dt$, we see that force is a bit more subtle than $m\mathbf{a}$. Indeed, if you aren't careful, serious mistakes are possible. Tempting as it is to explore one or two in some detail, we resist and turn to issues with greater import.

[9] When something doesn't change when it is operated upon (in this case, moved around), it is said to possess symmetry. Note that this is related to the fact that momentum is "conserved." In 1918 Emmy Noether, while working for Einstein, proved a very general and powerful theorem (now known as "Noether's theorem") showing that whenever a symmetry is present, there is an associated conservation law. Noether, as a woman, couldn't get a regular academic appointment in Germany, notwithstanding that she was a brilliant mathematician. When the faculty of Gottingen University considered her for an appointment, David Hilbert, one of the leading mathematicians of the day, chided his colleagues for their intolerance regarding Noether by allowing as how the faculty were not the members of a "bathing establishment."

must become a four-vector momentum in spacetime, and the length of the four-vector momentum in spacetime must be invariant in the absence of external forces.

In Newtonian physics the momentum of an object is defined as the product of its mass m and velocity $\mathbf{v}$, that is, $\mathbf{p} = m\mathbf{v}$. Mass, for Newton, was a measure of the "quantity of matter" of an object. In the early twentieth century, the concept of mass was expanded to encompass the notion that mass is the measure of the inertial resistance of entities to applied forces, that is, the m in $\mathbf{F} = m\mathbf{a}$, and m might include things hitherto not thought to be "matter." Mass, by the way, is also the "charge" of the gravitational field. Here, however, we are interested in the inertial aspect of mass. When we write momentum as a four-vector, the question is: What can we write for the timelike part of the four-vector that has the dimension of momentum? Well, it has to be a mass times a velocity, indeed, the fourth (timelike) component of the four-velocity times the mass. What is that fourth component of all velocities? The velocity of light, because it is the only velocity that is invariant (the same in all circumstances as measured by all observers). This makes the timelike component of the four-momentum equal to mc. The definition of the four-force, then, would seem to be the rate of change of four-momentum. What, or whose rate of change? After all, the rate of time depends on the motion of observers, and by the principle of relativity, none of them are preferred.

Well, it would seem that the only rate of time that all observers can agree upon is the rate of time in the rest frame of the object experiencing the force – that is, the "proper" time of the object acted on by the force.[10] So, the relativistic generalization of Newton's second law is: When an external force is applied to an object, the four-force is equal to the rate of change of the four-momentum with respect to the proper time of the object acted upon.

You may be wondering: What the devil happened to Mach's principle and the origin of inertia? What does all of this stuff about three- and four-vectors, forces, and momenta (and their rates of change) have to do with the origin of inertia? Well, inertia figures into momenta in the mass that multiplies the velocity. Mass, the measure of the "quantity of matter" for Newton, is the quantitative measure of the inertia of a body – its resistance to forces applied to change its state of motion. The more mass an object has, the smaller its acceleration for a given applied force. But what makes up mass? And what is its "origin"? From Einstein's first law, $E = mc^2$, we know that energy has something to do with mass. If we write Einstein's second law, $m = E/c^2$,[11] and we take note of the fact that SRT explicitly ignores gravity, then it would appear that we can define the mass of an object as the total (non-gravitational) energy of the object divided by the speed of light squared.

How does this relate to the timelike part of the four-momentum? Well, look at the timelike part of the four-momentum: mc. If you multiply this by c, you get mc^2. Dimensionally, this is an energy. And since relativistic mass depends on velocity, as the velocity of an object with some rest mass changes, its energy increases because its mass increases,

---

[10] As mentioned earlier, the term "proper" is always used when referring to a quantity measured in the instantaneous frame of rest of the object measured. The most common quantity, after time, designated as proper is mass – the restmass of an object is its proper mass.

[11] We continue to use Frank Wilczek's enumeration of Einstein's laws.

notwithstanding that c doesn't change. So, with this simple artifice we can transform the four-momentum vector into the energy-momentum four-vector.

In Newtonian physics, energy and momentum are separately conserved. In SRT it is the energy-momentum four-vector that is conserved. Einstein figured this out as an afterthought to his first work on SRT. And he didn't have the formalism and language of four-vectors to help him. That wasn't invented until a couple of years later – by one of his former teachers, Herman Minkowski.[12] Einstein had posed himself the question, "Does the inertia of a body depend on its energy content?" Indeed, that is the title of the paper that contains his second law: $m = E/c^2$. His first law doesn't even appear anywhere in the paper.

Distinguishing between Einstein's first and second laws, as they are the same equation in different arrangements, may seem a quibble to you. But as Frank Wilczek points out in his book *The Lightness of Being*, the way you put things can have profound consequences for the way you understand them. When you write $E = mc^2$, it's natural to notice that m is multiplied by, in everyday units like eith the centimeter-gram-second (cgs) or meter-kilogram-second (mks) system, an enormous number. Since m is normally taken to refer to the rest mass of an object, that means that rest mass contains an enormous amount of energy, and your thoughts turn to power plants and bombs that might be made using only a minuscule amount of mass. When you write, as Einstein did in 1905, $m = E/c^2$ completely different thoughts come to mind. Instead of ogling the enormous amount of energy present in small amounts of rest mass, you appreciate that all non-gravitational energy contributes to the inertial masses of things.

Non-gravitational? Why doesn't gravitational energy contribute to inertial mass? Well, it does. But only in special circumstances, in particular, in the form of gravity waves. There are other special circumstances where it doesn't. Gravitational potential energy due to the presence of nearby sources makes no contribution. But all these subtleties are part of GRT, and that's in the next section. For now we need only note that it is energy, not restmass alone, that is the origin of inertial mass. As Wilczek notes, more than 95% of the mass of normal matter arises from the energy contained in the rest massless gluons that bind the quarks in the neutrons and protons in the nuclei of the atoms that make it up.

A caveat should be added here. The foregoing comments about energy and mass only strictly apply to localized, isolated objects at rest in some local inertial frame of reference. The comments are also true in some other special circumstances. But in general things get more complicated when observers are moving with respect to the object observed and when other stuff is in the neighborhood that interacts with the object whose mass is being considered. Moving observers can be accommodated by stipulating that m is the relativistic mass. But nearby interacting entities can be trickier to deal with.

Summing up, the principle of relativity demands that the speed of light be "constant" so that it is impossible to identify (with local measurements) a preferred inertial frame of reference. The constancy of the speed of light leads to SRT, which in turn leads to

---

[12] Minkowski characterized Einstein the undergraduate student as a "lazy dog."

pseudo-Euclidean spacetime. When Newton's second law is put into a form that is consistent with SRT, the four-momentum (the proper rate of change of which is the four-force) multiplied by the object's four-velocity for zero spatial velocity, leads to $E = mc^2$. When this is written as Einstein's second law [$m = E/c^2$], it says that energy has inertia, in principle, even if the energy isn't associated with simple little massy particles that you can put on a balance and weigh. But there is no explanation why energy, be it massy particles or photons (particles of light) or gluons, has inertia. So we turn to general relativity theory to see if it sheds any light on the issue of the origin and nature of inertia.

GENERAL RELATIVITY

Einstein's first key insight – that the principle of relativity demanded that the speed of light be measured as the same by all observers, and that this required space and time to be conceived as spacetime – led to SRT. His second key insight – that Einstein called "the happiest thought of my life" – was his so-called "Equivalence Principle" (EP), the action of a gravity field that causes everything to "fall" in the direction of the field with the same acceleration irrespective of their masses and compositions, and this is equivalent to the behavior of everything in the absence of local gravity fields but located in an accelerating frame of reference – say, in a rocket ship accelerating in deep space.

Einstein realized that this equivalence could only be true if local inertial frames of reference – those in which Newton's first law is true – in the presence of a local concentration of matter like Earth are those that are in a state of "free fall." In Einstein's hands, the principle of relativity is extended to the case of accelerations, as the assertion that it is impossible to distinguish an acceleration in the absence of local matter concentrations from the action of a uniform (to sufficient accuracy) gravity field where one is forced out of a state of free fall without "looking out the window to see what's what."

It used to be said that accelerations could only be dealt with by employing general relativity, as SRT only dealt with relative velocities. As Misner, Thorne, and Wheeler made plain in Chap. 6 of their classic text on *Gravitation* (Freeman, San Francisco, 1973), accelerations are routinely dealt with using the techniques of SRT. The reason why general relativity is required is not to deal with accelerations. It is to deal with the geometry of spacetime in the presence of gravitating stuff – matter and energy and stress and mass currents. For the EP to be true, since for example stuff on all sides of Earth free fall toward its center, it must be the case that local concentrations of matter distort the geometry of spacetime rather than produce forces on objects in their vicinity. This eventually led Einstein to his General Relativity Theory where, in the words of John Wheeler, "spacetime tells matter how to move, and matter tells spacetime how to curve." "Matter," with its property of inertia and as the "charge" (or source) of gravity, does not simply produce a field in spacetime; the field is the distortion of spacetime itself. This is why GRT is called a "background independent" theory of gravity. It is this fact – that the field is not something in spacetime, but rather the distortion of spacetime itself – that makes possible the wormholes and warp drives that enable serious rapid spacetime "transport" – if we can figure out how to build them.

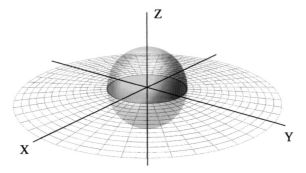

Fig. 1.5 Here we have a spherical massive object centered at the origin of coordinates in real three-dimensional space. The X, Y plane is shaded where it passes through the object. The mass of the object distorts the X, Y plane according to GRT as shown in the next figure

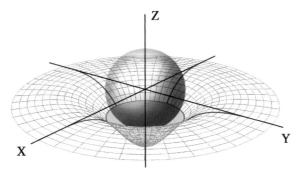

Fig. 1.6 The hyperspace diagram for the X, Y plane of the previous figure. The Z dimension here is now a hyperspace dimension that enables us to show how the X, Y surface in the previous figure is distorted by the presence of the spherical object

The customary visual rendition that is intended to show how this works is a "hyperspace embedding diagram." Consider the case of a simple spherical source of gravity, say, a star that is not changing in time. The warping that this source effects on space is the stretching of space in the radial direction. Space perpendicular to the radial direction is unaffected by the presence of the star. To show this, we consider a two-dimensional plane section through the center of the star, as shown in Fig. 1.5. We now use the third dimension, freed up by restricting consideration to the two-dimensional plane in Fig. 1.5, as a hyperspatial dimension – that is, a dimension that is not a real physical dimension – that allows us to show the distortion of the two-dimensional plane through the center of the star. This is illustrated in Fig. 1.6. Note that the radial stretching that distorts the two-dimensional plane through the center of the star has no affect at all on the circumferences of the circles in the plane centered on the center of the star. If the mass of the star is compacted into a smaller and smaller sphere, the radial distortion near its surface becomes larger and larger. If the compaction continues long enough, eventually the spacetime distortion in the radial direction becomes so large that nothing traveling in

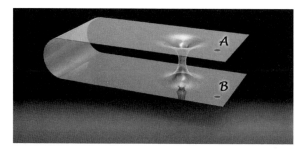

Fig. 1.7 One classic embedding diagram of a wormhole connecting two points *A* and *B*, which are distant in the normal two-dimensional space with the fold. But through the wormhole, they are much closer

the radial direction can make it past a circumferential surface called an "event horizon," and we have a "black hole."

When a massive object undergoes "gravitational collapse" and becomes a black hole, in the formation process a "wormhole" is fleetingly formed. This wormhole connects the spacetime of our universe with another universe where the space and time dimensions are interchanged. Such wormholes formed with positive mass only exist for an instant before they are "pinched" off and become a "singularity." Their existence is so brief that not even light can get through them before pinch-off takes place. Kip Thorne and his colleagues found in the 1980s that in order to keep a wormhole "throat" open, it must be stabilized by a very large amount of "exotic" matter – matter with negative mass, that is.

As most everyone now knows, wormholes are shortcuts through hyperspace between two locations in spacetime separated by arbitrarily long distances through normal spacetime. The now-famous embedding diagram of a wormhole is shown in Fig. 1.7. A simple line-drawing version of this diagram graced the pages of the Misner, Thorne, and Wheeler classic text on gravity, *Gravitation*. Indeed, Fig. 1.7 is a shaded version of their classic embedding diagram.

The length of the throat of the wormhole is exaggerated in this diagram. But it conveys the point: the distance through the wormhole is much shorter than the distance through the two-dimensional surface that represents normal spacetime. This exaggeration is especially pronounced in the case of an "absurdly benign" wormhole – a wormhole with a throat only a few meters long and with all of the flarings at each end of the throat restricted to at most a meter or two. A rendition of the appearance of such a wormhole is shown in Fig. 1.8.

Note that the wormhole here is a four-dimensional sphere, so the appearance of circularity is deceptive. It should be noted that the wormhole can connect both spatially and temporally distant events. That is, wormholes can be designed to be time machines connecting distant events in both the future and past.

Famously, Stephen Hawking has argued that the actual construction of such wormholes is prohibited by the laws of physics. But not all physicists share his conviction in this. Hawking's argument depends on making time machines with only smooth deformations of spacetime. That is, tearing spacetime to connect two distant events (and causing "topology change") is prohibited. With only smooth deformations allowed, you always end up at some point creating a "closed timelike curve" (CTC); and if even one measly photon starts

Fig. 1.8 This figure illustrates an absurdly benign wormhole connecting Earth and Jupiter's moon Titan as a commercial transit device

endlessly circulating along the CTC, its multiple copies build up infinite energy instantly in the wormhole throat, blowing it up. If we can tear spacetime, we should be able to avoid this disastrous situation, one would think! But back to the EP.

I'll bet you won't be surprised to learn that the EP has been a source of criticism and debate since Einstein introduced it and made it one of the cornerstones of GRT. The feature of the EP that many critics dislike is that it permits the local elimination of the local gravitational field by a simple transformation to a suitably chosen accelerating frame of reference, or, equivalently, *to a suitably chosen spacetime geometry*. This is only possible because everything responds to a gravitational field the same way. We know that Newtonian gravity displays this same characteristic (that Galileo discovered), and we can use Newtonian gravity to illustrate this point. Consider an object with (passive gravitational) mass m acted upon by Earth's gravity field. The force exerted by Earth is just:

$$\mathbf{F} = \frac{GMm}{R^3}\mathbf{R},$$

where M and R are the mass and radius of Earth. By Newton's second law, $\mathbf{F}$ is also equal to $m\mathbf{a}$, so:

$$\mathbf{F} = \frac{GMm}{R^3}\mathbf{R} = m\mathbf{a},$$

and since $\mathbf{R}$ and $\mathbf{a}$ point in the same direction, we can drop the vector notation and canceling m and write:

$$\frac{GM}{R^2} = a.$$

Note that this is true regardless of the value of m, and this is only possible if the Equivalence Principle is correct. That is, the passive gravitational mass that figures into the gravitational force and the inertial mass that figures into Newton's second law, in principle at least, need not necessarily be the same. Only if they are the same does the cancellation that shows all gravitational accelerations of various bodies to be the same carry through.

Gravity is the only interaction or force of the so-called four forces that satisfies the Equivalence Principle, making gravitation the unique universal interaction. Of the known interactions – gravity, electromagnetism, and the strong and weak forces – it is the only interaction that can be eliminated locally by a suitable choice of geometry. The local elimination of "apparent" forces is also possible for "fictitious" forces: "forces" that appear because of an infelicitous choice of coordinate frame of reference (for example, Coriolis forces), and all sorts of inertial reaction forces. (See Adler, Bazin and Schiffer's outstanding discussion of fictitious forces on pages 57 through 59 of their text, Introduction to General Relativity.) None of the other forces conveyed by fields has this property. They are all mediated by the exchange of "transfer" particles – photons, gluons, and the like – that pass through spacetime to convey forces. However, gravity can be accounted for by the warping of spacetime itself. If you are determined to believe that gravity is just another force, like all the others, you will likely want it to be mediated by "gravitons" that pass through spacetime. But no one has ever seen a graviton.

In no small part, much of the distaste for Einstein's version of the EP, the so-called Einstein Equivalence Principle (EEP), stems from the fact that it forbids the "localization" of gravitational energy (or, strictly speaking, energy-momentum). Gravity waves, considered over several wavelengths, are an exception to this prohibition. But that doesn't change the prohibition significantly. If a gravity field can be eliminated by a transformation to a suitably chosen accelerating frame of reference, or equivalently a suitable choice of geometry, then no energy can be associated with it locally, for in a sense it isn't really there in the first place.

If accelerations, in themselves, conferred energy on objects being accelerated, the situation might be different. Why? Because then they might produce energy equivalent to that which would be produced by the action of a gravity field on its sources – local gravitational potential energy. But accelerations per se don't produce energy in this way. Accelerations are related to changes in motion and resulting changes in the energies of objects.[13] But applying a force to an object in, say, Earth's gravity field to keep it from engaging in free-fall, a steady force of one "g" does not change the energy of the object, at least after stationary conditions have been achieved. So, for gravity to mimic accelerations as the EP stipulates, localization of gravitational (potential) energy must be forbidden. (See the Addendum to this chapter for Misner's, Thorne's, and Wheeler's comments in their book *Gravitation* on localization of gravitational energy and the EP.)

Einstein's critics based their attacks on the fact that the EP is only, strictly speaking, true for a "uniform" gravitational field – that is, the gravitational field that would be produced by a plane, semi-infinite mass distribution, something that cannot exist in reality.

---

[13] Changes in internal energies of accelerating objects may take place if the objects are extended and not rigid. As we will see later, this complication leads to the prediction of interesting transient effects.

For any realistic mass distribution, the field is not uniform, and the non-uniformity means that the EP is only approximately true in very small regions of spacetime. Indeed, they argue that no matter how small the region of spacetime under consideration is, "tidal" gravitational effects will be present and, in principle at least, measurable.

Tidal effects, of course, are a consequence of the non-uniformity of the field, so arguably their presence in real systems cannot be said to invalidate the EP. But that's not what the critics are after. What they want is to assert that in reality, gravitation is just another field, like all others.[14] Had his critics been successful, of course, Einstein's accomplishment would have been measurably diminished. Einstein stuck to his guns. An ideal uniform gravity field might be an unobtainable fiction in our reality, but it was clear to him that the E(E)P was correct notwithstanding the arguments of his critics. It is worth noting here that "idealization" has been used successfully in physics for hundreds of years to identify fundamental physical principles.

There is a very important point to be noted here. No matter how extreme the local distortion of spacetime produced by a local concentration of matter might be, the "constancy" of the speed of light at every point in spacetime remains true. That is, SRT is true at every point in spacetime in GRT. The extent of the spacetime around any given point where SRT is true may be infinitesimally small, but it is never of exactly zero extent. While SRT is true at every point – or, correctly "event" – in spacetime in GRT, the speed of light in GRT is no longer a "constant." That is, not all observers get the same number for the speed of light in a vacuum. All *local* observers still get the same number. But when distant observers measure the speed of light near a large local matter concentration, the speed they measure is less than the speed measured by the local observers. (Technically, distant observers measure the "coordinate" speed of light. It is not constant. The coordinate speed of light depends on the presence of gravity fields, and via the EEP, the coordinate speed of light for accelerating observers depends on their acceleration.)

This curious feature of GRT is especially obvious in the case of light from a distant source as it passes in the vicinity of a massive star. The light is deflected by the star, which warps the spacetime in its neighborhood. But a distant observer doesn't see the spacetime around the star warp. After all, empty spacetime is perfectly transparent. What is seen is the light moving along a path that appears curved, a path that results from the light appearing to slow down the closer it gets to the star.

If we are talking about some object with rest mass – mass you can measure on a balance in its proper frame of reference – the path followed is also curved, though a bit differently as

---

[14] Einstein's critics, it appears, have been quite happy to use unrealizable conditions when it suited their purposes in other situations. For example, they are quite content to assume that spacetime is Minkowskian at "asymptotic" infinity, or that spacetime in the absence of "matter" is globally Minkowskian. Actually, neither of these conditions can be realized. Their assumption is the merest speculation. Just because you can write down equations that model such conditions does not mean that reality actually is, or would be, that way. What we do know is that at cosmic scale, spacetime is spatially flat. And that condition corresponds to a mean "matter" density that, while small, is not zero. In fact, in Friedmann-Robertson-Walker cosmologies (which are homogeneous and isotropic) spatial flatness results from the presence of "crititcal" cosmic "matter" (everything that gravitates) density – about 2×10^{-29} g/cm^3. That's about one electron per 50 cm^3. Not very much stuff, to say the least.

objects with finite rest mass cannot reach the speed of light. These free-fall paths have a name: geodesics. They are found by solving Einstein's field equations of GRT for the particular distribution of sources of the local gravitational field. Because gravity warps spacetime so that things you measure depend on the direction in which you measure them, it turns out to be a tensor field – a more complicated thing than a scalar or vector field. So that you can see the formalism of GRT, now is the place to write down Einstein's field equations. They are:

$$G_{\mu\nu} = R_{\mu\nu} - \frac{1}{2} g_{\mu\nu} R = \frac{8\pi G}{c^4} T_{\mu\nu}.$$

$G_{\mu\nu}$ is the Einstein tensor (with indexes μ and ν that take on the values 1–4 of the dimensions of spacetime). $G_{\mu\nu}$ is constructed from the Ricci tensor, $R_{\mu\nu}$, a "contracted" form of the fourth rank Riemann curvature tensor, the "metric" $g_{\mu\nu}$ and the scalar curvature R. The Einstein tensor is equal to the coupling constant $8\pi G/c^4$, an exceedingly small number, times the "stress-energy-momentum" tensor $T_{\mu\nu}$. If you read the excerpt about energy localization in the gravity field, you'll know that $T_{\mu\nu}$ must be supplemented by a non-tensor quantity customarily identified as the pseudo-tensor of the gravitational field to make things work out right. Put in very general, schematic form, Einstein's field equations say:

GEOMETRY = MATTER SOURCES HERE AND THERE times a very small number.

For our purposes, we do not need to be worried about how all of this came to be. We need only note that GRT is to the best of everyone's knowledge the correct theory of gravity and inertia. Happily, tensor gravity has the property of symmetry, so several of the field components are paired, and only ten components have independent values. To find ten components you need ten equations, which is messier than scalar or vector theories. Often, however, it is possible to simplify things either by choosing simple circumstances, or by making simplifying approximations, to reduce the messiness. If this can't be done, and an analytic solution can't be found, there are always numerical techniques available. Numerical relativity nowadays is a major sub-field in its own right.

The most famous prediction of GRT is that of "black holes," or as they were known before John Wheeler gave them their catchy name, "frozen stars." These objects have all of their masses enclosed by their "event horizons." For a simple non-rotating spherical star, the radius of the event horizon, also sometimes called the "gravitational radius," is given by $R = 2GM/c^2$, where G is Newton's universal constant of gravitation, M the mass of the star, and c the speed of light in a vacuum.

As most everyone now knows, the event horizon of a black hole is a surface of "no return." Should you have the misfortune to fall to the event horizon, you will inexorably be sucked into the hole – and spagettified by tidal forces, too, as you approach the singularity at the center of the hole, where space and time cease to exist. Books have been written and movies made about black holes and the exploits of those in their vicinities. There is an important point about black holes, however, that sometimes doesn't get recognized. For distant observers, time stops at the event horizon.

So what? Well, this means that for us distant observers, we can never see anything fall into a black hole. Everything that has ever fallen toward a black hole, for us, just seems to pile up at the surface that is the event horizon. It never falls through. That's why, pre-Wheeler, they were called frozen stars. But what about observers who fall into a black hole? Time doesn't stop for them, does it? No, it doesn't. Indeed, you fall through the event horizon as if there were nothing there at all to stop you.

How can both stories be right? Well, as you fall towards the hole, the rate of time you detect for distant observers out in the universe far from the hole speeds up. And at the instant that you reach the event horizon, the rate of distant time becomes infinite and the whole history of the universe – whatever it may be – passes in that instant. So, an instant later, when you are inside the event horizon, the exterior universe is gone. Even if you could go back (you can't), there is no back to go to. For our purposes here, though, what is important is that for distant observers such as us, the measured speed of light at the event horizon is zero, because, for us, time stops there.

This is true if the mass of the black hole is positive. Should the mass of the hole be negative, however, time at the gravitational radius measured by distant observers would speed up. Indeed, at the gravitational radius, it would be infinitely fast. This means that if the mass of the hole is "exotic," stuff near the gravitational radius can appear to us to travel much, much faster than the speed of light. This odd behavior in the vicinity of negative mass stars (should they even exist) doesn't have much direct value for rapid spacetime transport. After all, you wouldn't want to hang around an exotic black hole so that you could age greatly before returning to much the same time as you left. But it is crucial to the nature of matter as it bears on the construction of stargates.

If the "bare" masses of elementary particles are exotic, they can appear to spin with surface velocities far in excess of the speed of light. And if a way can be found, using only "low" energy electromagnetic fields, to expose those bare masses, stargates may lie in our future. How all this works is dealt with in the last section of this book.

Now, all of this is very interesting, indeed, and in some cases downright weird. We normally don't think of space and time as deformable entities. Nonetheless, they are. And the drama of reality plays itself out in space and time that are locally uninfluenced, beyond the effects predicted by GRT, by the action taking place within them. The thing that distorts space and time, or more accurately, spacetime in GRT, is mass-energy. How the distortion occurs can be constructed with the Equivalence Principle and the principle of general covariance.

The principle of general covariance is the proposition that all physical laws should have the same form in all frames of reference – inertial or accelerated. Einstein noted early on that he was not happy about this as he thought the distribution of matter and its motions throughout the universe should account for inertia and thus be essential to a correct description of reality. The physical reason why this *must* be the case rests on Mach's principle, as Einstein suspected before he gave the principle its name in 1918. How this works involves subtleties that have made Mach's principle a topic of contention and confusion literally from the time Einstein introduced it to the present day.

Before turning to Mach's principle per se, a short digression on "fictitious forces," discussed by Adler, Bazin, and Schiffer. By definition, a "fictitious" force is one that can be eliminated by suitable choice of geometry (and this only works in the four-dimensional spacetime of relativity theory, not in the three absolute spatial dimensions of Newtonian physics). Three types of forces can be handled in this way: gravity, inertia, and "Coriolis" forces. The reason why is that they all, in a sense at least, satisfy the EP. They are "universal" in the sense that the masses of test objects and the like drop out of the equations of motion – as in the case of Newton's second law (equation of motion) and Newtonian gravity in the example worked out above – for all objects in a local region.

It has become customary in the relativity community to take this "fictitious" business more seriously than it deserves. It is the reason why people go around saying that there are no gravitational forces, and ignore inertial forces as meriting serious investigation and explanation. We'll be looking at this issue in some detail in the next chapter. But here it should be noted that some fictitious forces really are fictitious, and others are quite real.

Inertial forces – those that produce "frame dragging" in the modern canon – are quite real. When "frames" get "dragged" by the inductive effects of the motion of matter, both global and local, if you want your test object not to get "dragged" with the spacetime, you have to exert a force to keep that from happening. That force is the inertial reaction force to the action, the force of gravity disguised as frame dragging – and it is very real, not "fictitious."

There's a reason why both of these forces are "fictitious" in the universality sense. They are the same force. Before Einstein and Mach's principle, this did not seem so, for (local) gravity forces are ridiculously weak by comparison with, say, electromagnetic forces. Even after Einstein, most people couldn't take seriously that the two forces, gravity and inertial forces, could be one and the same. For example, electromagnetic forces induce inertial reaction forces as large as they are. And the electromagnetic force and its equal and opposite inertial reaction force are gigantic by comparison with gravitational forces produced by modest amounts of matter. How could the incredibly weak gravity force do that? Why, it might take the gravity action of the entire universe to produce a gravity force that big!

The fact of the matter is that notwithstanding their "fictitious" character, inertial forces don't get treated geometrically, whereas gravity does. The reason why is simple. Gravity is chiefly a phenomenon of astronomical-sized objects – planets, stars, galaxies, and the universe itself. It acts on everything in the vicinity of these types of objects – all the time. The visualization of this is the rubber sheet analogy often used in popularizations of GRT.

Inertial reaction forces, however, are not produced by planets, stars, and galaxies. It may be that they are produced by the universe, but they are normally inert – that is, dead. They only show up when a non-gravitational force acts on some object to accelerate it. These forces are, in a sense, episodic. They don't act on everything in some reasonably large region of spacetime all the time like gravity as we normally conceive it does. So the rubber sheet analogy simply doesn't work. Without a simple, easy to envision analogy like the rubber sheet, it's easy to ignore the possibility that inertial forces may be nothing more than gravity in a different guise.

A class of "fictitious" forces that really are fictitious exists: Coriolis forces. Coriolis forces are not real forces; they are "apparent" forces. Why? Because they appear to observers in accelerating (rotating) reference frames to act on test particles and other local objects that are in inertial (force-free) motion. The non-inertial motion of the observer causes them to appear. Since they satisfy the universality condition, though, they can be eliminated by suitable choice of geometry. But do not be fooled; they are not the same as gravity and inertia.

We now turn to Mach's principle, critically important to the making of starships and stargates.

ADDENDUM

The Misner, Thorne, and Wheeler discussion of localization of gravitational energy in their comprehensive textbook, Gravitation:

§20.4.   WHY THE ENERGY OF THE GRAVITATIONAL FIELD CANNOT BE LOCALIZED

Consider an element of 3-volume $d\Sigma_\nu$ and evaluate the contribution of the "gravitational field" in that element of 3-volume to the energy-momentum 4-vector, using in the calculation either the pseudotensor $t^{\mu\nu}$ or the pseudotensor $t^{\mu\nu}_{\text{L-L}}$ discussed in the last section. Thereby obtain

$$\boldsymbol{p} = \boldsymbol{e}_\mu t^{\mu\nu} d\Sigma_\nu$$

or

$$\boldsymbol{p} = \boldsymbol{e}_\mu t^{\mu\nu}_{\text{L-L}} d\Sigma_\nu.$$

Right? No, the question is wrong. The motivation is wrong. The result is wrong. The idea is wrong.

To ask for the amount of electromagnetic energy and momentum in an element of 3-volume makes sense. First, there is one and only one formula for this quantity. Second, and more important, this energy-momentum in principle "has weight." It curves space. It serves as a source term on the righthand side of Einstein's field equations. It produces a relative geodesic deviation of two nearby world lines that pass through the region of space in question. It is observable. Not one of these properties does "local gravitational energy-momentum" possess. There is no unique formula for it, but a multitude of quite distinct formulas. The two cited are only two among an infinity. Moreover, "local gravitational energy-momentum" has no weight. It does not curve space. It does not serve as a source term on the righthand side of Einstein's field equations. It does not produce any relative geodesic deviation of two nearby world lines that pass through the region of space in question. It is not observable.

Anybody who looks for a magic formula for "local gravitational energy-momentum" is looking for the right answer to the wrong question. Unhappily, enormous time and effort were devoted in the past to trying to "answer this question" before investigators realized the futility of the enterprise. Toward the end, above all mathematical arguments, one came to appreciate the quiet but rock-like strength of Einstein's equivalence principle. One can always find in any given locality a frame of reference in which all local "gravitational fields" (all Christoffel symbols; all $\Gamma^\alpha_{\mu\nu}$) disappear. No Γ's means no "gravitational field" and no local gravitational field means no "local gravitational energy-momentum."

Nobody can deny or wants to deny that gravitational forces make a contribution to the mass-energy of a gravitationally interacting system. The mass-energy of the Earth-moon system is less than the mass-energy that the system would have if the two objects were at infinite separation. The mass-energy of a neutron star is less than the mass-energy of the same number of baryons at infinite separation. Surrounding a region of empty space where there is a concentration of gravitational waves, there is a net attraction, betokening a positive net mass-energy in that region of space (see Chapter 35). At issue is not the existence of gravitational energy, but the localizability of gravitational energy. It is not localizable. The equivalence principle forbids.

Look at an old-fashioned potato, replete with warts and bumps. With an orange marking pen, mark on it a "North Pole" and an "equator". The length of the equator is very far from being equal to 2π times the distance from the North Pole to the equator. The explanation, "curvature," is simple, just as the explanation, "gravitation", for the deficit in mass of the earth-moon system (or deficit for the neutron star, or surplus for the region of space occupied by the gravitational waves) is simple. Yet it is not possible to ascribe the deficit in the length of the equator in the one case, or in mass in the other case, in any uniquely right way to different elements of the manifold (2-dimensional in the one case, 3-dimensional in the other). Look at a small region on the surface of the potato. The geometry there is locally flat. Look at any small region of space in any of the three gravitating systems. In an appropriate coordinate system it is free of gravitational field. The over-all effect one is looking at is a global effect, not a local effect. That is what the mathematics cries out. That is the lesson of the nonuniqueness of the $t^{\mu\nu}$!

2

Mach's Principle

So strongly did Einstein believe at that time in the relativity of inertia that in 1918 he stated as being on an equal footing three principles on which a satisfactory theory of gravitation should rest:

1. The principle of relativity as expressed by general covariance.
2. The principle of equivalence.
3. Mach's principle (the first time this term entered the literature):...that the $g_{\mu\nu}$ are completely determined by the mass of bodies, more generally by $T_{\mu\nu}$.

In 1922, Einstein noted that others were satisfied to proceed without this [third] criterion and added, "This contentedness will appear incomprehensible to a later generation however."

....It must be said that, as far as I can see, to this day Mach's principle has not brought physics decisively farther. It must also be said that the origin of inertia is and remains *the* most obscure subject in the theory of particles and fields. Mach's principle may therefore have a future – but not without the quantum theory.

–Abraham Pais, *Subtle is the Lord: the Science and the Life of Albert Einstein*, pp. 287–288.
(Quoted by permission of Oxford University Press, Oxford, 1982)

BACKGROUND

Recapitulating, we have seen that when the implications of the principle of relativity for space and time were understood in the early twentieth century, Einstein quickly apprehended that the quantity of interest in the matter of inertia was not (rest) mass *per se*, rather it was the total non-gravitational energy contained in an object (isolated and at rest). This followed from Einstein's second law, which says:

$$m = \frac{E}{c^2}, \tag{2.1}$$

where m is now understood as the total inertial mass, not just the rest mass of an object, and E is the total non-gravitational energy. If one restricts oneself to Special Relativity Theory (SRT), this is about all one can say about inertial mass. It was Einstein's hope that he could go farther in identifying the origin of inertia in General Relativity Theory (GRT), as is evident in the quote from Pais's biography of Einstein above.

J.F. Woodward, *Making Starships and Stargates: The Science of Interstellar Transport and Absurdly Benign Wormholes*, Springer Praxis Books, DOI 10.1007/978-1-4614-5623-0_2,
© James F. Woodward 2013

As we have seen in the previous chapter, Einstein didn't need "Mach's principle" to create GRT. Shortly after publishing his first papers on GRT, he did try to incorporate the principle into his theory. He did this by adding the now famous "cosmological constant" term to his field equations. Those equations, as noted in Chapter 1, without the cosmological constant term, are:

$$G_{\mu\nu} = R_{\mu\nu} - \frac{1}{2} g_{\mu\nu} = -\frac{8\pi G}{c^4} T_{\mu\nu}, \tag{2.2}$$

where $G_{\mu\nu}$ is the Einstein tensor that embodies the geometry of spacetime, $R_{\mu\nu}$ is the "contracted" Ricci tensor (obtained by "contraction" from the Riemann curvature tensor which has four "indexes," each of which can take on values 1–4 for the four dimensions of spacetime), $g_{\mu\nu}$ is the "metric" of spacetime, and $T_{\mu\nu}$ is the "stress-energy-momentum" tensor, that is, the sources of the gravitational field. The cosmological term gets added to $G_{\mu\nu}$. That is, $G_{\mu\nu} \rightarrow G_{\mu\nu} + \lambda g_{\mu\nu}$, where λ is the so-called cosmological constant. We need not worry about the details of these tensor equations. But it's worth remarking here that the coefficient of $T_{\mu\nu}$, with factors of Newton's constant of gravitation G in the numerator and the speed of light c to the fourth power in the denominator, is exceedingly small. This means that the sources of the field must be enormous to produce even modest bending of spacetime. That is why a Jupiter mass of exotic matter is required to make wormholes and warp drives.

Ostensibly, Einstein added the cosmological constant term to make static cosmological solutions possible by including a long-range repulsive force. But he also hoped that the inclusion of the cosmological constant term would render his field equations solutionless in the absence of matter. Willem deSitter quickly showed that Einstein's new equations had an expanding, asymptotically empty solution, one with full inertial structure. And a vacuum solution, too. So Einstein's attempt to include Mach's principle in this way was deemed a failure.

The chief reason for his failure seems to have been the way he defined the principle: that the inertial properties of objects in spacetime should be defined (or caused) by the distribution of matter (and its motions) in the universe. Put a little differently, Einstein wanted the sources of the gravitational field at the global scale to determine the inertia of local objects. He called this "the relativity of inertia." The problem Einstein encountered was that his GRT is a local field theory (like all other field theories), and the field equations of GRT admit global solutions that simply do not satisfy any reasonable formulation of, as he called it, Mach's principle. Even the addition of the "cosmological constant" term to his field equations didn't suffice to suppress the non-Machian solutions.

Alexander Friedmann and Georges Lemaitre worked out cosmological solutions for Einstein's field equations in the 1920s, but cosmology didn't really take off until Edwin Hubble, very late in the decade, showed that almost all galaxies were receding from Earth. Moreover, they obeyed a velocity-distance relationship that suggested that the universe is expanding. From the 1930s onward work on cosmology has progressed more or less steadily. The cosmological models initiated by Friedman, predicated on the homogeneity and isotropy of matter at the cosmic scale, were developed quickly by Robertson and Walker. So now cosmological models with homogeneity and isotropy are called

Friedmann, Robertson, Walker (FRW) cosmologies. One of them is of particular interest: the model wherein space is flat at cosmological scale.

Spatial flatness corresponds to "critical" cosmic matter density – 2×10^{-29} g per cubic centimeter – and has the unfortunate tendency to be unstable. Small deviations from this density lead to rapid evolution away from flatness. Since flatness is the observed fact of our experience and the universe is more than 10 billion years old, how we could be in a spatially flat universe so long after the primeval fireball was considered something of a problem. The advent of "inflationary" cosmologies 20 or so years ago is widely thought to have solved this problem. As we will see shortly, spatial flatness and critical cosmic matter density figure into the answer to the question of the origin of inertia. But we are getting ahead of the story.

MACH'S PRINCIPLE

As the Equivalence Principle makes clear, gravity defines local inertial frames of reference as those in a state of free fall in the vicinity of a local concentration of matter. Moreover, gravity is the only truly "universal" interaction in that gravity acts on everything. For these reasons Einstein was convinced that GRT should also account for inertial phenomena, for inertia, like gravity, is a universal property of matter, though it is normally "inert." (Good historical articles on his attempts to incorporate Mach's principle in GRT can be found in: *Mach's Principle: From Newton's Bucket to Quantum Gravity*, Brikhauser, Boston, 1995, edited by Julian Barbour and Herbert Pfister.)

Notwithstanding that Willem deSitter shot down his early efforts to build Mach's principle into GRT by adding the "cosmological constant" term to his field equations, Einstein persisted. When he gave a series of lectures on GRT at Princeton in 1921, he included extended remarks on the principle and the issue of inertia in GRT. (These remarks can be found in *The Meaning of Relativity*, 5th ed., Princeton University Press, Princeton, 1955, pp. 99–108). In his words:

> [T]he theory of relativity makes it appear probable that Mach was on the right road in his thought that inertia depends upon a mutual action of matter. For we shall show in the following that, according to our equations, inert masses do act upon each other in the sense of the relativity of inertia, even if only very feebly. What is to be expected along the line of Mach's thought?
>
> 1. The inertia of a body must increase when ponderable masses are piled up in its neighborhood.
> 2. A body must experience an accelerating force when neighbouring masses are accelerated, and, in fact, the force must be in the same direction as that acceleration.
> 3. A rotating hollow body must generate inside of itself a "Coriolis field," which deflects moving bodies in the sense of the rotation, and a radial centrifugal field as well.
>
> We shall now show that these three effects, which are to be expected in accordance with Mach's ideas, are actually present according to our theory, although their magnitude is so small that confirmation of them by laboratory experiments is not to be thought of. . . .

The first of Einstein's criteria is the idea that when "spectator" matter is present in the vicinity of some massive object, the spectator matter should change the gravitational potential energy of the object. And since $E = mc^2$, that gravitational potential energy should contribute to E and change the mass of the object.

It turns out that Einstein was wrong about this. Only non-gravitational energies contribute to E *when it is measured locally*. But the reason why E, locally measured, doesn't include gravity involves a subtlety about the nature of gravity and inertia that is easily missed. The second criterion is the prediction of, as it is now known, "linear accelerative frame dragging," though Einstein states it as the production of a force by the accelerating spectator matter on the body in question, rather than the dragging of local spacetime by the accelerating matter. This, when the action of the universe is considered, turns out to be the nub of Mach's principle. If the universe is accelerated in any direction, it rigidly drags inertial frames of reference along with it in the direction of the acceleration. Consequently, only accelerations relative to the universe are detectable; and inertia is "relative."

Einstein didn't consider the cosmological consequences of this term. But he showed that this term and its effects depends on gravity being at least a vector field theory (analogous to Maxwell's theory of electrodynamics). The effect is not to be found in Newtonian gravity, a scalar field theory (as the field equation can be written in terms of a scalar "potential" alone with the direction and magnitude of gravitational forces recovered using the "gradient operator"). The third criterion is just the Lens-Thirring effect and Gravity Probe B prediction.[1]

Solving the full tensor field equations of GRT exactly is notoriously difficult, so Einstein did a calculation in the "weak field" approximation (where the metric tensor $g_{\mu\nu}$ is approximated by $\eta_{\mu\nu} + h_{\mu\nu}$ with $\eta_{\mu\nu}$ the Minkowski tensor of the flat spacetime of SRT and $h_{\mu\nu}$ the tensor that represents the field) and put his results into vector formalism. Suffice it to say, he found results that seemed to support each of his three criteria. (The formal predictions can be found in an excerpt from a paper by Carl Brans on the localization of gravitational energy at the end of this chapter.) His predicted effects are indeed very small when one considers even quite large local concentrations of matter (other than black holes in the vicinity of event horizons, of course).

Why didn't Einstein see that the sort of force that, because of the universality of gravity, is equivalent to frame dragging in his second prediction could explain Mach's principle? At least part of the problem here seems to be that he wasn't thinking cosmologically when looking for predicted quantitative effects – and so little was understood about the structure and size of the universe in the 1920s that there was no plausible basis, other than the most general sorts of considerations, to make inferences about the action of cosmic matter on local objects.

Shortly after Einstein gave his Princeton lectures, he found out, through posthumously reported remarks made by Mach shortly before his death in 1916, that Mach had disavowed any association with Einstein's ideas on relativity and inertia.

---

[1] Initially conceived of by George Pugh and Leonard Schiff in the 1960s, Gravity Probe B was a collection of high precision gyroscopes flown in a satellite in polar orbit intended to detect the dragging of spacetime caused by the rotation of Earth. The project, which flew several years ago, spanned decades and cost nearly a billion dollars. One noted relativist, queried by the press on the launch of the satellite, was reported to have remarked, "never was so much spent to learn so little." The history of this project is yet to be written. But it will doubtless prove fascinating.

Einstein, not long thereafter, asserted that any correct cosmological model should be spatially closed so that its geometry (the left hand side of his field equations) would be completely determined by its sources (the right hand side of his field equations) without the stipulations of additional boundary conditions and abandoned further work on Mach's principle.

If you are an expert, you may also be thinking, Einstein's calculation was done in the weak field approximation where gravitational effects are small. In cosmological circumstances one can expect gravitational potentials to be very large; indeed, even as large as the square of the speed of light – as is the case near the event horizon of a black hole. Well yes. But the universe isn't like the region of spacetime near to the event horizon of a stellar mass black hole. The sort of curvature encountered there is simply absent in the universe considered at cosmic scale. At cosmic scale, the universe is spatially flat. And absent local concentrations of matter, spacetime looks Minkowskian, notwithstanding that the gravitational potential approaches the square of the speed of light. So using the weak field approximation to compute lowest order gravimagnetic effects is perfectly okay.

THE MACH'S PRINCIPLE REVIVAL

By the early 1950s, the cosmological situation had changed. Significant theoretical work on cosmology had taken place, for example, that of Roberston and Walker in the 1930s and 1940s. Thomas Gold, Herman Bondi, and Fred Hoyle had proposed "steady state" cosmology, and Walter Baade had shown that there were two populations of stars, dramatically increasing the age of the universe for FRW cosmological models. So when Dennis Sciama, one of the very few doctoral students trained by Paul Dirac, came along in the early 1950s, tackling the "problem of the origin of inertia" seemed a reasonable thing to do.

Sciama's approach was to ignore GRT and write down a vector theory of gravity analogous to Maxwell's theory of electrodynamics. He initially thought his vector theory different from GRT. But eventually it was found to be just an approximation to GRT. This, by the way, is an exceedingly important point. Sciama's calculations are *not* optional. They are the exact predictions of GRT when conditions make the vector approximation valid and the idealizations he adopted reasonable.

What Sciama noticed was that when you write out the equation for the gravity field that is the analog of the electric field in electrodynamics, in addition to the commonplace term involving the gradient of a scalar potential, there is a term that is the rate of change of the "vector potential." In electrodynamics, the vector potential is associated with the magnetic field, and the term involving the rate of change of the vector potential that appears in the equation for the electric field means that when the magnetic field changes, it contributes to the electric field, causing it to change, too. Sciama noted that in the analogous case for gravity, the rate of change of the vector potential leads to a term in the "gravelectric" field that depends on acceleration of an object relative to the (on average) uniform bulk of the matter in the universe. That is,

$$\mathbf{E}_g = -\nabla\phi - \frac{1}{c}\frac{\partial \mathbf{A}_g}{\partial t}.\tag{2.3}$$

where $\mathbf{E}_g$ is the gravelectric field strength, c the vacuum speed of light, and φ and $\mathbf{A}_g$ the scalar and three-vector gravitational potentials respectively produced by all of the "matter" in the causally connected part of the universe. Matter is in quotes because what counts as matter is not universally agreed upon. We take "matter" to be everything that gravitates. This includes things such as zero-restmass energetic radiation and "dark energy," which are sometimes excluded as matter. The "del" in front of the scalar potential is the "gradient" operator, which returns the rate of change of the potential in space and its direction. The relationship that allows one to write the change in $\mathbf{A}_g$ terms of the scalar potential and velocity is the fact that $\mathbf{A}_g$ is just the sum over all matter currents in the universe. That is,

$$\mathbf{A}_g = \frac{1}{c}\int_V \frac{\rho\mathbf{v}}{r}dV,\tag{2.4}$$

where ρ is the matter density in the volume element dV, $\mathbf{v}$ the relative velocity of the object and volume element, and r the radial distance to the volume element. The factor of c in the denominator appears because Gaussian units are employed.[2] Sciama assumed that gravity, like electromagnetism, propagates at speed c, so normally this integration would involve a messy calculation involving retarded Green's functions and other mathematical complications. But because of the extremely simple, idealized conditions Sciama imposed, he saw that he could sidestep all of that messiness by invoking a little trick.

Sciama noted that in the case of an object moving with velocity $\mathbf{v}$ with respect to the rest of the universe, one could change reference frame to the "instantaneous frame of rest" of the object; and in that frame the object is at rest and the rest of the universe moves past it – apparently rigidly – with velocity $-\mathbf{v}$. Since, in this special frame of reference everything in the universe, as detected by the object, is moving with the same velocity $-\mathbf{v}$ – the velocity in the integration of Eq. 2.4 can be removed from the integration, and Eq. 2.4 becomes:

$$\mathbf{A}_g = \frac{1}{c}\int_V \frac{\rho\mathbf{v}}{r}dV = \frac{\mathbf{v}}{c}\int_V \frac{\rho}{r}dV.\tag{2.5}$$

The result of this trick is to transform an integration over matter *current* densities into an integration over matter densities *per se*. Anyone familiar with elementary electrodynamics will instantly recognize this integration as that which gives the *scalar* potential of

---

[2] Nowadays in some quarters so-called SI units are used. They make the magnitudes of many things normally encountered in field theory unintuitively large or small. I use the traditional Gaussian units of field theory because there was a good reason why they were adopted decades ago by those who work in this area.

the field – but in this case, it returns the scalar potential of the gravitational field. As a result, for the simple case considered by Sciama, Eq. 2.5 becomes:

$$\mathbf{A}_g = \frac{1}{c} \int_V \frac{\rho \mathbf{v}}{r} dV = \frac{\mathbf{v}}{c} \int_V \frac{\rho}{r} dV \cong \frac{\mathbf{v}}{c} \frac{GM}{R} = \frac{\mathbf{v}\phi}{c}. \tag{2.6}$$

where we have taken r as the radial distance from the local object to a spherical volume element (of thickness dR), G is Newton's constant of gravitation, and M and R are the mass and radius of the universe respectively.

R was taken by Sciama as the radius of the "Hubble sphere," that is, the product of the speed of light and the age of the universe. A more accurate calculation would have employed the "particle horizon," the sphere centered on Earth within which signals traveling at the speed of light can reach Earth. The particle horizon encompasses considerably more material than the Hubble sphere. Sciama also neglected the expansion of the universe.

These issues notwithstanding, Sciama's work triggered an at times intense debate about the origin of inertia. Why? Because when we put the result of the integration in Eq. 2.6 back into Eq. 2.3, we get:

$$\mathbf{E}_g = -\nabla \phi - \frac{1}{c} \frac{\partial \mathbf{A}_g}{\partial t} = -\nabla \phi - \frac{\phi}{c^2} \frac{\partial \mathbf{v}}{\partial t}. \tag{2.7}$$

Now, we return to the consideration of our object moving with velocity $\mathbf{v}$ with respect to the homogenous and isotropic universe that we can envisage as moving *rigidly* with velocity $-\mathbf{v}$ past the object which is taken as (instantaneously) at rest. In this case the gradient of the scalar potential vanishes. And if $\mathbf{v}$ is constant or zero, so, too, does the second term – and there is no gravelectric field felt by the object.

However, if the object is accelerating with respect to the rest of the universe (due to the application of some suitable "external" force), then the second term does not vanish as $\partial \mathbf{v}/\partial t = \mathbf{a}$, the acceleration, is not zero. More importantly, from the point of view of the origin of inertia – and inertial reaction forces – if $\phi/c^2 = 1$, then the gravelectric field exactly produces the "equal and opposite" inertial reaction force the accelerating agent experiences. That is, *inertial reaction forces are exclusively gravitational in origin.* The reason why this was so intriguing is that the condition $\phi/c^2 = 1$ has special cosmological significance, as we will consider presently.

Clearly, Sciama's calculation is an approximation. In particular, it is a vector approximation to a field theory that was known to require tensor form in order to be completely general. And it is an idealization. Sciama's assumptions about the distribution and motion of the "matter" sources of the gravelectric field at the object considered are much simpler than reality, even in the early 1950s, was known to be. Nevertheless, Sciama's theory is *not* a "toy model." Toy models are created by physicists when they can't formulate their theory in tractable form in the full four dimensions of real spacetime. To make their theories tractable, they generate them with one or two spatial dimensions where the math is simple enough to be managed. Sciama's theory is four-dimensional. And the above calculation returns an answer for inertial reaction forces that is essentially correct despite the approximation and idealizations adopted. The part of Sciama's paper "On the Origin of

Inertia" where he calculates this expression is reproduced as Addendum #1 at the end of this chapter.

It is worth noting here that an important feature of inertial reaction forces is present in Eq. 2.7, and it was noted by Sciama. The two terms on the right hand side of the equation have different dependencies on distance. The scalar potential depends on the inverse first power of the distance. The gradient of the scalar potential, when you are far enough away from a body of arbitrary shape so that it can be approximated as a sphere, depends on the inverse second power of the distance. That is, Newtonian gravitational force exerted by a body on another sufficiently distant goes as the inverse square of the distance separating them.[3]

When you are calculating the effect of distant matter on a local object, inverse square dependence applies for the gradient of the scalar potential. And it drops off fairly quickly. The term arises from the time-derivative of the vector potential scales with the scalar potential, not its gradient. So the distance dependence of this term is inverse first power. When the distances involved in a situation are small, this difference between the terms may be unimportant. When the distances are large, the difference is crucial. The term arising from the vector potential dominates because it doesn't decrease nearly as rapidly as the Newtonian term does for large distances. This is the reason why the inertia of local objects is due almost exclusively to the action of distant matter.

The inverse first power of the distance dependence of the term from the vector potential that causes inertial forces also signals that the interaction is "radiative." That is, the interactions that arise from this term involve propagating disturbances in the gravity field. They do *not* arise from instantaneously communicated effects or the passive action of a pre-existing field. So inertial forces would seem to be gravity "radiation reaction" effects. This poses a problem, for an inertial reaction force appears at the instant an accelerating force is applied to an object. How can that be true if the inertial reaction force involves an active communication with chiefly the most distant matter in the universe, and communication with the stuff out there takes place at the speed of light?

If reaction forces were produced by the interaction with a passive, locally present pre-existing field, this would not be a problem. But that is not what is calculated in Sciama's treatment. The trick of using the instantaneous frame of rest where the universe very obviously appears to be moving rigidly past the accelerating object not only sidesteps a messy calculation involving Green's functions; it blurs the issue of instantaneity of reaction forces. This is arguably the most difficult aspect of coming to grips with the origin of inertia.

You may be wondering, if this sort of thing happens with gravity, why don't we see the same sort of behavior in electromagnetism? After all, if we accept Sciama's theory as the vector approximation to GRT that it is, they are both vector field theories with essentially

---

[3] Newton is routinely credited with the discovery of the inverse square law of universal gravitation. But his contemporary Robert Hooke claimed to have independently discovered the inverse square law before Newton made public his claim. Newton refused the presidency of the Royal Society until shortly after Hooke's death. Shortly thereafter, the Royal Society moved to new quarters, and Hooke's papers from the 1680s were lost in the move. Whether Hooke actually discovered the inverse square nature of gravity, absent his papers, is a matter of conjecture. It seems unlikely, though, that he discovered the universal nature of the interaction.

the same field equations. Ironically, as it turns out, the problems of the origin of inertia – in the form of electrical "self-energy" – and "radiation reaction" have plagued electrodynamics for years, too. It just hasn't been discussed much in recent years. But infinite"self-energies" of point particles was the motivation, for example, for the invention the "renormalization" program of quantum field theory, and of string theory.

We'll be looking at these issues in later chapters in some detail. Here we note that although the vector field formalisms for gravity and electromagnetism are essentially the same, this type of gravitational force from the action of cosmic matter does not arise in electrodynamics – because on average the universe is electric charge neutral, so cosmic electric charge currents sum to zero everywhere. More specifically, since on average there is as much negative electric charge as positive in any region of spacetime, the total charge density is zero. So, in the calculation of the vector potential – as in Eq. 2.5 – since ρ is zero, the integral for the potential vanishes. This means that in everyday electrodynamics you never have to deal with the action of distant electric charge and currents of any significance. But in gravity, you do.

Sciama's calculation is not optional. It is a prediction of GRT providing that $\phi / c^2 = 1$. Is $\phi / c^2 = 1$ true?

Yes. When is $\phi/c^2 = 1$? When "critical cosmic matter density" is reached, and space at the cosmic scale is flat. Sciama didn't know if this were true. Indeed, even in the 1950s it was thought that the amount of luminous matter in the universe was not sufficient to be "critical." So Sciama did not make a bald-faced claim that he could fully account for inertial reaction forces. But space at the cosmic scale sure looked pretty flat. And it was known that if cosmic scale space deviated from flatness, it would quickly evolve to far greater distortion. As the universe was at least billions of years old and still flat, most cosmologists assumed that space really was flat, and that critical cosmic matter density was obtained. And the fact that luminous matter was less than 10% of the critical value came to be called the "missing mass" problem.[4] Only after the turn of the century was space at the cosmic scale measured – by the Wilkinson Microwave Anisotropy Probe (WMAP) about a decade ago. So we know whether or not cosmic scale space is flat. It is.

You may be wondering, if we know that space at the cosmic scale is flat, why isn't it common knowledge that inertial reaction forces are caused by the gravitational interaction of local accelerating objects with chiefly cosmic matter? Well, two issues figure into the answer to this question. One is the consequence of an analysis done by Carl Brans in the early 1960s. (Excerpts from Brans' paper are to be found at the end of this chapter.) And the other, related to Brans' argument, is the business about there being no "real" gravitational forces. Brans showed that if the presence of "spectator" matter (concentrations of matter nearby to a laboratory that shields the stuff in it from all external influences except gravity, which cannot be shielded) were to change the gravitational potential energies of objects in the shielded laboratory, you could always

---

[4] Actually, the "missing mass" problem was first identified in the 1930s by Fritz Zwicky by applying the "virial theorem" to clusters of galaxies. The virial theorem says that on average, the kinetic and potential energies of galaxies in clusters should be the same. So, by measuring the motions of galaxies in a cluster, you can estimate the mass of the cluster. It leads to galaxy cluster mass estimates 10–100 times greater than the light emitted suggests is present. Only later was it extended to encompass cosmology, too.

tell whether you were in a gravity field or an accelerating lab in deep space by performing only local experiments.

In particular, the gravitationally induced changes in the masses of elementary particles in the lab would change their charge to mass ratios, and this would be locally detectable. No such changes in charge to mass ratios would occur in an accelerated reference frame in deep space. As a result, a gravity field could always be discriminated from an acceleration with local experiments. Since this would be a violation of the Equivalence Principle, Brans asserted that gravitational potential energy cannot be "localized." That is, the scalar gravitational potential must have exactly the same value, whatever it might be, everywhere in the laboratory, no matter where the lab is located or how it is accelerating. As Brans noted, this condition on gravitational potential energy reveals Einstein's first prediction quoted above as wrong. Evidently, it appears that the distribution of matter outside of the lab cannot have any identifiable effect on the contents of the lab. Mach's principle, however, would seem to suggest the opposite should be the case. And it was easy to infer that Mach's principle was not contained in pristine GRT.

The inference that Mach's principle is not contained in GRT, however, is mistaken. If you take account of the role of the vector potential in Sciama's gravelectric field equation,[5] it is clear that should spectator matter outside the lab be accelerated, it will have an effect on the contents of the lab, changing what are perceived to be the local inertial frames of reference. This is the action of Mach's principle. But as the accelerating spectator matter will act on all of the contents of the lab equally, for inertial forces are "fictitious," they produce the same acceleration irrespective of the mass of the objects acted upon. So, using local measurements in the lab it will not be discernible either as a force of gravity or a change in the acceleration of the lab. And it will not change the gravitational potential energies of the contents of the lab.

Brans' argument about the localizability of gravitational potential energy has an even more radical consequence – one found in the excerpt from Misner, Thorne, and Wheeler on energy localization in the gravitational field found in the previous chapter. If you can eliminate the action of the gravitational field point by point throughout the laboratory by a careful choice of geometry that, for us external observers, has the effect of setting inertial frames of reference into accelerated motion with respect to the walls, floor and ceiling of the lab, it seems reasonable to say that there is no gravitational field, in the usual sense of the word, present in the lab. This is what is meant when people say that GRT "geometrizes" the gravitational field. In this view there are no gravitational forces. Gravity merely distorts spacetime, and objects in inertial motion follow the geodesics of the distorted spacetime. The only real forces in this view are non-gravitational. Inertia, of course, is a real force. But if you believe that there aren't any real gravitational forces, then the origin of inertia remains "obscure" – as Abraham Pais remarked in the quote at the outset of this chapter – for it isn't a result of the electromagnetic, weak, or strong interactions (and can't be because they are not universal), and that leaves only gravity.

---

[5] Or Einstein's vector approximation equation for the force exerted by spectator matter that is accelerating on other local objects.

But we've excluded gravity because we know that there aren't any gravitational forces. And the origin of inertia remains a mystery.

There may not be any "real" gravitational forces in GRT, but there is "frame dragging." That is, in the conventional view, matter can exert a force on spacetime to produce frame dragging, but it can't act directly on the matter in the possibly dragged spacetime. If this sounds a bit convoluted, that's because it is. Let's illustrate this point.

About the time that Thorne and his graduate students were introducing the rest of us to traversable wormholes, a committee of the National Academy of Sciences was doing a decadal review of the state of physics, producing recommendations on the areas of physics that should be supported with real money. One of their recommendations was that Gravity Probe B should be supported because, allegedly, no other test of "gravitomagnetism" was contemplated, and this was an important, if difficult and expensive, test of GRT.

Ken Nordtvedt, a physicist with impeccable credentials who had proposed the "Nordtvedt effect,"[6] then being tested by ranging the distance of the Moon with a laser, but who had not been a member of the decadal survey committee, pointed out that the claim was just wrong. He noted that even in doing routine orbit calculations, unless care was taken to use special frames of reference, one had to take account of gravimagnetic effects to get reasonable results. Using "parameterized post Newtonian" (PPN) formulation of gravity, a formalism that he and others had developed as a tool to investigate a variety of theories of gravity some 20 years earlier, he showed explicitly how this came about.

In the course of his treatment of orbital motion, Nordtvedt drew attention to the fact that gravity predicts that linearly accelerated objects should drag the spacetime in their environs along with themselves since the gravitational vector potential does not vanish.[7] Nordtvedt's 1988 paper on the "Existence of the Gravitomagnetic Interaction" where he discussed all this is excerpted in Addendum #3 at the end of this chapter. In effect, he recovered the same basic result as Einstein and Sciama, only where they had talked about gravitational forces acting on local objects, Nordtvedt put this in terms of "frame dragging."[8]

Are they the same thing? Well, yes, of course they are. The reason why you may find this confusing is because in the case of everything except gravity, one talks about the sources of fields, the fields the sources create, and the actions of fields *in* spacetime on other sources. That is, spacetime is a *background* in which sources and fields exist and interact. In GRT spacetime itself is the field. There is no background spacetime in which the gravitational field exists and acts. Since there is no background spacetime, GRT is called a "background independent" theory.

---

[6] The Nordtvedt effect proposes that gravitational potential energies do contribute to the mass-energy of things and predicts (small) deviations from the predictions of GRT that would follow. Such effects have not been observed.

[7] He also predicted that the masses of things should vary as they are accelerated, an effect of the sort that we'll be looking at in the next chapter.

[8] Nordtvedt considered only a rigid sphere of uniform density of modest dimensions. He did not extend the argument to the case where the sphere is the entire universe, as did Sciama.

It is this background independence that makes gravity and GRT fundamentally different from all other fields. And it is the reason why "frame dragging" is fully equivalent to the action of a gravitational force. If you want to preserve the configuration of a system before some nearby objects are accelerated, when the nearby objects begin to accelerate you have to exert a force that counteracts the effect of the frame dragging produced by the acceleration of the nearby objects. When you do that, what do you feel? An inertial reaction force – the force produced by the action of the dragged spacetime, which is produced by the gravitational action of the accelerated nearby objects. By interposing frame dragging we've made it appear that no gravitational force is acting. But of course gravity is acting, notwithstanding that we've introduced the intermediary of frame dragging to make it appear otherwise.

When only nearby objects are accelerated to produce frame dragging, as Einstein noted for the equivalent force he expected, the predicted effects are quite small. When it is the universe that is accelerated, it is the full normal inertial reaction force that is felt if you constrain some object to not accelerate with the universe. Why the difference? Because when the entire universe is "rigidly" accelerated, the interior spacetime is rigidly dragged with it, whereas nearby objects, even with very large masses, produce only small, partial dragging.

You may be thinking, yeah, right, rigidly accelerating the whole universe. That would be a neat trick. Getting the timing right would be an insuperable task. The fact of the matter, nonetheless, is that you can do this. We all do. All the time. All we have to do is accelerate a local object. Your fist or foot, for example. The principle of relativity requires that such local accelerations be equivalent to considering the local object as at rest with the whole universe being accelerated in the opposite direction. And the calculation using the PPN formalism for frame dragging (with GRT values for the coefficients in the equation assumed) bears this out. At the end of his paper on gravimagnetism Nordtvedt showed that a sphere of radius R and mass M subjected to an acceleration $\mathbf{a}$ drags the inertial space within it as:

$$\delta \mathbf{a}(\mathbf{r}, t) = -\left(2 + 2\gamma + \frac{\alpha_1}{2}\right) \frac{U(\mathbf{r}, t)}{c^2} \mathbf{a} \qquad (2.8)$$

where the PPN coefficients have the values $\gamma = 2$ and $\alpha_1 = 0$ for the case of GRT and $U(\mathbf{r}, t)$ is the Newtonian scalar potential, that is, $U = GM/R$. So we have four times ϕ (changing back to the notation of Sciama's work on Mach's principle) equal to c^2 to make $\delta \mathbf{a} = \mathbf{a}$ in Eq. 2.8; that is, if the universe is accelerated in any direction, spacetime is rigidly dragged with it, making the acceleration locally undetectable.

You may be concerned by the difference of a factor of 4 between the Nordtvedt result and Sciama's calculation. Factors of 2 and 4 are often encountered when doing calculations in GRT and comparing them with calculations done with approximations in, in effect, flat spacetime. In this case, resolution of the discrepancy was recently provided by Sultana and Kazanas, who did a detailed calculation of the contributions to the scalar potential using the features of modern "precision" cosmology (including things like dark matter and dark energy, and using the particle horizon rather than the Hubble sphere), but merely postulating the "Sciama force," which, of course, did not include the factor of 4 recovered in Nordtvedt's calculation. They, in their relativistically correct calculation, found ϕ to have only a quarter of the required value to make the coefficient of the acceleration equal to one. Using the general relativistic calculation, with its factor of 4 in the coefficient, makes the full coefficient of the acceleration almost exactly equal to one – as expected if Mach's principle is true.[9]

You might think that having established the equivalence of frame dragging by the universe and the action of inertial forces, we'd be done with the issue of inertia. Alas, such optimism is premature. A few issues remain to be dealt with. Chief among them is that if $\phi = GM/R$, since at least R is changing (because of the expansion of the universe), it would seem that $\phi = c^2$ must just be an accident of our present epoch. However, if the laws of physics are to be true everywhere and during every time period, and inertial reaction forces are gravitational, then it must be the case that $\phi = c^2$ everywhere and at all times if Newton's third law of mechanics is to be universally valid.

Well, we know that the principle of relativity requires that c, when it is locally measured, has this property – it is a "locally measured invariant." So, perhaps it is not much of a stretch to accept that ϕ is a locally measured invariant, too. After all, GM/R has dimensions of velocity squared. No fudging is needed to get that to work out right. But there is an even more fundamental and important reason to accept the locally measured invariance of ϕ: it is the central feature of the "Einstein Equivalence Principle" (EEP) that is required to construct GRT. As is universally known, the EEP prohibits the "localization" of gravitational potential energy. That is, it requires that whenever you make a local determination of the total scalar gravitational potential, you get the same number, whatever it may happen to be (but we know in fact to be equal to c^2). Note that this does *not* mean that the gravitational potential must everywhere have the same value, for distant observers may measure different values at different places – just as they do for the speed of light when it is present in the gravity fields of local objects. Indeed, this is not an accident, because ϕ and c are related, one being the square of the other.

Should you be inclined to blow all of this off as some sort of sophistry, keep in mind that there is a compelling argument for the EEP and the locally measured invariance of ϕ – the one constructed by Carl Brans in 1962 that we've already invoked. If you view the gravitational field as an entity that is present in a (presumably flat) background spacetime – as opposed to the chief property of spacetime itself (as it is in GRT) – it is easy to believe that gravitational potential energies should be "localizable" – that is, gravitational potentials should have effects that can be detected by local measurements. Brans pointed out that were this true, it would be a violation of the principle of relativity as contained in the Equivalence Principle. Why? Because, as mentioned above, you would always, with some appropriate local experiment, be able to distinguish a gravitational field from accelerated frames of reference.

---

[9] See: J. Sultana and D. Kazanas, arXiv:1104.1306v1 (astro-ph.CO, later published in the *Journal of Modern Physics D*). They find that the "Sciama" force is one quarter of that needed for an exact inertial reaction force. The factor of 4 discrepancy arises from the fact that Sultana and Kazanas simply assumed the "Sciama" force without deriving it from GRT, and Sciama's calculation is not exactly equivalent to a general relativistic calculation like Nordtvedt's. The difference is the factor of 4 that when multiplied times their result returns 1 almost exactly.

Brans' way was to measure the charge to mass ratios of elementary particles. An even simpler, cruder way to make the discrimination between gravity field and accelerated reference frame is to drop stuff. You won't be able to tell the difference between a gravity field and accelerated frame of reference by the way things "fall" since they all "fall" with the same acceleration in both cases, irrespective of their masses or compositions. But you will be able to tell by how big a dent in the floor they make – because their masses are presumably different when gravity is present, versus when it is not, and bigger masses make bigger dents. Brans' argument makes clear that the EEP must be correct if the principle of relativity is correct – and that Einstein was wrong in 1921 when he assumed that the piling up of spectator matter would change the masses of local objects. Notwithstanding that the non-localizability of gravitational potential energies, however, the fact that inertial reaction forces are independent of time and place requires that the masses of things be equal to their total gravitational potential energies. That is, $E = mc^2$ and $E_{grav} = m\phi$, so if $E = E_{grav}$ and $\phi = c^2$ as Mach's principle demands, we have a simple identity.

ANOTHER EXAMPLE

To bring home the full import of the foregoing discussion of GRT and Mach's principle, we briefly consider a slightly more complicated example than that used so far. Instead of considering a test body in an otherwise uniform universe, we look at the behavior of a test object (with negligible mass) in the vicinity of Earth. In Newtonian physics we say that the mass of Earth produces a gravitational field in its vicinity that exerts a force on the test object. If the test object is unconstrained, it falls toward the center of Earth with an acceleration of one "gee." We can arrest this motion by applying an upward force with equal magnitude, balancing the "force" of gravity. The agent applying the upward balancing force, of course, experiences the downward force which he or she attributes to Earth's gravity. This is the commonplace explanation of these circumstances that even relativists intuitively recognize.

The general relativistic explanation of the circumstances of our test body in proximity to Earth, however, is *fundamentally* different. Earth does *not* produce a gravity field that acts to produce a force on the test body. Earth does produce a local distortion of spacetime (which is the gravity field), changing the local inertial structure of spacetime from the otherwise flat character it would have (as measured by the WMAP project). As a result, if our test body engages in unconstrained motion, it responds inertially and finds itself in a state of free fall. Despite the fact that the test body appears to us to be accelerating, and we intuitively assume that accelerations are the consequence of the application of forces, *no forces act on the falling test body*.

What happens, then, when we apply a constraining force to the test body to stop its free fall acceleration? Does this somehow turn on Earth's gravity force to balance the constraining force we have applied? No. You can't turn gravity off and on (yet). *The balancing force that you feel is the inertial reaction force that arises in response to the "arresting" force that you have applied to the test object.* Your arresting force has actually produced acceleration of the test object – with respect to local inertial frames of reference that are in free fall. The force that we normally ascribe to the gravitational action of Earth, which is quite real, is not produced by Earth. It is produced chiefly by the distant matter in the universe. The reason why we associate it with the action of Earth is because Earth determines the extent of the local distortion of inertial spacetime, and thus the amount of acceleration required to arrest the inertial motion of objects in the vicinity of Earth's surface.

One may ask: is it really necessary to adopt this arguably very odd way of looking at the circumstances that seem to make such intuitive sense when viewed from the Newtonian point of view? That is, can we in some sense accept GRT, but take the above description as an "equivalent representation" to the Newtonian viewpoint with its objective gravity field that produces forces on nearby objects? No. The representations are in no sense equivalent. The reason why is the EEP. The geometrization of the gravitational field in GRT depends on the complete indistinguishability of accelerated reference frames from the local action of gravity fields.

There are those who argue that the presence of tidal effects in all but (unphysical) uniform gravity fields always allow us to distinguish gravity fields from accelerated reference frames, but this is a red herring. We can always choose our local Lorentz frame sufficiently small so as to reduce tidal effects to insignificant levels, making the two types of frames indistinguishable. Were gravitational potential energies localizable, however, we would be faced with a real violation of the indistinguishability condition that would vitiate field geometrization. Using either Brans' charge to mass ratios, or the cruder dents criterion, no matter how small we make the region considered, we can always make determinations that tell us whether we are dealing with a gravity field or an accelerated reference frame, because, unlike tidal forces, charge to mass ratios and dents don't depend on the size of the region considered. They are so-called "first" or "lowest" order effects.

The foregoing considerations are sufficient in themselves to reject attempts to "objectify" static gravity fields. But they are attended by an even stronger argument. If local gravitational potential energies really did contribute to locally observable phenomena, then $\phi/c^2 = 1$ everywhere and at all times would *not* in general be true. Consequently, inertial reaction forces would not always equal "external" applied forces, and Newton's third law would be false. That would open the way to violations of the conservation of energy and momentum. If you're trying to make revolutionary spacecraft, you may not think this necessarily bad. It is.

As we have now seen, the principle of relativity has present within it a collection of interlocking principles – one of which is Mach's principle, which says both that inertial reaction forces are the gravitational action of everything in the universe, and the inertia of objects is just their total gravitational potential energy (divided by c^2). Objects are to be understood as including everything that gravitates (including things we do not yet understand in detail like dark matter and dark energy). Are these principles ones that can be individually rejected if we don't like them without screwing up everything else? No. If the principle of relativity is correct, then the EEP and Mach's principle follow inexorably. If either the EEP or Mach's principle is false, then so, too, is the principle of relativity – and Newton's laws of mechanics. That's a pretty high price to pay for rejecting a principle you may not care for.

Two issues remain to be addressed in a little detail. One is the instantaneity of inertial reaction forces. The other is how Mach's principle squares with traditional gravity wave physics. We address inertial reaction forces and how they relate to gravity wave physics first.

INERTIAL REACTION FORCES AND GRAVITY WAVE PHYSICS

It has been known since Einstein created GRT in 1915 that his theory predicted propagating disturbances in the gravitational field, that is, it predicted "gravitational waves." The whole business of gravity waves and how they are generated by and interact with matter sources, however, was at times quite contentious. Should you want to know the details of how all of this developed, Dan Kennefick has written an outstanding history of the subject: *Traveling at the Speed of Thought: Einstein and the Quest for Gravitational Waves.*

Most, if not all, of the issues of debate were settled many years ago now. One of the issues was the manner in which the prediction is calculated. As noted above, exact solutions of the full non-linear Einstein field equations are few and far between. One of the standard techniques for dealing with this is to invoke the "weak field approximation," where you assume that the Einstein tensor (describing the geometry of spacetime) can be written as the "Minkowski" metric of flat spacetime with an added "perturbation" metric field that accounts for gravity, as mentioned earlier in this chapter. Since the flat spacetime metric in this approach is effectively a "background" spacetime unaffected by the presence of matter and gravity fields, Einstein's theory is effectively "linearized" by this procedure. With a few further assumptions, Einstein's field equations can be put in a form that closely resemble Maxwell's equations for the electromagnetic field – as Einstein himself did in his discussion of Mach's principle mentioned above, and Sciama and Nordtvedt (among many others) subsequently did.

Solutions of Maxwell's equations have been explored in great detail in the roughly century and a half since their creation. The standard techniques include classification according to the disposition of the sources of the fields and their behavior (how they move). This leads to what is called a "multipole expansion" of the field, each component of the field being related to a particular aspect of the distribution and motion of its sources. The simplest part of the field in this decomposition is the so-called "monopole" component, where the sources can be viewed as consisting of a single "charge" located at one point in spacetime.

In electromagnetism the next least complicated source distribution is the so-called "dipole" component. Electrical charges come in two varieties: positive and negative, and the dipole component of a multipole expansion consists of the part that can be characterized by a positive charge located at one point and a negative charge located somewhere else in spacetime. The measure of this charge distribution is called its dipole "moment," defined as the product of the charges times the separation distance between them. If the dipole moment of the dipole component of the field is made to change, the changes in the surrounding field are found to propagate away from the charges at the speed of light. The propagating disturbance in the field is the "radiation" field. Non-propagating fields are called "induction" fields, as they are induced by the presence of sources of the field and do not depend on their moments changing.

The next term in the multipole expansion for source distributions and their associated field components is the so-called "quadrupole" term. It is the part of the field that takes into account the simplest charge distribution for sources of the same sign (positive or negative in the case of electromagnetism) that cannot be covered by the monopole term. It corresponds to two charges of the same sign separated, like the dipole distribution, by some distance in spacetime. Just as there is a dipole moment, so, too, is there a quadrupole moment. And if the quadrupole moment changes, like the dipole term, a propagating disturbance in the field is produced.

Since there are no negative masses (yet), and the vector approximation of GRT is a vector theory analogous to Maxwell's equations for electrodynamics, it is found that the "lowest order" radiative component of the gravitational field is that produced by sources with time-varying quadrupole moments. An example is a dumbbell spinning about the axis of symmetry that passes perpendicularly through the bar separating the bells. Another more fashionable example is a pair of black holes in orbit around each other. An example that does not involve spinning stuff is two masses separated by a spring that are set into oscillatory motion along their line of centers. Even in the case of orbiting black holes, the amount of momenergy involved in the gravitational radiation is exceedingly minute. (This is the stuff being sought with the Laser Interferometer Gravitational wave Observatory, with a price tag now approaching a gigabuck.) Laboratory scale gravitational quadrupoles, even operating at very high frequencies, produce hopelessly undetectable amounts of gravitational radiation.[10]

What does all this have to do with inertial reaction forces? Well, as Sciama was at pains to point out, his calculation of those forces show two things: one, they depend on the acceleration of sources; and two, their dependence on distance in his gravelectric field equation goes as the inverse first power, not inverse square. These are the well-known signatures of radiative interactions. It would seem then that inertial reaction forces should involve radiation, and that they should be called radiation reaction forces. But there is a problem. The quadrupole radiation given off by an accelerating massive object is incredibly minute. And the monopole component of the field in electrodynamics is non-radiating. How can this be squared with the fact that inertial reaction forces are, by comparison, enormous, decades of orders of magnitude larger than quadrupole radiation reaction? To answer this question we must first tackle the instantaneity of inertial reaction forces.

---

[10] The field strength of gravitational radiation depends on the frequency at which it is emitted. Gravitational waves, all other things held constant, depend on the fifth power of the emission frequency. This strong frequency dependence has led some to speculate that very high frequency gravitational waves might be used for propulsive purposes. Since the momenergy in gravity waves produced by human scale sources is so hopelessly minute, even allowing for unrealistically high frequency sources, gravity waves hold out no promise of practical scale effects.

THE INSTANTANEITY OF INERTIAL REACTION FORCES

The immediate fact of inertial reaction forces is that they respond to applied forces instantaneously. Why? Well, if you believe, as Newton and legions after him have, that inertia is an inherent property of material objects needing no further explanation, then this question needs no answer. The problem with this view, of course, is the fact noted famously by Mach that inertial frames of reference seem to be those in inertial motion with respect to the "fixed stars." Today we would say inertial motion with respect to the local cosmic frame of rest, and that, remarkably, isn't rotating. This suggests that the stuff out there has something to do with inertia. But it is so far away, typically billions of light-years distant. How can that produce instantaneous effects?

The easy answer to this question is to assert that the distant stuff produces a gravity field, which we know to *be* spacetime in GRT, here, and when we try to accelerate anything in spacetime, spacetime pushes back. Since the local spacetime is the gravity field of the distant stuff, obviously we should expect local inertia to be related to the distant stuff. This is the "local pre-existing field" argument.

Sounds good, doesn't it? It is, however, a flawed view of things, as was made evident by Sciama's argument back in the early 1950s. As we've noted already, Sciama used a little trick to avoid a tedious calculation involving Lienard-Wiechert potentials, Green's functions, and a lot of associated mathematical machinery. To calculate the effect of very distant matter on a local accelerating body, he noted that from the perspective of the local body, the entire universe appears to be accelerating rigidly in the opposite direction. The apparent rigid motion provides the justification for removing the velocity from the integral for the vector potential. Sciama, of course, knew that this was just a trick to avoid a messy integration, for, as already mentioned, he was quick to point out that distance dependence of the scalar gravitational potential was inverse *first* power, rather than the inverse *second* power of Newtonian gravity. Those familiar with the process of radiation immediately recognize the inverse first power as the signature of a radiative interaction. What Sciama's calculation (and those of Einstein, Nordtvedt, and others) shows is that inertial reaction forces are conveyed by a radiative process. Inertial forces are not the simple passive action of a pre-existing field that acts when local objects are accelerated.

A way to visualize what's going on here is to consider what happens to the spacetime surrounding a local object that is given a quick impulsive acceleration. Before the acceleration, its gravity field is symmetrically centered on it. The same is true shortly after the impulse. But the impulse displaces the center of symmetry of the field from the prior center of symmetry. That produces a "kink" in the gravity field, like that shown in Fig. 2.1. The radiative nature of the interaction means that the kink induced in the field by the impulsive acceleration[11] propagates outward from the object during the acceleration at the speed of light.

It is the production of the kink in the field by the source, not the field itself, that produces the inertial reaction force on the source and accelerating agent. In electrodynamics, this is known as the problem of "radiation reaction." Should you trouble yourself to

---

[11] The technical term for such an acceleration is a "jerk."

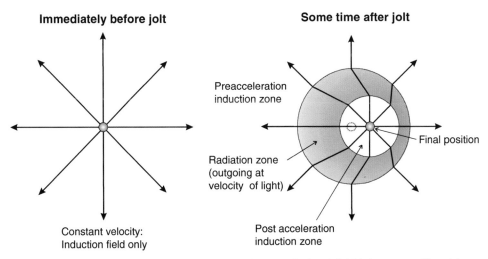

Fig. 2.1 The "kink" diagram. When a source of the gravitational field is in a state of inertial motion, it carries its field, represented by the *lines* radiating from the source's location, along with it without distortion. If the source is sharply accelerated, and then decelerated, so that it moves after the "jolt" as it did before, a "kink" is introduced into the field lines. The kink does not move to infinity at once. It propagates outward at the speed of light

read up on this in, say, Feynman's *Lectures on Physics*, or pretty much any advanced text on electrodynamics, you'll find that this is a messy problem with some very curious features, for example, "pre-acceleration," where an object starts to accelerate *before* the force producing the acceleration acts (as Dirac showed in a classic paper on electromagnetic radiation reaction published in 1938). All those problems carry over to the gravity case if inertial reaction forces are forces of radiative reaction – as seems to be the case now that the WMAP results are known.

Now, there are two problems here. The first is that the kink in the field is normally taken as due to the monopole term in the multipole expansion, and it is allegedly non-radiative. We will deal with this issue presently. The second problem is that if the coupling between the test object and the distant matter in the universe is carried by the kink in the field propagating at the speed of light, it will take billions of years for the kink to reach the distant matter, and billions of years for a return signal to get back to the accelerating object. Inertial reaction forces, however, are instantaneous. Push something and it pushes back immediately. How can the distant matter in the universe act instantly on an object when it is accelerated by an external force without violating the speed limit, c, of SRT?

ACTION AT A DISTANCE AND "ADVANCED" WAVES

The simplest, most elegant way to deal with the problems just mentioned was worked out for electrodynamics by John Wheeler and Richard Feynman in the 1940s. Their theory, intended to deal with the problems attending classical electron theory (infinite self-energies,[12] radiation reaction, and so forth), goes by the name "action-at-a-distance" or "absorber" electrodynamics. It is a scheme designed to account for seemingly instantaneous radiation reaction forces that are produced by an interaction with a distant "absorber." To do this, Wheeler and Feynman noted that the propagating solutions to "classical" wave equations can either be "retarded" – that is, propagate forward in time – or "advanced" – that is, propagate backward in time.

Physically and mathematically, there is no discernible difference between the two classes of solutions. Since it appears to us that waves propagate into the future, we just ignore the solutions that propagate backward in time. After all, we do not appear to be constantly buffeted by waves coming back from the future.

The business of advanced waves can be a bit confusing, so we make a brief foray into this topic to ensure that we are all on the same page. The usual story about the role of time in the laws of physics is that the laws of physics possess a property called "time reversal symmetry." That is, you can replace the time t with $-t$ everywhere in your equations, and the processes described by the time-reversed equations are just as valid as the original equations. Another way this is sometimes illustrated is to film some process running forward in time, and then point out that if the film is run backward, the processes depicted also obey the laws of physics, albeit the time-reversed laws.

The fact that the laws of physics are time-reversal invariant has led to endless speculations on "the arrow of time," and how time could be asymmetric given the symmetry of the underlying laws. Philosophers, and physicists with a philosophical bent, seem to be those most prone to delving into the mysteries of time. We'll be concerned here with a much more mundane problem: How exactly do advanced waves work?

A commonplace example used to illustrate advanced waves is the spreading of ripples on a pond when a rock is thrown into the middle. When the rock hits the water, it sets up a series of waves that propagate from the point of impact in symmetrical circles toward the shoreline. If we make a film of this sequence of events and run it backward, we will see the waves forming near the shoreline, and then moving in concentric circles of decreasing diameters toward the center. And when the waves arrive at the center, the rock will emerge from the water as though thrust from the depths by the waves. This wave behavior is illustrated in Fig. 2.2 as sequences of time-lapsed pictures of waves, with time proceeding from left to right. The normal view of things is shown in the upper strip of pictures, and the reversed in the lower strip.

The problem with this picture is that when we run the movie backward to supposedly reverse the direction of time, what we really do – since we can only run the movie forward in time, regardless of which end of the movie we start with – is run the waves *backward in space* as the movie runs *forward in time*. A true advanced wave starts in the future at the shoreline and propagates backward in time toward the center of the pond, something we

---

[12] Self energy in electrodynamics arises because the parts of an electric charge repel the other parts of the charge, and work must be done to compress the parts into a compact structure. The energy expended to affect the assembly is stored in the field of the charge. When the electron was discovered by J. J. Thomson in 1897, it was not long until H. A. Lorentz and others suggested that the electron's mass might be nothing more than the energy stored in its electric field (divided by c^2). They used this conjecture to calculate the so-called "classical electron radius" that turns out to be about 10^{-13} cm. But should you assume that the size of the electron is zero, the energy of assembly turns out to be infinite.

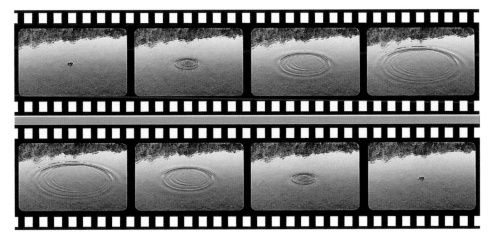

Fig. 2.2 The *top set of frames*, reading left to right, show waves propagating forward in time and space as they spread from a rock being thrown into a pond. When people talk about "advanced" waves, they often remark that waves propagating backward in time are those seen by running a movie of the waves in reverse, producing the sequence of pictures in the *bottom row*. However, the *bottom row* shows waves propagating backward in space as time goes forward

cannot actually see from the present. So, when we watch a movie running backward, we must imagine that we are running backward in time, notwithstanding that we are actually "moving" forward in time.

What we do see, moving forward in time, when and advanced wave comes back from the future is a wave that appears to be propagating away from the impact of the rock toward the shoreline of the pond. That is, the advanced wave looks exactly like a retarded wave. As long as the advanced wave coming back from the future didn't propagate farther into the past than the rock hitting the water that initiated all of the waves, neither you nor I could tell whether the waves in the pond had any advanced component. So, using retarded and advanced waves to get distant objects to "instantaneously" affect local objects becomes finding a solution for wave action that cancels the advanced waves at the source (the rock hitting the water) to keep them from traveling farther into the past.

What Wheeler and Feynman noted was that if a forward in time propagating wave in the electromagnetic field was eventually absorbed by enough material out there in the distant universe, and as it was absorbed it produced an "advanced" wave propagating backward in time, all of the contributions from all of the parts of the absorber would just get back to the source at exactly the right time to produce the apparent force of radiative reaction. And as they passed the origin of the waves into the past, if the waves were half advanced and half retarded, they would cancel out the "advanced" wave propagating from the source into the past. So future events would not indiscriminately screw up the past (and our present). But the half-advanced waves coming back from the future provide a way for arbitrarily distant objects to affect events in the present seemingly instantaneously. In the case of gravity, this allows the whole universe to act on any object that's accelerated by an external (non-gravitational) force with an equal and opposite force. This solution to the problems of radiation reaction is so neat it almost has the appearance of a cheap tourist trick, too good to be true. But it actually works.

Some exceedingly important features of action-at-a-distance electrodynamics must be mentioned, as they figure critically into the understanding of inertial reaction forces when the theory is extended to include gravity. Of these, far and away the most important is the fact that *there is no radiation as understood in conventional electrodynamics in the action-at-a-distance version*. It has not been mentioned yet, but in addition to acceleration dependence and inverse first power of the distance dependence of the "amplitude" or field strength[13] of the radiation field, there is another condition that a radiation field must satisfy: it must have a "freely propagating" non-vanishing energy density as it approaches "asymptotic infinity." This property gives the field "independent degrees of freedom."

What this means, in simple physical language, is that once a radiation field has been launched by the acceleration of some charges of the field, the radiation is "decoupled" from both the source (which can no longer affect it) *and the sinks (just sources soaking up the field), if any, that ultimately absorb it.* Note that the launching of the radiation does *not* depend on it ever being absorbed by sinks out there somewhere in the future. That's what "freely propagating at asymptotic infinity" means. Note, too, that there are no classical radiation fields in action-at-a-distance electrodynamics, for no electromagnetic disturbances (that might be considered radiation in classical theory) are ever launched without the circumstances of their eventual absorption being established *before* they are launched. That is, there are no field "modes" with "independent degrees of freedom," no loose radiation that might make it to "asymptotic infinity."

Why is this the case? Because the theory only works if the eventual absorption of all disturbances is guaranteed so that the requisite "advanced" disturbances, needed to combine with the "retarded" disturbances, are present to yield the world as we see it. What this means it that if your field theory is an action-at-a-distance theory, you can have "monopole" propagating disturbances in the field that carry energy and momentum – as the "kink" diagram suggests ought to be possible – and that they can have the acceleration and inverse first power of the distance characteristics of classical radiation, but they will not be considered "radiation" by those ignorant of action-at-a-distance theory.

You may ask at this point, how can such radically different results be obtained from action-at-a-distance and classical field theory? The answer is really quite simple. Michael Faraday, the pre-eminent experimental physicist of the nineteenth century, hated action-at-a-distance. In his day, it was the chief feature of Newtonian gravitation, and even Newton himself had thought that instantaneous action of gravity over arbitrarily large distances stupid.[14] Indeed, Newton's famous "hypotheses non fingo" [I make no hypotheses {about

---

[13] The "amplitude" (for an oscillatory field) or "field strength" (the magnitude of the scalar potential or field vector) is not the same as the "intensity" of the field. The intensity is proportional to the square of the field strength. So, a field whose strength decreases as $1/r$ has an intensity that decreases as $1/r^2$, as does electromagnetic radiation (light), for example. When the intensity decreases at this rate, some energy just barely makes it to "asymptotic infinity." If the intensity decreases faster than $1/r^2$, as it does for any field whose strength decreases more quickly than $1/r$, then no freely propagating energy makes it to asymptotic infinity.

[14] As Faraday discovered in the early 1840s when Newton's "third letter to Bentley" was first published. Hitherto, Newton's true views on action-at-a-distance were not generally known. After reading Newton's letter, it is said that Faraday became positively boorish regaling everyone with the news that Newton rejected action-at-a-distance.

the mechanism of gravity}] remark was his response to critics who assailed him about action-at-a-distance.

Faraday, despite repeated attempts, never found a way to rid gravity of action-at-a-distance. But he invented the field concept for electrodynamics to head off a similar fate for electrical and magnetic phenomena. Maxwell incorporated Faraday's concept into his elaboration of the equations of electrodynamics. If you look at reality through "local" eyes, this approach makes eminent good sense. After all, you can wiggle some electric charges and launch an electromagnetic wave without giving any thought at all to what eventually happens to the wave. For all you know, it may well end up propagating freely at asymptotic infinity. If all you know about the totality of reality is emerging astronomical knowledge of the galaxy, as was the case through the early twentieth century, this is perfectly reasonable. But when you know more about cosmology, the know-nothing strictly local view is not so obviously reasonable.

How can the classical "local" view be squared with the action-at-a-distance picture? Well, we can't just take some source distribution with a quadrupole moment, say, a dumbbell, create a time varying quadrupole moment by spinning the dumbbell or making the masses of the dumbbell accelerate along their line of centers with respect to each other. That will just give us back the idiotically small radiation calculated by gravity wave physicists. What's missing in the dumbbell picture? The rest of the universe. How can we include it? By taking note of the fact that it acts seemingly instantaneously, so we can imagine that some non-negligible part of the whole universe is located in very close proximity to one (or the other) of our dumbbell masses.

The dumbbell mass, if you will, anchors the local system in the universe. And this anchoring mass must be present in any real system in order to accelerate the primary mass to produce the "monopole" kink in the field depicted in Fig. 2.1. That is, the idealization of a single mass that is accelerated is unrealizable, as there must always be a second reaction mass against which the accelerating agent acts to produce the acceleration of the primary mass. So all real accelerations necessarily involve quadrupoles.[15] But when we are talking about the monopole kink in the field of one of the masses, the second mass of the equivalent quadrupole is a significant part of the mass of the universe. We can consider the mass of the universe effectively present at the second dumbbell mass because of the instantaneous action-at-a-distance character of inertial effects. The radiation produced by this quadrupole is decades of orders of magnitude larger than that for the local dumbbell quadrupole taken by itself. The reaction to the quadrupole radiation produced by the effective universe-dumbbell system is the inertial reaction force that acts on the dumbbell mass being accelerated.

There are obvious problems with carrying through a calculation of the sort just sketched. Concentrating a large fraction of the mass of the universe at a point in proximity to anything will recreate the initial singularity, and so on. But the point nonetheless

---

[15] Two exceptions to this rule should be noted. First, a spherical object whose parts are undergoing a uniform radial acceleration does not radiate as the quadrupole moment is and remains zero. While such an expansion changes the radial tension in the field, it produces no "kink" in the field of the sort shown in Fig. 2.1. Second, there are those who hope to find a way to couple an object directly to the distant matter in the universe and produce accelerations without the need for an anchoring local mass. Such speculations are sometimes referred to as "field effect" propulsion. Hope springs eternal.

remains that if you insist on doing a standard quadrupole calculation, you've got to get the right quadrupole if you expect to get reasonable results. When you are considering inertial reaction forces, the right quadrupole always includes the effective universe, and it acts immediately and as if it were very, very nearby.

Wheeler and Feynman's elegant solution to the problem of radiation reaction is the only apparent way to get seemingly instantaneous reaction forces that depend on distant matter *without screwing up the dictum of the principle of relativity that limits signal propagation velocities to the speed of light.* Feynman may have harbored similar views, for he devoted the first part of his Nobel address to absorber electrodynamics.[16] In electrodynamics you can hold either view, for the two are fully equivalent. But when you come to grips with Mach's principle, you find that this is the only convincing way to deal with inertial reaction forces while preserving the finite signal velocity required by the principle of relativity.

When Mach's principle was hotly debated in the 1960s, Fred Hoyle and Jayant Narlikar figured this out and wrote papers and a book on the subject. No one paid much attention, it seems.[17] Their contemporaries may have been influenced by Hoyle's support for the "steady state" cosmology, which was then losing credibility. Wheeler's last book in the mid-1990s was an attempt to evade action-at-a-distance by invoking "constraint" equations on "initial data" that have instantaneous propagation (because they are "elliptic" rather than "hyperbolic"). Wheeler had abandoned the action-at-a-distance theory that he and Feynman had developed 50 years earlier. However, this should be evaluated keeping in mind that the *propagating kink in the field* is the field response to the acceleration of sources. Inertia is not just the action of the pre-existing gravity field on sources as they accelerate.

In the 1980s, John Cramer adapted the Wheeler-Feynman theory to quantum mechanics to explain "entanglement," another instance of seemingly instantaneous signal propagation that is customarily explained away in less than completely convincing ways. Cramer's "transactional interpretation" of quantum mechanics has not yet attracted widespread adherents. The culture of "shut up and calculate" has softened over the years. But serious examination of alternate interpretations of quantum mechanics has yet to make it into the mainstream of physics pedagogy.

Before Hoyle, Narlikar, and Cramer were others who saw the writing on the wall. Herman Weyl, the father of "gauge theory," famously remarked shortly after the first of the Wheeler-Feynman papers on action-at-a-distance electrodynamics, "Reality simply *is*, it does not *happen*." And Olivier Costa de Beauregard made early attempts to apply it to quantum theory.

The reason why the action-at-a-distance view of radiation reaction meets such stiff resistance is captured in Weyl's remark just quoted. The passage of time is an illusion.

---

[16] When I read it as a grad student in the 1960s, I thought he was nuts. But Feynman knew what he was doing. Frank Wilczek recounts (in *The Lightness of Being*, pp. 83–84) a conversation with Feynman in 1982 about fields: "... He had hoped that by formulating his theory directly in terms of paths of particles in space-time – Feynman graphs – he would avoid the field concept and construct something essentially new. For a while, he thought he had. Why did he want to get rid of fields? 'I had a slogan, ... The vacuum doesn't weigh anything [dramatic pause] *because nothing's there!* ...'" Feynman initially thought that his path integral approach captured the chief feature of the action at a distance theory: no freely propagating radiation in spacetime.

[17] Paul Davies, author of many popular books on physics, however, recounts in his *About Time* that it was attendance at one of Hoyle's lectures on this topic that set him on his early research career.

Indeed, "persistent illusion" was exactly the way Einstein characterized our notions of past, present, and future and the passage of time to the relatives of his lifelong friend Michel Besso after Besso's death, shortly before his own. The past and the future are really out there. Really. Not probably. You may think that this must all be a lot of nonsense dreamed up by people who don't have enough real work to fill their time. But let me point out that if absurdly benign wormholes are ever to be built and actually work, then this worldview *must* be correct. The past and the future must really "already" be out there. How can you travel to a past or future that doesn't "already" exist?

THE "RELATIONAL" AND "PHYSICAL" VERSIONS OF MACH'S PRINCIPLE

Should you find the forgoing confusing and contentious, you'll doubtless be disappointed to learn that we haven't yet covered the full range of arguments involving Mach's principle. As arguments about Mach's principle developed over the decades of the 1950s, 1960s, and 1970s, two distinct ways of "interpreting" the principle emerged. One came to be called the "relationalist" view, and the other we shall call the "physical" view.

Serious arguments about Mach's principle ceased to be fashionable in the mid-1970s. A few hardy souls wrote about the principle in the late 1970s and 1980s, but no one paid them much mind. Mach's principle became fashionable again in the early 1990s, and Julian Barbour and Herbert Pfister organized a conference of experts in the field held in Tübingen in the summer of 1993. The proceedings of the conference were published as volume six of the *Einstein Studies* series with the title: *Mach's Principle: From Newton's Bucket to Quantum Gravity* (Birkhauser, Boston, 1994). This is an outstanding book, not least because the questions, comments, and dialog were published, as well as the technical papers presented.

Both the relationalist and physical positions on Mach's principle were on display at the conference. Many of the attendees seem to have been convinced relationalists. The essence of the relationalist position is that all discussion of the motion of massive objects should be related to other massive objects; that relating the motion of objects to spacetime itself is not legitimate. This probably doesn't sound very much like our discussion of Mach's principle here. That's because it isn't. The relationalist approach says nothing at all about the origin of inertial reaction forces. The physical view of Mach's principle, however, does. After the conference, one of the leading critics of Mach's principle, Wolfgang Rindler, wrote a paper alleging that Mach's principle was false, for it led to the prediction of the motion of satellites in orbit around planets that is not observed – that is, the motion was in the opposite direction from that predicted by GRT. It was 3 years before Herman Bondi and Joseph Samuel's response to Rindler was published. They pointed out that while Rindler's argument was correct, it was based on the relationalist interpretation of Mach's principle. They argued that the physical interpretation that they took to be exemplified by GRT and Sciama's model for inertia gave correct predictions. Therefore, Mach's principle could not be dismissed as incorrect on the basis of satellite motion, as Rindler had hoped to do. It seems that Einstein was right in 1922, and Pais in 1982, when they remarked that Mach's principle was a missing piece of the puzzle of the origin of inertia. We should now know better. After all, the WMAP results show that *as a matter of fact* space is flat, and it is certainly not empty, so if the principle of relativity, introduced by

Galileo, is right, then Mach's principle is correct, too. And we should simply drop all of the arguments and assumptions that distract us from this conclusion.[18]

MACH'S PRINCIPLE, STARSHIPS, AND STARGATES

You may be thinking that all of this Mach's principle stuff is just too confusing and contentious to take seriously. There must be another way – one with simple principles that no one argues about – to make starships and stargates. Sorry. No such luck. The only way to build starships and stargates is by making traversable absurdly benign wormholes. That can only happen when we understand the role of inertia in gravity. It might seem to you, if this is true, that we are doomed never to build such devices. It's now a 100 years since Einstein first tried to model Mach's ideas in a vector theory of gravity, and we seem no closer to getting a version of Mach's principle that might collect a consensus.

The problem here is that Mach's principle has been understood from the first days of general relativity to be essentially a cosmological problem. Look at Einstein's statement of the principle in the quote at the beginning of this chapter. The geometry must be fully specified in terms of the sources – that is, no solutions of the field equations should exist when there are no sources, or when other "non-Machian" conditions (like rotation of the universe) exist. The fact of the matter is that non-Machian, self-consistent solutions of Einstein's equations do exist. This has led some to the view that the principle should be taken to be a boundary condition on the cosmological solutions of Einstein's equations. But even this approach yields equivocal results.

Let's look at an example of what we're talking about. In the years before Alan Guth and others proposed the cosmological models containing the process of "inflation," one of the outstanding issues of cosmology was the so-called "flatness" problem. The then prevailing preferred cosmological models – Friedman-Robertson-Walker (FRW) cosmologies – could be classified as "open" [expands forever] or "closed" [expands to some finite radius and then collapses] separated by a model that expands forever, but tends to zero expansion at temporal asymptotic infinity. The separating model is characterized by spatial flatness (and "critical" cosmic matter density) at all times. Even then (and now more so given the WMAP results), the universe looked very flat at cosmological scale. As noted above, the problem with the spatially flat model is that it is unstable. The slightest deviation from exact flatness produces very rapid evolution away from flatness – but the universe has been around for billions of years. The inflationary scenario invented by Guth and others, in fact, was intended to address precisely this problem.

---

[18] In this connection, Paul Davies relates an apposite story: "... I ventured: "What is the origin of the random phase assumption?" To my astonishment and dismay, [David] Bohm merely shrugged and muttered: "Who knows?"

"But you can't make much progress in physics without making that assumption," I protested.

"In my opinion," replied Bohm, "progress in science is usually made by *dropping* assumptions!"

This seemed like a humiliating put-down at the time, but I have always remembered these words of David Bohm. History shows he is right. ...

Most cosmologists accept the inflationary model. But it doesn't have the status of a paradigm, not yet anyway. Other cosmological models are offered for other reasons. And there is a camp that argues that the consensus cosmology is wrong for other reasons. Hoping for a consensus to emerge on Mach's principle in such circumstances is simply not realistic. Frankly, the technical details of fashionable cosmological models are not important here. If we want to build starships and stargates, do we need to wait until cosmologists decide on some model and then see if it includes Mach's principle? No! Whatever that model, if it is ever found, turns out to be, it will be one with spatial flatness. Why? Because spatial flatness is *measured* to be the fact of our reality. Spatial flatness in FRW cosmologies guarantees "critical" cosmic matter density obtains, and that guarantees $\phi = c^2$. We know that for the EEP, Mach's principle, and Newton's third law to be true, this condition must be true everywhere and at every time in local measurements. And this *must* be true no matter what cosmological model you choose to believe in.

Now, building starships and stargates is not a matter of cosmology. It is a matter of using the law of gravity and inertia at the local level. We want to find a way to manipulate stuff we can lay our hands on and figure out how to make it produce effects that will make it possible to effectively induce outrageous amounts of exotic matter. We may have to pay attention to cosmological scale effects in some circumstances. But whether the fashionable cosmological model is explicitly Machian is really irrelevant to what we are up to. So we accept the physical version of Mach's principle – the assertion that inertial reaction forces are gravitational, and mass is just the total gravitational potential energy divided by the square of the speed of light – and ask: does the principle lead to any effects that we might be able to use to make starships and stargates? We address the answer to this question in the next chapter. Here, to sum up, we note that one way to suppress the confusion surrounding Mach's principle is to codify the principle in the form of a simple law or two. Imagine trying to do mechanics without Newton's laws, or electrodynamics without Maxwell's equations, or relativity without Einstein's laws. Therefore let's propose the adoption of the Mach-Einstein-Sciama laws of inertia:

First law: $\phi = c^2$ locally always; or, inertial reaction forces are due to the gravitational action of causally connected "matter", where matter is understood as everything that gravitates.

Second law: $m = E/\phi$, or the mass of an entity (isolated and at rest) is equal to its non-gravitational energy divided by the locally measured total gravitational potential.

A zeroth law might be added: Inertial reaction forces are instantaneous. But that is arguably belaboring the obvious. The first and second laws, in contradistinction, are not obvious. We will use these laws, and mostly ignore cosmology, to derive some interesting local effects that may make stargates possible. Cosmology will only come back into our consideration after those effects have been derived and some experimental work aimed at detecting them has been presented.

ADDENDA

Addendum #1: On the Origin of Inertia Article

It is convenient to begin by calculating the potential at a test-particle that is at rest in a universe containing no irregularities. Since our field equations have the same form as Maxwell's, we can use electrodynamic formulae to calculate the potential, and to bring out the analogy with electrodynamics we use a similar notation and terminology, but we emphasize that in this paper we shall be concerned with purely gravitational phenomena.

Retardation effects are taken to arise in the same way as in electrodynamics, so that the contribution of any region of the universe to the potential at a point P at time t is computed by ascribing to that region just the properties that are observed at P at time t.

We thus have for the scalar potential (8)

$$\Phi = -\int_V \frac{\rho}{r} dV. \tag{1}$$

We use the minus sign in (1) because inertial mass then turns out to be positive, but in fact either sign can be used (Section 4(vii)). The vector potential **A** vanishes by symmetry.

We shall assume that matter receding with velocity greater than that of light makes no contribution to the potential, so that the integral in (1) is taken over the spherical volume of radius $c\tau$. An assumption of this sort is necessary since we have naïvely extrapolated the Hubble law without considering relativistic effects, and should give the correct order of magnitude. A relativistic treatment is given in II.

Since the density is supposed uniform, (1) gives

$$\Phi = -2\pi\rho c^2\tau^2. \tag{2}$$

Owing to our assumptions, the numerical factor 2π is only approximate.

We now calculate the potentials for the simple case when the particle moves relative to the smoothed-out universe with the small rectilinear velocity $-\mathbf{v}(t)$. In the rest-frame of the particle the universe moves rectilinearly with velocity $\mathbf{v}(t)$. Now at time t there will be observable at the particle, in addition to the Hubble effect, a Doppler shift corresponding to $\mathbf{v}(t)$ from all parts of the universe. Hence, in computing the potential in the rest-frame of the particle at time t, we must ascribe to every region of the universe the velocity that is observed at time t, that is, $\mathbf{v}(t) + \mathbf{r}/\tau$.

Neglecting terms of order v^2/c^2, we have

$$\Phi = -2\pi\rho c^2\tau^2$$

as before. The vector potential no longer vanishes, but has the value

$$\mathbf{A} = -\int_V \frac{\mathbf{v}\rho}{cr} dV. \tag{3}$$

Since **v** is independent of r, we can take it outside the integral. We then obtain

$$\mathbf{A} = \frac{\Phi}{c}\mathbf{v}(t).$$

Since the change of ρ with time is very small, the gravelectric part of the field is approximately

$$\mathbf{E} = -\operatorname{grad}\Phi - \frac{1}{c}\frac{\partial\mathbf{A}}{\partial t}$$
$$= -\frac{\Phi}{c^2}\frac{\partial\mathbf{v}}{\partial t},$$

while the gravomagnetic field is

$$\mathbf{H} = \operatorname{curl}\mathbf{A} = 0.$$

"On the Origin of Inertia" by D.W. Sciama, *Monthly Notices of the Royal Astronomical Society*, vol. 113, pp. 34–42. Reprinted under Wiley's fair dealing policy.

Addendum #2: Brans on Gravitational Energy Localization

PHYSICAL REVIEW   VOLUME 125, NUMBER 1   JANUARY 1, 1962

Mach's Principle and the Locally Measured Gravitational Constant in General Relativity*

CARL H. BRANS†

Palmer Physical Laboratory, Princeton University, Princeton, New Jersey

(Received August 14, 1961)

It has been conjectured that a "Mach's principle" might lead to a dependence of the local Newtonian gravitational constant, K, on universe structure, $K^{-1} \sim M/R$. Einstein and others have suggested that general relativity predicts such a result. A closer analysis, however, including the carrying out of the geodesic equations to second order, seems to indicate that this is not true and that the apparent "Mach's principle" terms involving total universe structure are really only coordinate effects. Further, the measure of gravitating mass obtained in a local, proper Newtonian gravitational experiment is compared in a coordinate-free way to an experimentally measurable inertial mass and found to be related to it in a way independent of the rest of the universe. A generalization of these results is given. It is based on the fact that in general relativity the only way the universe can influence experiments done in an electrically shielded laboratory is through the metric and that this can be "transformed away" to any degree of accuracy for a sufficiently small laboratory. Consequences of this are summarized in Dicke's "strong principle of equivalence." It is noted, however, that there are other statements which might be called "Mach's principles" which are satisfied in general relativity.

I. INTRODUCTION

THE principal idea which guided Einstein in formulating the general theory of relativity was the local equivalence of gravitational and inertial effects, that is, the equivalence of a uniform gravitational force field and a constant acceleration of the reference frame. Another idea relating gravity and inertia is Mach's principle. This is less precisely formulated but suggests that the inertial properties of a body are determined by the distribution of matter in the universe. Since the gravitational field interacts with *all* matter, one could hope to see the Mach principle relationship between inertial and distant matter described in terms of the gravitational field. To state this in a way independent of units, consider the ratio of the inertial mass of a body to its active gravitational mass.[1]

In particular, let us see that this ratio might be in a static universe consisting only of a mass shell of radius R and inertial mass M together with a relatively small body of inertial mass m at its center. If we probe the gravitational field of m with a small test particle, we might expect from the Eötvös experiment that the acceleration of the test particle is independent of its mass. It certainly depends, however, on m and r and conceivably on M and R. The fact that the Newtonian theory of gravity is valid to a high degree of accuracy suggests that for $m \ll M$, $r \ll R$, the acceleration is

$$a = -[m/r^2 F(M,R)], \qquad (1.1)$$

where F is a function of dimensions mass over length (velocity of light $c = 1$). Dimensional analysis then suggests

$$F = AM/R, \qquad (1.2)$$

where A is a constant dimensionless number. For a more general type of universe with masses m_a at distances r_a from some point x, this might be extended to

$$F(x) = A \sum_a m_a / r_a. \qquad (1.3)$$

Until recently, experimental determinations of F from (1.1) were possible only on the earth. The value found is not inconsistent with (1.3), a positive value of A in the neighborhood of 10^0 or 10^1, and present astronomical knowledge of m_a and r_a. It is clear that in a uniform universe, $m_a \sim r_a^2$, so that the dominant contribution to the sum on the right side of (1.3) comes from distant matter and the resulting $F(x)$ is fairly constant in space and time. This also is consistent with present observations.

A comparison of (1.1) with the standard classical Newtonian theory of gravity shows that F^{-1} plays the role of Newton's "universal gravitational constant." However, if (1.3) is true, this number is not a universal constant but depends on the distribution of mass in the universe about the point where it is measured. To investigate possible resulting changes in value of this number, it is convenient to introduce a standard value and refer variations to it. Specifically, let $K_0/8\pi$ be defined as the presently observed terrestrial value of $F(x)^{-1}$. $K_0/8\pi$ is thus a constant number of dimensions length over mass. Then rewrite (1.1) as

$$a = -\frac{K_0}{8\pi r^2}\left(\frac{8\pi m}{K_0 F}\right). \qquad (1.4)$$

This equation is identical with Newton's if the quantity in parentheses, $m_g \equiv 8\pi m/K_0 F$, is taken to be the active gravitational mass[1] associated with m. Notice that by definition of K_0, this gives $m_g = m$ at the present time on earth. However, if (1.3) is true, a Cavendish-type experiment interpreted in the context of a Newtonian theory with fixed gravitational constant $K_0/8\pi$ would give a measurement of active gravitational mass m_g

* Based on part of a multilithed Ph.D. thesis submitted to Princeton University.
† National Science Foundation Predoctoral Fellow, 1957–1960. Now at Loyola University, New Orleans, Louisiana.
[1] H. Bondi, Revs. Modern Phys. **29**, 423 (1957).

MACH'S PRINCIPLE IN GENERAL RELATIVITY

yielding a ratio

$$\frac{m}{m_g} = A \sum_a \frac{K_0 m_a}{8\pi r_a}, \tag{1.5}$$

which would not necessarily always be unity.

Einstein[2] claims to find such a result in general relativity. In order to study this problem, consider the creation of relatively small masses m_a', at distances r_a', from the present standard laboratory in which, prior to the creation of m_a', $m/m_g = 1$ by definition of K_0. With m_a' present, however, (1.5) then yields

$$\frac{m}{m_g} = 1 + A \sum_{\text{"new matter"}} \frac{K_0 m_a'}{8\pi r_a'}. \tag{1.6}$$

If it is assumed that each $K_0 m_a'/8\pi r_a'$ is small compared to unity, the weak-field equations might be used to check (1.6). Einstein does this and arrives at

$$\left(1 + \sum_{\text{"new matter"}} \frac{K_0 m_a}{8\pi r_a}\right) a \cong -\frac{K_0 m}{8\pi r^2}. \tag{1.7}$$

Thus

$$\frac{m}{m_g} = 1 + \sum_{\text{"new matter"}} \frac{K_0 m_a}{8\pi r_a}, \tag{1.8}$$

which is identical with (1.6) if $A = 1$. Einstein argued from this that since some matter contributes to the ratio, m/m_g, all the universe probably does (Sec. II). There has been some discussion[3] of what the numerical coefficient A of the sum in the right side of (1.8) should be, and indeed the first approximation procedure seems inadequate to resolve this. Consequently, the equations of motion through second order will be applied to this problem in Sec. II.

This result (1.8), or its corrected form (2.11), is clearly coordinate dependent, however. Hence the relationship between its numerical description of the path of a particle and the actually observed path is not defined without further analysis. The usual interpretation of general relativity is based on the identification of the *invariant* theoretical measure of an interval, proper time, with time experimentally measured in some fundamental way, e.g., on an atomic clock. An invariant measure of distance and thus acceleration can be obtained from this by setting the velocity of light equal to one. When this is done, the invariant description of the path of a test particle relative to a central mass is found to be approximately Newtonian with coefficients independent of the rest of the universe. (See Sec. III.)

However, the number m appearing in the left side of (1.5) has not yet been related to an experimentally measured inertial mass. To remedy this, a description of a process for invariantly studying the acceleration of charged bodies in a known electric field is given. The

resultant ratio of "force" to acceleration is defined as the inertial mass. For a simple theory of matter m_{inert} is found to be just the m appearing in (1.5) (Sec. IV). This procedure assumes given standards of charge and time interval.

The independence of the relationship between the two numbers, m_g and m_{inert}, from the rest of the universe is more generally true than the above special case might indicate. In fact, assume that the space in the neighborhood of an electrically shielded laboratory is sufficiently flat that in a certain coordinate system the differences between the metric components and those of the Minkowskian, together with the first two derivatives of these differences, are negligible over the laboratory. Then, according to general relativity, if small masses, charged or uncharged, are introduced into the laboratory, the description of their motions and interactions in this coordinate system is independent of the rest of the universe. This is due to the fact that once the laboratory is shielded, the only way the rest of the universe could influence it, according to general relativity, is through the metric. If this is sensibly flat within, its influence can be transformed away by a coordinate transformation, thus eliminating any effects from the rest of the universe. This is Dicke's "strong principle of equivalence."[4] (See Sec. V.)

There are, however, other statements which might be considered Mach's principles. These are based on the fact that in general relativity gravitational and inertial forces have the same formal origin. (See Sec. V.)

II. EINSTEIN'S RESULTS

Gravity and general relativity being largely concerned with the interaction between masses as masses, Einstein was naturally interested in whether or not Mach's principle as discussed in Sec. I above was satisfied in general relativity. Specifically, is the attraction and resultant relative motion of two gravitating bodies influenced by the rest of the universe?

Einstein investigated this in the weak-field approximation.[2] The metric he found to represent the gravitational field due to a distribution of small masses corresponding to a "density" σ and having small velocities, dx^i/ds, can be written as

$$g_{00} = 1 - \frac{K}{4\pi} \int \frac{\sigma dV}{r},$$

$$g_{0i} = \frac{K}{2\pi} \int \frac{\sigma(dx^i/ds)}{r} dV, \tag{2.1}$$

$$g_{ij} = -\delta_{ij}\left(1 + \frac{K}{4\pi} \int \frac{\sigma dV}{r}\right),$$

[2] A. Einstein, *The Meaning of Relativity* (Princeton University Press, Princeton, New Jersey, 1955), 5th ed., pp. 99–108.
[3] W. Davidson, Monthly Notices Roy. Astron. Soc. 117, 212 (1957).

[4] R. H. Dicke, Science 129, 621 (1959). See also Revs. Modern Phys. 29, 355 (1957); J. Wash. Acad. Sci. 48, 213 (1958); Am. J. Phys. 28, 344 (1960).

CARL H. BRANS

on replacing Einstein's imaginary time x^4 by the real $x^0 = -ix^4$. Here K is just the constant introduced in the Einstein field equations and thus not yet related to K_0 or other observed numbers. Equation (2.1) is correct only to first order in $K\int\sigma dV/r$, and dx^i/ds. The geodesic equation for a test particle in this field becomes

$$\frac{d}{dx^0}[(1+\bar\sigma)\mathbf{v}] = \nabla\bar\sigma + \frac{\partial \mathbf{A}}{\partial x^0} + (\nabla\times\mathbf{A})\times\mathbf{v}, \quad (2.2)$$

where

$$\mathbf{v} \equiv d\mathbf{x}/ds,$$

$$\bar\sigma \equiv \frac{K}{8\pi}\int\frac{\sigma dV}{r}, \quad (2.3)$$

$$\mathbf{A} \equiv \frac{K}{2\pi}\int\frac{\sigma\mathbf{v}}{r}dV.$$

For simplicity, consider the application of these results to the case of the motion of a test particle near a small mass m at rest at the origin, all inside a static, spherical shell of mass M_s and radius R_s,[5] (2.2) now becomes

$$\frac{d}{dx^0}\left[\left(1+\frac{KM_s}{8\pi R_s}+\frac{Km}{8\pi r}\right)v^i\right] = \frac{Km}{8\pi}\frac{\partial}{\partial x^i}\left(\frac{1}{r}\right). \quad (2.4)$$

Thus, $(1+KM_s/8\pi R_s+Km/8\pi r)$ times the coordinate acceleration of the test particle is just the Newtonian term, to this approximation. Einstein interpreted this by saying that the "inert mass is proportional to $1+\bar\sigma$,"[2] or in (2.4) to $1+(K/8\pi)(M_s/R_s+m/r)$. However, an equivalent statement, more convenient for this discussion and in keeping with that of Sec. I, can be made. Specifically, dividing (2.4) by $[1+(K/8\pi)\times(M_s/R_s+m/r)]$ gives, for v_i instantaneously zero,

$$\frac{d}{dx^0}v^i = \frac{Km}{8\pi[1+(K/8\pi)(M_s/R_s+m/r)]}\frac{\partial}{\partial x^i}\left(\frac{1}{r}\right). \quad (2.5)$$

This, in keeping with Einstein's interpretation above, would suggest that the locally measured Newtonian active gravitational mass of m is

$$m_a = m/[1+(K/8\pi)(M_s/R_s+m/r)], \quad (2.6)$$

or that the effective, locally measured Newtonian gravitational constant is

$$K_E = K/[1+(K/8\pi)(M_s/R_s+m/r)]. \quad (2.7)$$

If this is true, a comparison of (2.6) with (1.5) would show that a Mach's principle in the sense of Sec. I would be satisfied in general relativity, since the number K_E in (2.7) measuring the attraction of m for

---

[5] This example, while admittedly rather specialized, is sufficient to illustrate the ideas under consideration. It should also be noted that here $KM_s/R_s \ll 1$ so that this does not correspond to the total "universe mass shell" discussed in Sec. I and for which $KM/R \sim 1$.

test particles would depend on the mass distribution M_s/R_s in the rest of the universe. To clarify the relation of (2.6) and (2.7) to the discussion in Sec. I, it is necessary to consider M_s and m as small additions to a background universe [i.e., as the m_a' were in the discussion preceding (1.6) above]. For the background universe assume that K has been chosen equal to K_0. Thus, (2.6) will coincide with (1.6) if $A=1$ in the latter.

V. SUMMARY AND GENERALIZATION

This section will be mainly concerned with investigating some of the consequences of the fact that in general relativity the entire gravitational interaction between masses is carried by the metric tensor which can be "transformed away" to any desired degree of accuracy over a sufficiently small neighborhood of any point. This fact leads naturally to the following definition relating a standard physical laboratory to a mathematical "coordinate patch." A locally almost Minkowskian coordinate system is one in which test particles of any velocity experience no observable acceleration when there is no matter or radiation present in the laboratory. The description of experiments done in a standard physical laboratory is assumed to correspond to the mathematical description given by such a coordinate system.

Using this definition, Dicke's[4] strong principle of equivalence can be defined as the assertion that as far as inertial and gravitational effects are concerned, the numerical content of experiments described in a locally almost Minkowskian coordinate system is independent of any characteristics of the mass distribution in the rest of the universe. It is important to realize that this is a definite extension of such results of the Eötvös experiment as generalized in the weak principle, i.e., the assertion that the acceleration of a test particle instantaneously at rest relative to a small gravitating body is independent of the mass of the test particle in the limit as this mass goes to zero. In other words, the Eötvös experiment suggests that the acceleration effects of an external gravitating body on a sufficiently small laboratory can be at least approximately eliminated by allowing the laboratory to fall "freely" since it seems to imply that all parts of the laboratory would fall with very little, if any, relative acceleration. However, it contains nothing to suggest that the only effect of the gravitating body on the laboratory is accelerative, which is the basis for the strong principle.

A sketch of an argument generalizing the results of Sec. IV and suggesting the validity of a strong principle in general relativity follows.

Consider a region having space-time dimensions, in arbitrary but fixed units, bounded by a number ϵ. This region is to represent the space contained in a laboratory in which standard experiments are to be performed. Let the matter tensor in the laboratory be represented by λT_L (here and in the following, to avoid unnecessary clutter, tensor indices will be suppressed when no confusion will arise), where λ is a positive number. Further, let the matter tensor for the rest of the universe be T_U and assume that $T_U = 0$ within the laboratory, while $T_L = 0$ outside it. The total matter tensor is thus $T_U + \lambda T_L$ everywhere. The purpose of the following discussion is then to show that under certain conditions the influence of the "rest of the universe" on real, proper experiments done in such a laboratory can be made arbitrarily small by making ϵ sufficiently small. The crux of the argument is the fact that the observable outcome of such experiments cannot depend on the purely mathematical choice of coordinate systems in which the calculations are performed.

To this end, let ρ (again suppressing indices) stand for all the matter variables other than the metric, ρ_L referring to matter in the laboratory, and ρ_U to all other matter. Thus, λT_L is a function of ρ_L and T_U is a function of ρ_U. Assume the variables satisfy "equations of motion"

$$f(\rho, g, g') = 0, \tag{5.1}$$

where g' stands for all first derivatives of g. Further, let the metric, g, be written as the sum of two parts $^0g + \gamma(\lambda)$, with ^0g independent of λ and where $\lim \gamma(\lambda) = 0$ as $\lambda \to 0$.

Let $^0\rho$ represent the functional form of ρ when $\lambda = 0$. Hence, when there is no matter within the laboratory, $\lambda = 0$, and $^0\rho$ and ^0g satisfy

$$f(^0\rho, {}^0g, {}^0g') = 0, \tag{5.2}$$

$$S(^0g) - T_U(^0\rho) = 0 \quad (S^{\alpha\beta} \equiv R^{\alpha\beta} - \tfrac{1}{2}g^{\alpha\beta}R). \tag{5.3}$$

If ρ_{null} represents the form of ρ corresponding to the vacuum and η is the Minkowski metric, then it will be assumed that

$$f(\rho_{\text{null}}, \eta, 0) = 0, \tag{5.4}$$

$$T_U(\rho_{\text{null}}) = 0. \tag{5.5}$$

The two most important assumptions will now be made. Within the space of the laboratory it is assumed

that (1) $^0\rho = \rho_{\text{null}}$, and (2) the differences between $^0g + \gamma(\lambda)$ and η together with the first two derivatives of $^0g + \gamma(\lambda)$ go continuously to zero with ϵ and λ. The first assumption is simply that when $\lambda = 0$ there is really no matter or fields within the laboratory. In other words, it ensures that when the tensor for matter in the laboratory, λT_L, is zero, the matter variables, ρ_L, actually correspond to the vacuum. This assumption is probably unnecessary for the ordinary descriptions of matter. The second assumption may seem strong in its requirement on the second derivatives of the metric. However, it will be used in the argument following Eq. (5.13).

Similarly, let $^0\rho_L$ and $\eta + {}^0\gamma$ be the matter variables and metric describing the situation inside the laboratory in the absence of any matter outside, i.e., when $\rho_U = \rho_{\text{null}}$. Thus, by definition,

$$f(^0\rho_L, \eta + {}^0\gamma, {}^0\gamma') = 0, \tag{5.6}$$

$$S(\eta + {}^0\gamma) - \lambda T_L(^0\rho_L) = 0. \tag{5.7}$$

Finally, the full field equations can be written

$$f(\rho, {}^0g + \gamma, {}^0g' + \gamma') = 0, \tag{5.8}$$

$$S(^0g + \gamma) - \lambda T_L(\rho_L) - T_U(\rho_U) = 0. \tag{5.9}$$

In particular, within the laboratory,

$$f(\rho_L, {}^0g + \gamma, {}^0g' + \gamma') = 0, \tag{5.10}$$

$$S(^0g + \gamma) - \lambda T_L(\rho_L) = 0. \tag{5.11}$$

However, by assumption, within the laboratory ^0g differs from η, and its first two derivatives from zero, only by numbers which go to zero as $\epsilon \to 0$. Thus (5.10) and (5.11) can be rewritten as

$$f(\rho_L, \eta + \gamma, \gamma') = H, \tag{5.12}$$

$$S(\eta + \gamma) - \lambda T_L(\rho_L) = E, \tag{5.13}$$

where $H \to 0$ as $\epsilon \to 0$ and $E \to 0$ as $\epsilon \to 0$. Notice that since S depends on the second derivatives of $^0g + \gamma$, it is sufficient that these vanish as $\epsilon \to 0$ for $E \to 0$ as $\epsilon \to 0$. Actually, this condition may not also be necessary, but this point is irrelevant to the main argument.

The final result is thus that the variables, ρ_L and $\eta + \gamma$, satisfy, within the laboratory, Eqs. (5.12) and (5.13) which differ from those, (5.6) and (5.7), satisfied by the corresponding variables in the absence of matter in the rest of the universe only by functions H and E which can be made arbitrarily small by making ϵ sufficiently small.

Thus, it seems reasonable to expect that for each λ, the solutions with matter in the rest of the universe, $^0g + \gamma$ and ρ_L, and those with matter only in the laboratory, $\eta + {}^0\gamma$ and $^0\rho_L$, can be brought arbitrarily close together by making the laboratory sufficiently small. Further, the outcome of proper, local experiments done in such a laboratory can depend only on the behavior of the metric and matter variables within it. Thus, the

CARL H. BRANS

results of such experiments can be made as nearly independent of the matter in the rest of the universe as desired by making ϵ sufficiently small.

Of course, the definition of quantities to be measured[11] and local laws to be tested within the laboratory may require $\lambda \to 0$. It might then be thought that for λ small enough the effects of the matter in the rest of the universe would become comparable to those of matter within the laboratory, vitiating the above argument. To prevent this, a lower limit for λ is demanded. This limit could be determined by the lower bound of available experimental accuracy for the measurements requiring $\lambda \to 0$. That is, values of λ below this limit would not produce observable differences in measurements. For this fixed λ, ϵ can then be determined as above.

There are, however, other statements which might possibly be called "Mach's principles" which are valid in general relativity. For example, inertial and gravitational forces have a common formal origin in general relativity. Specifically, for a test particle of mass m and velocity w^β,

$$F^\mu \equiv -m\Gamma_{\alpha\beta}{}^\mu w^\alpha w^\beta \qquad (5.14)$$

might be identified with the gravitational force acting on m. On the other hand, this quantity transforms just as an inertial force should, i.e., in going to a relatively accelerated system, the acceleration enters F^μ linearly. For example, in a coordinate system rotating relatively to a Lorentz system in a flat space, F^μ as defined in (5.14) contains the centrifugal and Coriolis forces experienced by particles in this rotating system.

---

[11] For example, inertial mass. See Eq. (4.23) and the discussion following it.

Thus, F^μ might also be identified with "inertial force" acting on m. Inertial coordinate systems would then be those in which F^μ vanishes or equivalently, those in which "free" uncharged test particles are unaccelerated. This coincides with the definition of locally almost Minkowskian coordinate systems above. Another way of saying this is that the locally almost Minkowskian or inertial coordinate systems are those in which the total gravitational force vanishes.

If suitable boundary conditions could then be exhibited for a general type of universe, the Einstein equation would predict the over-all state of motion of inertial frames relative to the total mass distribution in the universe. This statement alone has been mentioned as a "Mach's principle."[12] However, once it is required that fundamental, standard experiments be done within such frames, the rest of the universe cannot, in general relativity, influence their results.

Another paper[13] will discuss modifications of general relativity violating the strong principle of equivalence by the introduction of a variable gravitational "constant" determined through field equations by the mass distribution in the universe.

ACKNOWLEDGMENTS

The author is grateful to R. H. Dicke and C. W. Misner for many helpful discussions of these problems. He also wishes to thank B. Block, W. Morgan, and J. Peebles for their many helpful comments.

---

[12] For a general discussion see F. A. E. Pirani, Helv. Phys. Acta, Suppl. IV, 198 (1956). Actually, Pirani's "Mach's principle" is stronger than that mentioned above. His requires that inertial systems be nonrotating relative to some average mass density in the universe.

[13] C. H. Brans and R. H. Dicke, Phys. Rev. 124, 925 (1961).

Addenda #3: Excerpt from Nordtvedt

International Journal of Theoretical Physics, Vol. 27, No. 11, 1988

Existence of the Gravitomagnetic Interaction

Ken Nordtvedt[1]

Received September 25, 1987

The point of view expressed in the literature that gravitomagnetism has not yet been observed or measured is not entirely correct. Observations of gravitational phenomena are reviewed in which the gravitomagnetic interaction—a post-Newtonian gravitational force between moving matter—has participated and which has been measured to 1 part in 1000. Gravitomagnetism is shown to be ubiquitous in gravitational phenomena and is a necessary ingredient in the equations of motion, without which the most basic gravitational dynamical effects (including Newtonian gravity) could not be consistently calculated by different inertial observers.

1. INTRODUCTION

In the overview *Physics Through the 1960s*, the National Academy of Sciences (1986) review of opportunities for experimental tests of general relativity, they declare that "At present there is no experimental evidence arguing for or against the existence of the gravitomagnetic effects predicted by general relativity. This fundamental part of the theory remains untested." Similar points of view have been expressed elsewhere in promotion of various experiments designed to "see" gravitomagnetism.

In this paper I make two points on this issue, which together lead to a position contrary to the viewpoint summarized by the above statement.

1. The gravitomagnetic interaction is a consequence of the gravitational vector potential. This vector potential pays a crucial, unavoidable role in gravitation; without the gravitational vector potential the simplest gravitational phenomena—the Newtonian-order Keplerian orbit and the deflection of light by a central body—cannot be consistently calculated in two or more inertial frames of observation. Gravitation without the vector potential is an incomplete, ambiguous theory in the most fundamental sense.

[1]Physics Department, Montana State University, Bozeman, Montana 59717.

0020-7748/88/1100-1395$06.00/0 © 1988 Plenum Publishing Corporation

5. DRAGGING OF INERTIAL FRAMES AND MACH'S IDEAS

What seems to have especially caught the interest of physicists in searching for the spin–spin interaction in gravity is that this would seem to be a manifestation of ideas of Mach, who a century ago believed that inertia was caused, in some sense, by the universe's matter distribution. Lense and Thirring later showed that, indeed, in general relativity rotating matter would drag the inertial frame around at a slow rate which fell off with distance from the rotating matter,

$$\Omega = \frac{G}{c^3}\left(\frac{\mathbf{J}-3\mathbf{J}\cdot\hat{\mathbf{r}}\hat{\mathbf{r}}}{r^3}\right) \tag{16}$$

$\mathbf{J}$ is the angular momentum of the spinning body and $\mathbf{r}$ is the distance to the point of space in question, $\Omega(\mathbf{r})$ is the rotation rate and rotation axis for the inertial space at that point of space which is induced by the spinning source. Equation (16) follows from (12) with choice of PPN coefficients appropriate to general relativity, and the identification

$$\Omega = -\frac{c}{2}\nabla\times\mathbf{h}$$

Looking at the general case, one can ask what is the complete effect of the gravitational vector potential in dragging inertial frames? This question can be addressed by calculating the contribution of $\mathbf{h}$ in establishing the geodesic coordinate frames (inertial frames). The general formula

$$[x^\gamma - x^\gamma_{(0)}]' = [x^\gamma - x^\gamma_{(0)}] + \tfrac{1}{2}\Gamma^\gamma_{\alpha\beta}[x^\alpha - x^\alpha_{(0)}][x^\beta - x^\beta_{(0)}] \tag{17}$$

in which $\Gamma^\gamma_{\alpha\beta}$ are the Christoffel symbols produced from first derivatives of the gravitational metric field, gives the transformation from original space-time coordinates x^γ to inertial (geodesic) coordinates $x^{\gamma'}$ in the vicinity of any chosen space-time point $x^\gamma(0)$. Examining solely the vector potential (g_{0i}) contribution to (17) yields

$$[\mathbf{r}-\mathbf{r}_{(0)}]' = [\mathbf{r}-\mathbf{r}_{(0)}] - c\left[\frac{1}{2}\frac{\partial\mathbf{h}}{\partial t}(t-t_0)^2 + \left(\frac{\nabla\times\mathbf{h}}{2}\right)\times(\mathbf{r}-\mathbf{r}_{(0)})(t-t_0)\right] \tag{18}$$

The gravitational vector potential produces in this general case a "dragging" of inertial space at each locality with both an acceleration of the inertial frame at rate

$$\mathbf{a}(\mathbf{r}, t) = -c\,\partial\mathbf{h}/\partial t \tag{19a}$$

and a rotation of the inertial frame at angular rate and axis

$$\Omega(\mathbf{r}, t) = -\tfrac{1}{2}c\nabla\times\mathbf{h} \tag{19b}$$

If we return to the problem of light deflection by a body moving at speed w and employ the vector potential given by (7), we find that (19a) gives no contribution to the light ray deflection; however, (19b) produces a rotational dragging of inertial frames at a rate

$$\Omega(r, t) = (1 + \gamma)\frac{GMDw}{c^2}\frac{1}{|\mathbf{r} - \mathbf{w}t|^3}$$

and in a counterclockwise sense. The time integral of this rotation rate over the entire trajectory of the light ray produces the total deflection or rotation angle

$$\delta\theta = -\frac{2w}{c}\theta_0$$

which is what is needed to obtain agreement with (5) as discussed in Section 2.

The periastron precession of the binary pulsar orbit discussed previously received contributions of inertial frame dragging from both (19a) and (19b). The situation can be viewed this way; part of the motion of the two bodies in the binary pulsar results from the "Coriolis" acceleration that each body experiences because the motion of the other body is producing rotational dragging of the inertial frame at the locality of each body in question.

Finally, the accelerated celestial body mentioned previously drags the inertial frames through (19a), with the resulting acceleration of inertial space being

$$\delta\mathbf{a}(\mathbf{r}, t) = -\left(2 + 2\gamma + \frac{\alpha_1}{2}\right)\frac{U(\mathbf{r}, t)}{c^2}\mathbf{a}$$

in which $U(\mathbf{r})$ is the Newtonian potential function of that body's mass distribution and $\mathbf{a}$ is the body's acceleration.

6. CONCLUSION

The gravitomagnetic interaction—the post-Newtonian gravitational interaction between moving masses—has been observed and measured in a number of different phenomena. The strength of this interaction is now known to an accuracy of 1 part in 1000. The gravitomagnetic interaction is also required in order to have a complete and consistent theory of gravity at all: even static source gravitational effects when viewed in another inertial frame require the gravitomagnetic interaction in order for basic consistency of a theory's equations of motion. Just as in electromagnetic theory, there is no absolute separation of "electric" and "magnetic" effects; such a division is inertial frame dependent.

Ken Nordtvedt, "Existence of the Gravitomagnetic Interaction," *International Journal of Theoretical Physics*, vol. 27, pp. 1395–1404. Reprinted with permission of Springer Verlag.

3

Mach Effects

SOURCES AND FIELDS

Gravitation, including the Mach-Einstein-Sciama laws of inertia, and by extension Mach effects, involve the interrelationship between a field and its sources – notwithstanding that the field in the case of gravitation, at least in General Relativity Theory (GRT), is spacetime itself. So, what we want to look at is the "field equations" for gravity. We'll find that we do not need to examine the full tensor field equations of Einstein, for we are looking for hopefully fairly large effects – and they, if present, may be expected to occur "at Newtonian order." That is, anything big enough to be of any use in revolutionary propulsion is likely to be something present at the scale of Newtonian effects that has been missed in earlier work.

The customary way to write a field equation is to collect all of the terms involving the field(s) of interest on the left hand side of the equation(s), and arrange things so that the terms involving the sources of the field appear on the right hand side. Usually, this is a pretty much straightforward matter. Armed with one's field equations, one usually stipulates a particular arrangement of the sources and then solves for the corresponding field quantities. The field(s) are then allowed to act on the sources where the fields are calculated, and so on.

Standard techniques have been worked out since the time of Newton for doing these computational procedures. We will be using standard procedures, but instead of asking the question, what are the fields for a given arrangement of sources, we will be asking a somewhat different question. When a source of the gravitational field is acted upon by an external force – we know that the action of the gravitational field is to produce the inertial reaction force experienced by the agent applying the external force – what effect does the action of the gravitational force on the object being accelerated have on the *source*?

This may seem a silly question. How could a field acting on a source in any circumstances change the source? But if we hope to manipulate inertia, this question is one worth asking. For, in the last analysis, all we can ever do is apply forces to things and hope to be able to produce effects that enable us to do what we want to do.

J.F. Woodward, *Making Starships and Stargates: The Science of Interstellar Transport and Absurdly Benign Wormholes*, Springer Praxis Books, DOI 10.1007/978-1-4614-5623-0_3, © James F. Woodward 2013

Since the advent of GRT, those scientists interested in gravity have pretty much ignored Newtonian gravity. After all it is, at best, just an approximation to Einstein's correct theory of gravity. Engineers, however, have worked with Newtonian gravity all along. If you are doing orbit calculations for, say, a spacecraft on an interplanetary mission, the corrections to Newtonian mechanics from GRT are so utterly minuscule as to be irrelevant for practical purposes. Why engage in lengthy, tedious, and complicated calculations using GRT and increase the risk of miscalculation when the Newtonian gravity approximation is more than sufficient? True, the same cannot be said in the case of GPS calculations because the timing involved is more than precise enough to make GRT corrections essential. For your vacation trip, or shopping downtown, being off by as much as a 100 m or more is likely inconsequential. But if you are trying to blast the bunker of some tin-horned dictator, getting the position right to less than a few meters does make a difference (unless you are using a tactical nuke, in which case being off by up to half a kilometer probably won't matter).

RELATIVISTIC NEWTONIAN GRAVITY

The relativistic version of Newtonian gravity gets mentioned in texts, but the field equations for relativistic Newtonian gravity do not get written out in standard vector notation as a general rule. Why bother writing down something that's a crude approximation? Nonetheless, George Luchak did so in the early 1950s, when constructing a formalism for the Schuster-Blackett conjecture. The Schuster-Blackett conjecture asserts that rotating, electrically neutral, massive objects generate magnetic fields. Were this true, it would couple gravity and electromagnetism in a novel way.

Luchak was chiefly interested in the anomalous coupling terms that get added to Maxwell's equations if the conjecture is true, so relativistic Newtonian gravity was a good enough approximation for him. Accordingly, contrary to established custom, instead of using the four dimensions of spacetime for tensor gravity and adding a fifth dimension to accommodate electromagnetism, he wrote down the equations of electromagnetism in the four dimensions of spacetime, and in the fifth dimension he wrote out a vector formalism for the scalar Newtonian approximation of relativistic gravity. Along with the curl of the gravity field **F** being zero, he got two other equations, one of which being of sufficient interest to be worth writing out explicitly:

$$\nabla \bullet \mathbf{F} + \frac{1}{c}\frac{\partial q}{\partial t} = -4\pi\rho. \tag{3.1}$$

ρ is the matter density source of the field **F**, and q is the rate at which gravitational forces do work on a unit volume. (The other equation relates the gradient of q to the time rate of change of **F**.)[1] The term in q in this equation appears because changes in gravity now propagate at the speed of light. It comes from the relativistic generalization of force,

---

[1] For formalphiles, the equations are: $\nabla \times \mathbf{F} = 0$ and $\nabla \bullet q + \frac{1}{c}\frac{\partial \mathbf{F}}{\partial t} = 0$.

namely, that force is the rate of change in proper time of the four-momentum, as discussed in Chap. 1. The time-like part of the four-force is the rate of change of mc, and since c is a constant in SRT (and strictly speaking, a locally measured invariant), that is just the rate of change of m, which is the rate of change of E/c^2. If m were a constant, too, this term would be zero. But in general m is not a constant.

A serious mistake is possible at this point. It is often assumed that the rest masses of objects are constants. So, the relativistic mass m can be taken as the rest mass, usually written as m_o, multiplied by the appropriate "Lorentz factor," an expression that appears ubiquitously in Special Relativity Theory (SRT) equal to one divided by the square root of one minus the square of the velocity divided by the square of c.[2] The symbol capital Greek gamma is commonly used to designate the Lorentz factor. (Sometimes the small Greek gamma is used, too.) When v approaches c, the Lorentz factor, and concomitantly the relativistic mass, approaches infinity. But the proper mass is unaffected. If the proper mass m_o really is a constant, then the rate of change of mc is just the rate of change of the Lorentz factor. As Wolfgang Rindler points out in section 35 of his outstanding book on SRT, *Introduction to Special Relativity*, this is a mistake. It may be that in a particular situation rest mass can be taken as a constant. In general, however, this is simply not true. In a situation as simple as the elastic collision of two objects, during the impact as energy is stored in elastic stresses, the rest masses of the colliding objects change. (The germane part of Rindler's treatment is reproduced at the end of this chapter as Addendum #1.) This turns out to be crucial to the prediction of Mach effects.

Now, Luchak's relativisitic Newtonian gravity equation looks very much like a standard classical field equation where the d'Alembertian operator (which involves taking spatial and temporal rates of change of a field quantity) acting on a field is equal to its sources. That is, it looks like a classical wave equation for the field with sources. It's the time dependent term in q that messes this up because q is not **F**. q, however, by definition, is the rate at which the field does work on sources, that is, the rate at which the energy of the sources changes due to the action of the field. So the term in q turns out to be the rate of change of the rate of change of the energy in a volume due to the action of the field on its sources. That is, it is the second time-derivative of the energy density.

This, the second time-derivative (of the field), is the correct form for the time-dependent term in the d'Alembertian of a field. The problem here is that the energy density isn't the right thing to be acted upon by the second time-derivative if the equation is to be a classical wave equation. It should be the field itself, or a potential of the field that is acted on by the second time-derivative.

The interesting aspect of this equation is the ambiguity of whether the time-dependent term should be treated as a field quantity, and left on the left hand side of the equation, or if it can be transferred to the right hand side and treated as a source of the field. Mathematically, where the time-dependent term appears is a matter of choice, for subtracting a term from both sides of an equation leaves the equation as valid as the pre-subtraction equation.

---

[2] More formalism: $\Gamma = \dfrac{1}{\sqrt{1 - \frac{v^2}{c^2}}}$

Physically speaking, whether something gets treated as a field, or a source of the field, is not a simple matter of formal convenience. q is not $\mathbf{F}$, so transferring the term in q to the source side wouldn't obviously involve treating a field as a source. But q may contain a quantity that should be treated as a field, not a source. In the matter of rapid spacetime transport, this question has some significance because if the time-dependent term can be treated as a source of the gravitational field, then there is a real prospect of being able to manipulate inertia, if only transiently. So it is worth exploring to find out if Luchak's equation for relativistic Newtonian gravity can be transformed into one with transient source terms.

FIRST PRINCIPLES

The way to resolve the issues involved here is to go back to first principles and see how the field equation is evolved from the definition of relativistic momentum and force. When this is done taking cognizance of Mach's principle, it turns out that it is possible to recover not only Luchak's field equation but also a classical wave equation for the gravitational potential – an equation that, in addition to the relativistically invariant d'Alembertian of the potential on the (left) field side, has transient source terms of the sort that Luchak's equation suggests might be possible. *But without Mach's principle in the form of the formal statement of his laws of inertia, this is impossible.*

The procedure is straightforward. You assume that inertial reaction forces are produced by the gravitational action of the matter in the universe that acts through a field. The field strength that acts on an accelerating body – written as a four-vector – is just the inertial reaction four-force divided by the mass of the body. That is the derivative with respect to proper time of the four-momentum divided by the mass of the body. To put this into densities, the numerator and denominator of the "source" terms get divided by the volume of the object. In order to get a field equation of standard form from the four-force per unit mass density, you apply Gauss' "divergence theorem."[3] You take the four-divergence of the field strength. Invoking Mach's principle judiciously, the field and source "variables" can be separated, and a standard field equation is obtained.

All of the foregoing is presented here with the aid of 20-20 hindsight. With that hindsight, it all looks pretty straightforward and simple. Actually wading through all of the considerations for the first time, though, was a good deal more tortuous. Only by going back to the basics, the relativistic definition of momentum and force and constructing the argument from first principles made it possible to have any confidence in the results.

---

[3] The "divergence" operation computes the rate at which a vector field is changing at some location by taking the "scalar product" of the gradient operator with the vector field. Gauss showed that if you sum this operation over a volume, you get the total of the sources of the field in the volume as the sum is equal to the total net flux of the field through the enclosing surface. As an aside, when the "curl" of the field vanishes, as it does for Newtonian gravity, the field can be written as the gradient of a scalar potential, and the divergence of the gradient of the potential is written as the square of the gradient operator, as below.

To be sure that nothing had been screwed up, the calculation was taken to a couple of general relativist friends: Ron Crowley and Stephen Goode, both colleagues at California State University at Fullerton. If you are doing "speculative" work like this, you are a fool should you pass up the chance to get the opinion of first-class, brutally honest professionals. In this case, seeking critical professional evaluation had an unintended consequence of real significance. When Luchak did his calculation, since he wasn't looking for small relativistic effects, he made an approximation: he suppressed all terms of "higher order" in the quantity v/c. By the way, in these sorts of circumstances this is a customary practice. When I did the first principles reconstruction of the Newtonian order field equation taking account of Mach's principle, since I was only interested in getting a transient source term that might get us some purchase on inertia, I made the same approximation.

When I took the derivation to Ron for criticism, almost immediately he came across the assumed approximation. Ron had spent some time in earlier years doing experimental work,[4] but he was really a theorist at heart. Theoreticians, in general, hate to see calculations done with approximations if, in their judgment, the approximations employed are not needed to get to a result.

When he came upon the approximation, Ron asked, "Why did you do that?" I explained that I didn't need the higher order terms to get the result I was interested in. He was contemptuous. He thought my approach either lazy or foolish, or both, and told me that he wouldn't go through the rest of the calculation until I had either done the calculation without the approximation, or could show that the approximation was essential to get any result at all. It turns out that the calculation can be done without the approximation in about three pages of algebra, even if your handwriting isn't small.[5] The exact calculation produces additional terms, normally very small, that do not appear in the Luchak level approximation. In particular, it yields a transient term that is always negative. This is the term that holds out the promise of being able to make starships and stargates (that is, absurdly benign wormholes).[6]

---

[4] I first met Ron on my first formal day on campus at CSUF in the fall of 1972. Our mutual interest in gravity was discovered almost immediately. But Ron had taken up work on experiments related to "critical" phenomena. This involved trying to get moderately complicated thermal systems involving vacua working. He was trying to use different melting point solders for different joints in the apparatus, hoping to be able to selectively melt particular joints by getting the right temperature for the particular joint he wanted to melt. The melting points he had chosen were about 50° apart – in a system made mostly of brass. Even an inexperienced experimentalist could see that this wouldn't work. Not long after, Ron returned to theory, especially after a sabbatical working with Kip Thorne's group at Cal Tech.

[5] Ron didn't object to my specializing to the instantaneous frame of rest of the accelerating object so as to suppress a bunch of relativistic terms of no particular physical significance for, while large accelerations might be present, relativistic velocities are not expected.

[6] Until this term turned up in the early 1990s, as an experimentalist interested in making things go fast, I had little interest in wormholes and all that. Indeed, Ron and I had been at the Pacific Coast Gravity meeting in the spring of 1989 at Cal Tech and watched Kip Thorne be told by several speakers that traversable wormholes were physically impossible for one reason or another. Had I been asked to choose sides, I probably would have sided with Thorne's critics. Wormholes enabled time travel, and as far as I was concerned at the time, time travel was just silly. That attitude changed when the Mach effect calculation was done exactly.

The field equation for the gravitational field acting on an accelerating body to produce the inertial reaction force the body communicates to the accelerating agent looks like:

$$\nabla^2 \phi - \frac{1}{c^2}\frac{\partial^2 \phi}{\partial t^2} = 4\pi G \rho_0 + \frac{\phi}{\rho_0 c^2}\frac{\partial^2 \rho_0}{\partial t^2} - \left(\frac{\phi}{\rho_0 c^2}\right)^2\left(\frac{\partial \rho_0}{\partial t}\right)^2 - \frac{1}{c^4}\left(\frac{\partial \phi}{\partial t}\right)^2, \qquad (3.2)$$

or, equivalently (since $\rho_0 = E_0/c^2$ according to Einstein's second law, expressed in densities),

$$\nabla^2 \phi - \frac{1}{c^2}\frac{\partial^2 \phi}{\partial t^2} = 4\pi G \rho_0 + \frac{\phi}{\rho_0 c^4}\frac{\partial^2 E_0}{\partial t^2} - \left(\frac{\phi}{\rho_0 c^4}\right)^2\left(\frac{\partial E_0}{\partial t}\right)^2 - \frac{1}{c^4}\left(\frac{\partial \phi}{\partial t}\right)^2. \qquad (3.3)$$

ρ_0 is the proper matter density (where "matter" is understood as everything that gravitates) and E_0 is the proper energy density. The left hand sides of these equations are just the d'Alembertian "operator" acting on the scalar gravitational potential ϕ. "Mach effects" are the transient source terms involving the proper matter or energy density on the right hand sides.

The equations are classical wave equations for the scalar gravitational potential ϕ, and notwithstanding the special circumstances invoked in their creation (the action of the gravity field on an accelerating object), they are general and correct, for when all the time derivatives are set equal to zero, Poisson's equation for the potential results. That is, we get back Newton's law of gravity in differential form with sources. When are the transient source terms zero? When the accelerating object considered in the derivation *does not absorb "internal" energy during the acceleration*. That is, if our accelerating body is not deformed by the acceleration, these terms are zero. This means that in situations like elementary particle interactions, you shouldn't see any Mach effects, for elementary particles per se are not deformed in their interactions, though they may be created or destroyed. The derivation of these effects was first published in 1995 in. "Making the Universe Safe for Historians: Time Travel and the Laws of Physics." The title is a takeoff on a joke made by Stephen Hawking in a paper on his "chronology protection conjecture" that he had published a few years earlier. According to Hawking, the prohibition of wormhole time machines makes it impossible to travel to the past to check up on reconstructions of the past by historians. One of the unintended consequences of this choice of title is that it has the acronym MUSH.

A small digression at this point is warranted to head off possible confusion. You may be wondering, especially after all of the fuss about ϕ and c being "locally measured invariants" in the previous chapter, how the derivatives of ϕ in these wave equations can have any meaning. After all, if ϕ has the same value everywhere and at all times, how can it be changing in either space or time?

The thing to keep in mind is "locally measured." As measured by a particular observer, c and ϕ have their invariant values wherever he or she is located. But everywhere else, the values measured may be quite different from the local invariant values. And if there is any variation, the derivatives do not vanish.

Let's look at a concrete example. Back around 1960, a few years after the discovery of the Mössbauer effect (recoilless emission and absorption of gamma rays by radioactive iron and cobalt), Pound and Rebka used the effect – which permits timing to an accuracy of a part in 10^{17} s – to measure the gravitational redshift in a "tower" about 22.5 m high on Harvard's campus. The gravitational redshift results because time runs slower in a stronger gravitational field, so an emitter at the bottom of the tower produces gamma rays that have a different frequency from those emitted and absorbed at the top of the tower. Pound and Rebka measured this shift for a source at the top of the tower by using a moving iron absorber at the bottom of the tower. The motion of the absorber produces a Doppler frequency shift that compensates for the higher frequency of the source at the top of the tower. From the speed of the absorber, the value of the frequency shift can be calculated.

Since time runs slower at the bottom of the tower, the speed of light there, measured by someone at the top of the tower, is also smaller. And since $\phi = c^2$, the value of ϕ at the bottom of the tower measured by the person at the top is also different from the local invariant value. Obviously, the derivative of ϕ in the direction of the vertical in the tower does not vanish. But if you measure the value of c, a proxy for ϕ, with, for example, a cavity resonator, you will get exactly the local invariant value everywhere in the tower. From all this you can infer that the locally measured value of ϕ is the same everywhere in the tower, notwithstanding that it has a non-vanishing derivative everywhere in the tower.

Alas, a further small digression is needed at this point. If you've read the relativity literature, you've likely found that everyone is very careful to distinguish inertial frames of reference from those associated with accelerations and gravity. Indeed, it used to be said that to deal with accelerations, general relativity was required because special relativity only dealt with inertial motions. Misner, Thorne, and Wheeler pretty much put an end to such nonsense by showing (in Chap. 6 of their classic text *Gravitation*, Freeman, San Francisco, 1973) that accelerations are routinely dealt with using the techniques of special relativity. The reason why general relativity is required is that local matter concentrations deform spacetime and that, in turn, requires non-Euclidean geometry to connect up local regions.

Nonetheless, it is well known that funny things happen to time and the speed of light in regions where gravity is significant. At the horizon of a black hole, time stops and the speed of light goes to zero as measured by distant observers. But what happens to these quantities when they are measured by a local observer in a strong gravity field, or equivalently undergoing a strong acceleration? Does the speed of light measured by them differ from the value measured in a coincident inertial frame of reference? That is, in terms of our cavity resonator mentioned above that computes the speed of light as the product of the frequency and wavelength for the cavity at resonance, does it read out 3×10^{10} cm/s irrespective of whether it is in inertial motion or accelerating?

This question can be answered by noting that in a local measurement, the size of the cavity resonator is unchanged by acceleration since acceleration is dealt with by making a series of Lorentz transformations to a succession of instantaneous frames of rest. This is especially obvious if we orient the cavity resonator perpendicular to the direction of the acceleration and any velocity it may have. Lengths in that direction are uninfluenced by either acceleration or velocity (since the velocity in each successive instantaneous frame of rest is zero).

What about time measured in an accelerating reference frame? That's the thing that stops at the horizon of a black hole for distant observers. Well, it doesn't stop for local accelerating observers. Indeed, for accelerating observers and their time measuring apparatus, no change in the rate of time results from the acceleration. How do we know? It has been measured, by Champney, Isaak and Kahn. ("An 'Aether Drift' Experiment Based on the Mössbauer Effect", *Physics Letters* **7**, 241–243 [1963].) The Mössbauer effect enables timing with an accuracy of a part in 10^{17}, so very accurate timing measurements can be made with it. What Champney and his co-authors did was mount a Mössbauer source on one tip of a rotor and a Mössbauer absorber on the opposite tip of the rotor, the rotor being about 8 cm from tip to tip. They then spun the rotor up to high speeds, ultimately a bit over 1,200 Hz, producing accelerations at the tips of roughly 30 Earth gravities. Accelerations and gravity fields produce "red shifts," as in the Pound-Rebka experiment mentioned above. But Champney, et al., found no red shift with their rotor.

Now, the Mössbauer source is a clock ticking at the frequency of the gamma rays it emits. And the absorber is a detector of the clock rate (frequency) of the source as it appears at the location of the absorber. Since the source and absorber share the same acceleration (albeit in opposite directions), if the rate of locally measured time is unaffected by accelerations, then no red shift should be detected in this experiment, and none was. When this is coupled with the fact that the length of a cavity resonator that determines the resonant wavelength is unaffected by accelerations, it follows immediately that the product of the resonant frequency and wavelength – the local speed of light – is unaffected by accelerations and gravity fields.

Who cares? Well, everyone should. What this means is that the speed of light is fully deserving of being included in the so-called Planck units, for it really is a locally measured constant – it has the same numerical value for all observers, no matter who, where, or when they are, *when they carry out a local measurement*. It is more than simply the ratio of the distance traveled to time of travel for a special kind of radiation. The gravitational origin of inertial reaction forces and the validity of Newton's law of action and reaction could have been guaranteed only by requiring that $\phi = c^2$, irrespective of the numerical value of c, which might be different in accelerating systems. But it isn't different. It is the fundamental "constant" (locally measured) that couples the two long-range interactions, gravity and electromagnetism, and makes them of one piece. The intimate relationship between gravity and electromagnetism will find further support when we get to Chap. 7, where we look into the structure of matter at the most elementary level.

For those of you who want to see the details of the derivation of Mach effects, a version with all of the algebra spelled out, is excerpted at the end of this chapter. It was published in 2004 as part of "Flux Capacitors and the Origin of Inertia." All of the line-by-line algebra was included because a number of mathematically capable people have reported difficulty reconstructing the details of the derivation published in MUSH years earlier. You don't need to know how to derive these equations, however, to see some of their implications. The terms that are of interest to us are the transient source terms on the right hand sides. We can separate them out from the other terms in the field equation, getting for the time-dependent proper source density:

$$\delta \rho_0(t) \approx \frac{1}{4\pi G} \left[\frac{\phi}{\rho_0 c^4} \frac{\partial^2 E_o}{\partial t^2} - \left(\frac{\phi}{\rho_0 c^4} \right)^2 \left(\frac{\partial E_0}{\partial t} \right)^2 \right], \tag{3.4}$$

where the last term on the right hand side in the field equation has been dropped, as it is always minuscule.[7] The factor of $1/4\pi G$ appears here because the parenthetical terms started out on the field (left hand) side of the derived field equation.[8]

If we integrate the contributions of this transient proper matter density over, say, a capacitor being charged or discharged *as it is being accelerated*, we will get for the transient total proper mass fluctuation, written δm_0:

$$\delta m_0 = \frac{1}{4\pi G} \left[\frac{1}{\rho_0 c^2} \frac{\partial P}{\partial t} - \left(\frac{1}{\rho_0 c^2} \right)^2 \frac{P^2}{V} \right], \tag{3.5}$$

where P is the instantaneous power delivered to the capacitor and V the volume of the dielectric.[9] If the applied power is sinusoidal at some frequency, then $\partial P/\partial t$ scales linearly with the frequency. So operating at elevated frequency is desirable. Keep in mind here that the capacitor must be accelerating for these terms to be non-vanishing. You can't just charge and discharge capacitors and have these transient effects be produced in them. None of the equations that we have written down for Mach effects, however, show the needed acceleration explicitly. And it is possible to forget that the effects only occur in accelerating objects. If you forget, you can waste a lot of time and effort on experiments doomed to ambiguity and failure.

Writing out the explicit acceleration dependence of the Mach effects is not difficult. We need to write the first and second time-derivatives of the proper energy in terms of the acceleration of the object. All we need note is that the work done by the accelerating force

---

[7] The term in question, $\frac{1}{c^4} \left(\frac{\partial \phi}{\partial t} \right)^2$, contains only the scalar potential, treated as a source however so that the d'Alembertian alone appears on the left hand side of the equation. While the potential is enormous, being equal to c^2, the time-derivative of the potential is always quite small because things out there in the distant universe, as viewed here on Earth, don't happen very quickly. So we have a small quantity multiplied by c^{-4}. The product is an utterly minuscule quantity. No propulsive advantage is to be found in this term.

[8] Since G, a small number, appears in the denominator, this factor dramatically increases the magnitude of the parenthetical terms. Note that the same thing happens when the cosmological term in Einstein's equations is treated as a source, rather than field quantity.

[9] The instantaneous power in a charging/discharging capacitor is just the product of the voltage across the capacitor and the current flowing to/from the capacitor. That is, $P = iV$. Since V is also used to designate volume, care must be taken to correctly identify which V is involved in the situation you are considering.

acting on the object is the scalar product $\mathbf{F} \bullet d\mathbf{s} = dE$, and the rate of change of work is the rate of change of energy, so:

$$\frac{\partial E_0}{\partial t} = \frac{\partial}{\partial t}(\mathbf{F} \bullet d\mathbf{s}) = \mathbf{F} \bullet \mathbf{v} = m\mathbf{a} \bullet \mathbf{v}. \tag{3.6}$$

It is important to note here that $\mathbf{v}$ is not the velocity normally encountered in elementary calculations of work and energy because we are talking about *proper*, that is, internal energy, changes, rather than the kinetic energy acquired as the result of the acceleration of a rigid object by an external force. For the same reason, the derivatives should be taken in the instantaneous rest frame of the object. The $\mathbf{v}$ here is the typical velocity of the parts of the object as it is compressed by the external force (while the object in bulk in the instantaneous rest frame has zero velocity). If the object is incompressible then $\mathbf{v}$ is zero and there are no internal energy changes. Concomitantly, there are no Mach effects. If $\mathbf{v}$ is not zero, it will likely be smaller than the bulk velocity acquired by the object due to the action of the force over time (unless the object is externally constrained as it is compressed). The second time-derivative of E_0 now is:

$$\frac{\partial^2 E_0}{\partial t^2} = \frac{\partial}{\partial t}(m\mathbf{a} \bullet \mathbf{v}) = m\mathbf{a} \bullet \frac{\partial \mathbf{v}}{\partial t} + \mathbf{v} \bullet \frac{\partial}{\partial t}(m\mathbf{a}) = m\mathbf{a} \bullet \frac{\partial \mathbf{v}}{\partial t} + \mathbf{v} \bullet \left(m\frac{\partial \mathbf{a}}{\partial t} + \mathbf{a}\frac{\partial m}{\partial t} \right). \tag{3.7}$$

Equations 3.6 and 3.7 can be used to explicitly display acceleration dependence of the effects via substitution in Eq. 3.4, but only when two considerations are taken account of. First, E_0 in Eq. 3.4 is the proper energy *density* because it follows from a field equation expressed in terms of densities, whereas E_0 in Eqs. 3.6 and 3.7 is the proper energy of the entire object being accelerated, so Eq. 3.4 must effectively be integrated over the whole accelerated object, making E_0 the total proper energy, before the substitutions can be carried out.

The second consideration that must be kept in mind is that the accelerating force can produce both changes in internal energy of the object accelerated and changes in its bulk velocity that do not contribute to internal energy changes. Only the part of the accelerating force that produces internal energy changes contributes to Mach effects. That is why $\partial \mathbf{v}/\partial t$ is written explicitly in Eq. 3.7, as it is only part of the total acceleration of the object. We can take account the fact that $\mathbf{v}$ in Eqs. 3.6 and 3.7 is only the part of the total $\mathbf{v}$ for the extended macroscopic object being accelerated by writing $\mathbf{v}_{int} = \eta\mathbf{v}$ and replacing $\mathbf{v}$ with $\mathbf{v}_{int}$ in the above equations. This leads to:

$$\frac{\partial^2 E_0}{\partial t^2} = \frac{\partial}{\partial t}(m\mathbf{a} \bullet \eta\mathbf{v}) = \eta m\mathbf{a}^2 + \eta\mathbf{v} \bullet \frac{\partial}{\partial t}(m\mathbf{a}) = \eta m\mathbf{a}^2 + \eta\mathbf{v} \bullet \left(m\frac{\partial \mathbf{a}}{\partial t} + \mathbf{a}\frac{\partial m}{\partial t} \right), \tag{3.8}$$

with $0 \leq \eta \leq 1$. As long as η can be taken to be a constant, the RHS of Eq. 3.8 obtains and things are fairly simple. But in general, η will be a function of time, making matters more complicated and solutions more complex.

The first thing we note about the equation for δm_0 is that the second term on the right hand side is always negative since the volume V of the dielectric cannot be negative and all other factors are squared. But we then notice that the coefficient of P^2/V is normally very small since there is a factor of c^2 in the denominator, and the coefficient is squared, making the factor $1/c^4$ – an exceedingly small number. Indeed, since this coefficient is the square of the coefficient of the first term involving the power, it will normally be many orders of magnitude smaller than the first term. However, the coefficients also contain a factor of the proper matter density in their denominators.

While c^2 may always be very large, ρ_0 is a variable that, at least in principle, can become zero and negative. When ρ_0 approaches zero, the coefficients become very large, and because the coefficient of the second term is the square of that for the first term, it blows up much more quickly than that for the first term. And the second term dominates the sources of the field. At least transiently, in principle, we should be able to induce significant amounts of "exotic" matter. This is the first necessary step toward building starships and stargates.

You might think that any prediction of an effect that might make the construction of wormholes and warp drives possible would have been actively sought. But in fact, as noted above, it was an unanticipated accident, though its significance was understood when it was found. Because the second effect is normally so small, it was ignored for years. The possibility that it might be produced in small-scale laboratory circumstances wasn't appreciated until Ron Crowley, Stephen Goode, and myself, the faculty members on Thomas Mahood's Master's degree committee at CSUF in 1999, got into an argument about it during Tom's thesis defense. Ron and Stephen were convinced that seeing the effects of the second transient term should be possible in laboratory circumstances. I wasn't. But they were right. Even then, it was a year until evidence for this effect was sought – after John Cramer, a prominent physicist from the University of Washington, suggested in a meeting at Lockheed's Lightspeed facility in Fort Worth, Texas, that it should be done.[10] When evidence for the effect was found (see: "The Technical End of Mach's Principle" in: *Mach's Principle and the Origin of Inertia*, eds. M. Sachs and A. R. Roy, Apeiron, Montreal, 2003, pp. 19–36) a decade ago, the evidence was not followed up on because production of the effect depends on "just so" conditions that are very difficult to reliably produce. The lure of simple systems involving only the first term Mach effect was too great. Producing the first-term Mach effect, however, turns out to be much more challenging than was appreciated at that time. If present at all, the first-term Mach effect also depends on "just so" conditions that are difficult to control.

---

[10] The Lockheed-Martin Lightspeed connection proved helpful in another regard. When questions arose about heating in some devices being tested between 1999 and 2002, Jim Peoples, manager of Millennium Projects, had a nice far infrared camera (with an expensive fluorite lens) shipped out for our use for several months.

LABORATORY SCALE PREDICTIONS

The predicted effects in simple electronic systems employing capacitors and inductors, for the leading term in Eq. 3.5 at any rate, are surprisingly large. Even at fairly low frequencies, it is reasonable to expect to see manifestations of the leading effect. Larger mass fluctuations are expected at higher frequencies, at least for the first term of Eq. 3.5, for if P is sinusoidal, then $\partial P/\partial t$ scales linearly with the frequency. But to be detected with a relatively "slow" weigh system, a way to "rectify" the mass fluctuation must be found so that a time-averaged, stationary force in one direction can be produced and measured.

The mass fluctuation itself, of course, cannot be "rectified," but its physical effect can be rectified by adding two components to the capacitor in which a mass fluctuation is driven. This element is identified as FM (fluctuating mass) in Fig. 3.1. The two additional components are an electromechanical actuator (customarily made of lead-zirconium-titanate, so-called PZT), designated A (actuator) in Fig. 3.1, and a "reaction mass" (RM) located at the end of the actuator opposite the fluctuating mass (FM) element, as shown in Fig. 3.1.

The principle of operation is simple. A voltage signal is applied to the FM element and PZT actuator so that the FM element periodically gains and loses mass. A second voltage signal is applied to the PZT actuator. The actuator voltage signal must have a component at the power frequency of the FM voltage signal, that is, twice the frequency of the signal applied to the FM. *And it must also have a component at the FM signal frequency to produce the acceleration of the FM required for a Mach effect to be produced.* The relative phase of the two signals is then adjusted so that, say, the PZT actuator is expanding (at the power frequency) when the FM element is more massive and contracting when it is less massive. The inertial reaction force that the FM element exerts on the PZT actuator is communicated through the actuator to the RM.

Evidently, the reaction force on the RM during the expansion part of the PZT actuator cycle will be greater than the reaction force during the contraction part of the cycle. So, the

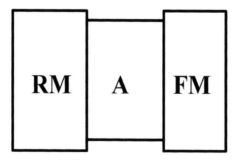

Fig. 3.1 A schematic diagram of a Mach effect "impulse engine." RM stands for reaction mass, A for actuator and FM for fluctuating mass. The leading Mach effect is used to cause the FM to periodically change its mass. The actuator then pushes "heavy" and pulls "light" to produce a stationary force on the reaction mass

time-averaged force on the RM will not be zero. Viewed from the "field" perspective, the device has set up a momentum flux in the "gravinertial" field – that is, the gravitational field understood as the cause of inertial reaction forces – coupling the FM to the chiefly distant matter in the universe that causes the acceleration of the mechanical system of Fig. 3.1.

Formal analysis of this system is especially simple in the approximation where the mass of the RM is taken as effectively infinite, and the capacitor undergoes an excursion $\delta\,l = \delta\,l_0 \cos(2\omega\,t)$ due to the action of the PZT actuator with respect to the RM. We obtain for the time-averaged reaction force on the RM:

$$\langle F \rangle = -4\omega^2 \delta\,l_0 \delta m \, \sin(2\omega t) \, \sin(2\omega t + \phi), \tag{3.9}$$

where φ is the phase angle between the PZT excursion and the mass fluctuation. Further algebra yields:

$$\langle F \rangle = -2\omega^2 \delta l_0 \delta m \, \cos \phi \tag{3.10}$$

as the only term that survives the time-averaging process. Evidently, stationary forces can be obtained from mass fluctuations in this way.

Much of the past decade was devoted to testing schemes that held out promise of being able to avoid the subtleties and complexities that seem to attend even the simplest systems where one might look for Mach effects. Although much has been learned, and many improvements to the experimental apparatus in use have been made, progress has been slow. Early in the decade, some time was devoted to discussion of the application of Newton's second law in devices of the type in Fig. 3.1. Even seasoned professionals occasionally make mistakes when applying the second law, even in simple circumstances.[11] The problem is the interpretation of the time rate of change of momentum definition of force. The issue is not one involving SRT, though it does involve relative motion. Newton's second law written out formally is:

$$\mathbf{F} = d\mathbf{p}/dt = d(m\mathbf{v})/dt = m\mathbf{a} + \mathbf{v}dm/dt. \tag{3.11}$$

The problem is the $\mathbf{v}dm/dt$ term. Unless you are careful, it is easy to make mistakes, treating this term as a force in assessing the forces acting in a system. The *only* physically meaningful contributions that can be attributed to this term are ones for which $\mathbf{v}$ has an invariant meaning – for example, the velocity of the just ejected exhaust plume of a rocket

---

[11] As friend and colleague Keith Wanser has pointed out, physics pedagogy journals periodically print articles on the subtleties involved in the interpretation of Newton's second law of mechanics. The irony is that sometimes the authors of these articles, already sensitized to the issues involved, themselves get things wrong (and the mistakes pass the reviewers' notice, too).

with respect to the rocket. That all observers can agree on. In rocket type situations, this is the *only* velocity with invariant meaning. In general, in the case of Mach effect devices such as that in Fig. 3.1, this term does not contribute to the inertial reaction force on the RM because it does not represent a force *on* the FM that is communicated through the PZT actuator to the RM. This is easily shown by noting that in the instantaneous frame of rest of the capacitor (FM) vdm/dt vanishes as v in that frame is zero. Since the vdm/dt "force" that purportedly acts on the FM is zero in this inertial frame of reference, it must also be zero in all other frames of reference, and it follows that a vdm/dt "force" does not act *on* the FM, and thence through the PZT on the RM.

So, in the case of Mach effect thrust generators, the vdm/dt term plays a role analogous to the rocket situation: it represents the changing momentum flux in the gravinertial field that conveys that momentum to the distant future universe and preserves local momentum conservation in the present. Since the instantaneity of inertial reaction forces implies that this is an "action-at-a-distance" interaction, the seeming retarded field "kink" involved is actually half due to an advanced wave returning from the future. So what seems to be momentum being dumped to the future is also momentum from the future getting dumped in the present. That is, in the words of my then-graduate student Tom Mahood on a sign that graced the door to my lab for many years, "Tomorrow's momentum today!"[12] The confusion over momentum conservation became so troublesome a decade ago that I appended a short piece to "Flux Capacitors" to deal with it.

What is important here is that there is no "new" physics in any of the above considerations. You don't have to believe anything hopelessly weird to discover Mach effects. All you have to do is ask a simple set of questions in the right order and apply standard techniques in the answers to those questions. Actually, asking the right questions in the right order is almost always the hardest part of any investigation. In this case, the answers to those questions are amenable to experimental test, so you can find out if the predicted effects really exist. If you want to make stargates, you do have to believe some weird stuff about the nature of elementary particles. That, however, is another story, told in the third section of this book.

ADDENDA

Addendum #1: Excerpt from Rindler

---

[12] The rest of the sign read, "Starfleet Labs" and was adorned with a circled rocket nozzle with a strike bar through it, and a rendition of a wormhole.

Our first 'guess' in Section 26 for the definition of four-force was

$$F = m_0 A = m_0 \frac{dU}{d\tau}, \qquad (?) \tag{35.2}$$

modelled on the first form of (35.1). But the 'correct' definition, consistent not only with other important 'half-laws' about force but also with our emphasis on **P** in the basic axiom (26.1), is modelled on the *second* form of (35.1):

$$F = \frac{dP}{d\tau} = \frac{d}{d\tau}(m_0 U). \tag{35.3}$$

Both (35.2) and (35.3) are manifestly tensorial definitions. But they are equivalent *only* if the rest mass is constant, which is by no means always the case. For example, if two particles collide elastically, their rest masses *during* collision will vary, but that would be precisely when we might be interested in the elastic forces acting on them.

If we define the (*relativistic*) *three-force* by

$$f = \frac{dp}{dt} = \frac{d(mu)}{dt}, \tag{35.4}$$

where **p**, m are, of course, the *relativistic* momentum and mass of the particle on which **f** acts, we can write (35.3) in the forms [cf. (23.2), (26.2), and (27.2)]

$$F = \gamma(u)\frac{d}{dt}(mc, p) = \gamma(u)\left(c\frac{dm}{dt}, f\right) = \gamma(u)\left(\frac{1}{c}\frac{dE}{dt}, f\right). \tag{35.5}$$

Thus the *power* dE/dt, i.e. the rate at which the force transfers energy to the particle, is the natural partner of **f** in the formation of the relevant four-vector.

Introduction to Special Relativity by Wolfgang Rindler, (2nd edition, Oxford University Press, 1991) pp. 90–91. By permission of Oxford University Press.

Addendum #2: Flux Capacitors and the Origin of Inertia

1. Introduction

Over a century has passed since Ernst Mach conjectured that the cause of inertia should somehow be causally related to the presence of the vast bulk of the matter (his "fixed stars") in the universe. Einstein translated this conjecture into "Mach's principle" (his words) and attempted to incorporate a version of it into general relativity theory (GRT) by introducing the "cosmological constant" term into his field equations for gravity.[1] Einstein ultimately abandoned his attempts to incorporate Mach's principle into GRT. But in the early 1950s Dennis Sciama revived interest in the "origin of inertia".[2] Mach's principle can be stated in very many ways. (Bondi and Samuel in a recent article list twelve versions, and their list is not exhaustive.[3]) Rather than try to express Mach's principle with great subtlety, Sciama, in 1964, adopted a simple (and elegant) statement:[4]

> *Inertial forces are exerted by matter, not by absolute space. In this form the principle contains two ideas:*
>
> (1) *Inertial forces have a dynamical rather than a kinematical origin, and so must be derived from a field theory [or possibly an action-at-a-distance theory in the sense of J.A. Wheeler and R.P. Feynman. . . .]*
> (2) *The whole of the inertial field must be due to sources, so that in solving the inertial field equations the boundary conditions must be chosen appropriately.*

Taking into account the fact that the field produced by the chiefly distant matter in the universe must display the same universal coupling to matter as gravity to properly account for inertial reaction forces, the essence of Mach's principle can be put into yet more succinct form: **Inertial reaction forces are the consequence of the gravitational action of the matter located in the causally connected part of the universe on objects therein accelerated by "external" forces.**

. . . .

2. Transient Mach Effects:

. . . The predicted phenomena in question arise from considering the effect of an "external" accelerating force on a massive test particle. Instead of assuming that such an acceleration will lead to the launching of a (ridiculously minuscule) gravitational wave and asking about the propagation of that wave, one assumes that the inertial reaction force the accelerating agent experiences is caused by the action of, in Sciama's words, "the radiation field of the universe" and then asks, given the field strength as the inertial reaction force per unit mass, what is the local source charge density at the test particle? The answer is obtained by taking the four-divergence of the field strength at the test particle. The field equation that results from these operations is:

$$\nabla^2 \phi - \frac{1}{\rho_0 c^2} \frac{\partial^2 E_0}{\partial t^2} + \left(\frac{1}{\rho_0 c^2}\right)^2 \left(\frac{\partial E_0}{\partial t}\right)^2 = 4\pi G \rho_0. \tag{3}$$

In this equation ϕ is the scalar potential of the gravitational field, ρ_0 the local proper matter density, E_0 the local proper energy density, c the vacuum speed of light, and G Newton's constant of gravitation. This equation looks very much like a wave equation. However, the space-like part (the Laplacian) involves a scalar potential, whereas the time-like part (the time-derivatives) involves the proper rest energy density. (A full derivation of the Mach effects discussed here is given in Appendix A.)

Equation (3) can be put into the form of a standard classical wave equation by using Mach's principle to "separate variables", for Mach's principle implies more than the statement above involving the origin of inertial reaction forces. Indeed, Mach's principle actually implies that the origin of mass is the gravitational interaction. In particular, the inertial masses of material objects are a consequence of their potential energy that arises from their gravitational interaction with the rest of the matter in the causally connected part of the universe. That is, in terms of densities,

$$E_g = \rho\phi, \tag{4}$$

where E_g is the local gravitational potential energy density, ρ the local "quantity of matter" density, and ϕ the total gravitational potential at that point. (Note that it follows from Sciama's analysis that $\phi \equiv c^2$, so Equation (4) is nothing more than the well-known relationship between mass and energy that follows from special relativity theory if E_g is taken to be the total local energy density.) Using this form of Mach's principle, we can write:

$$E_0 = \rho_0\phi, \tag{5}$$

and this expression can be used in Equation (3) to affect the separation of variables. After some straight-forward algebra (recounted in Appendix A) we find that:

$$\nabla^2\phi - \frac{1}{c^2}\frac{\partial^2\phi}{\partial t^2} = 4\pi G\rho_0 + \frac{\phi}{\rho_0 c^2}\frac{\partial^2\rho_0}{\partial t^2} - \left(\frac{\phi}{\rho_0 c^2}\right)^2\left(\frac{\partial\rho_0}{\partial t}\right)^2 - \frac{1}{c^4}\left(\frac{\partial\phi}{\partial t}\right)^2, \tag{6}$$

or, equivalently,

$$\nabla^2\phi - \frac{1}{c^2}\frac{\partial^2\phi}{\partial t^2} = 4\pi G\rho_0 + \frac{\phi}{\rho_0 c^4}\frac{\partial^2 E_0}{\partial t^2} - \left(\frac{\phi}{\rho_0 c^4}\right)^2\left(\frac{\partial E_0}{\partial t}\right)^2 - \frac{1}{c^4}\left(\frac{\partial\phi}{\partial t}\right)^2. \tag{7}$$

This is a classical wave equation for the gravitational potential ϕ, and notwithstanding the special circumstances invoked in its creation, it is general and correct, for when all the time derivatives are set equal to zero, Poisson's equation for the potential results. That is, we get back Newton's law of gravity in differential form.

Some of the implications of this equation [either (6) or (7)] have been addressed elsewhere.[7,8] Here we note that the transient source terms on the RHS can be written:

$$\delta\rho_0(t) \approx \frac{1}{4\pi G}\left[\frac{\phi}{\rho_0 c^4}\frac{\partial^2 E_o}{\partial t^2} - \left(\frac{\phi}{\rho_0 c^4}\right)^2\left(\frac{\partial E_0}{\partial t}\right)^2\right], \tag{8}$$

or, taking account of the fact that $\phi/c^2 = 1$,

$$\delta\rho_0(t) \approx \frac{1}{4\pi G}\left[\frac{1}{\rho_0 c^2}\frac{\partial^2 E_o}{\partial t^2} - \left(\frac{1}{\rho_0 c^2}\right)^2\left(\frac{\partial E_0}{\partial t}\right)^2\right], \qquad (9)$$

where the last term in Equations (6) and (7) has been dropped as it is always minuscule. It is in the transient proper matter density effects – the RHSs of Equations (8) and (9) – that we seek evidence to demonstrate that the origin of inertia, as conjectured by Mach, Einstein, Sciama, and others, is in fact the gravitational interaction between all of the causally connected parts of the universe.

Appendix A [from *Fux Capacitors and the Origin of Inertia*]

Armed with the definition of Mach's principle presented in the body of this paper, we tackle the detailed derivation of Eq. A2.5 above (which was first obtained in complete form in *Making the Universe Safe for Historians*, see select bibliography). The correct gravitational field equation, of course, is Einstein's field equation of GRT, and the vector approximation to that equation is a set of Maxwell-like field equations. But for our purposes we are less interested in the field *per se* than we are in the *sources* of the field, for it is they that carry mass, and thus inertia. In GRT, and in its vector approximation, the sources of the field are *stipulated*. What we want to know, however, is: Does Mach's principle tell us anything interesting about the nature of the sources of the field? To answer this question, it turns out, we do not need either the machinery of GRT or its vector approximation with their stipulated sources. We only need the relativistically invariant (i.e., Lorentz invariant) generalization of Newtonian gravity, for that is all that is necessary to recover the transient matter terms found in Eq. A2.5.

Why does this work? Because inertia is already implicitly built into Newtonian mechanics. The reason why it is possible to ignore the explicit contribution of the distant matter in the universe to local gravity is because of the *universality* of the gravitational interaction (crudely, it affects everything the same way, in proportion to its mass), as pointed out by Sciama and noted here, and so that contribution can always be eliminated by a coordinate (i.e., gauge) transformation, as noted by Brans.[15] (As an aside, this is the reason why gravitational energy is "non-localizable" in GRT, a well-known consequence of the Equivalence Principle in that theory.) Moreover, by demanding the Lorentz invariance we insure that correct time-dependence is built into our simplest possible approximation to the field equation(s) of GRT.

To derive one considers a "test particle" (one with sufficiently small mass that it does not itself contribute directly to the field being investigated) in a universe of uniform matter density. We act on the test particle by, say, attaching an electric charge to it and placing it between the plates of a capacitor that can be charged with suitable external apparatus. That is, we accelerate the test particle by applying an external force. The acceleration, via Newton's third law, produces an inertial reaction force in the test particle that acts on the accelerating agent. In view of the Machian nature of GRT and Sciama's analysis of the origin of inertia, we see that the inertial reaction force produced in these circumstances is just the action of the gravitational field of the chiefly distant matter in the universe on the test particle as it is accelerated. So we can write the field strength of the gravitational action on the test particle as the inertial reaction force it experiences divided by the mass of the test particle (since a field strength is a force per unit charge, the "charge" in this case being mass). Actually, the standard forms of field equations are expressed in terms of charge densities, so one has to do a volumetric division to get the force per unit mass expression into standard form.

There are two critically important points to take into account here. The first is that the mass density that enters the field equation so constructed is the *matter density of the test particle, not the matter density of the uniformly distributed cosmic matter that causes the inertial reaction force*. The second point is that in order to satisfy the Lorentz invariance, this calculation is done using the *four-vectors of relativistic spacetime*, not the three-vectors of classical space and time. Formally, we make two assumptions:

1. Inertial reaction forces in objects subjected to accelerations are produced by the interaction of the accelerated objects with a field – they are not the immediate consequence *only* of some inherent property of the object. And from GRT and Sciama's vector approximation argument, we know that the field in question is the gravitational field generated by the rest of the matter in the universe.
2. Any acceptable physical theory must be locally Lorentz invariant; that is, in sufficiently small regions of spacetime special relativity theory (SRT) must obtain.

We then ask: In the simplest of all possible circumstances – the acceleration of a test particle in a universe of otherwise constant matter density – what, in the simplest possible approximation, is the field equation for inertial forces implied by these propositions? SRT allows us to stipulate the inertial reaction force **F** on our test particle stimulated by the external accelerating force $\mathbf{F}_{ext}$ as:

$$\mathbf{F} = -\mathbf{F}_{ext} = -\frac{d\mathbf{P}}{d\tau}, \tag{A1}$$

with

$$\mathbf{P} = (\gamma m_0 c, \mathbf{p}), \tag{A2}$$

$$\gamma = \frac{1}{\sqrt{1 - \frac{v^2}{c^2}}}, \tag{A3}$$

where bold capital letters denote four-vectors and bold lower case letters denote three-vectors, **P** and **p** are the four- and three-momenta of the test particle respectively, τ is the proper time of the test particle, v the instantaneous velocity of the test particle with respect to us, and c the *vacuum* speed of light. *Note that the minus sign has been introduced in Eq. A1 because it is the inertial reaction force, which acts in the direction opposite to the acceleration produced by the external force, that is being expressed. One could adopt* *another sign convention here; but to do so would mean that other sign conventions introduced below would have to be altered to maintain consistency.*

We specialize to the instantaneous frame of rest of the test particle. In this frame we can ignore the difference between coordinate and proper time, and γs (since they are all equal to one). We will not recover a generally valid field equation in this way, but that is not our objective. In the frame of instantaneous rest of the test particle Eq. A1 becomes:

$$\mathbf{F} = -\frac{d\mathbf{P}}{d\tau} = -\left(\frac{\partial m_0 c}{\partial t}, \mathbf{f}\right), \tag{A4}$$

with,

$$\mathbf{f} = \frac{d\mathbf{p}}{dt}. \tag{A5}$$

Since we seek the equation for the field (i.e., force per unit mass) that produces $\mathbf{F}$, we normalize $\mathbf{F}$ by dividing by m_0. Defining $f = \mathbf{f}/m_0$, we get,

$$F = \frac{\mathbf{F}}{m_0} = -\left(\frac{c}{m_0}\frac{\partial m_0}{\partial t}, f\right). \tag{A6}$$

To recover a field equation of standard form we let the test particle have some small extension and a proper matter density ρ_0. (That is, operationally, we divide the numerator and the denominator of the time-like factor of F by a unit volume.) Equation A6 then is:

$$F = -\left(\frac{c}{\rho_0}\frac{\partial \rho_0}{\partial t}, f\right). \tag{A7}$$

From SRT we know that $\rho_0 = E_0/c^2$, E_0 being the proper energy density, so we may write:

$$F = -\left(\frac{1}{\rho_0 c}\frac{\partial E_0}{\partial t}, f\right). \tag{A8}$$

With an equation that gives the gravitational field strength that causes the inertial reaction force experienced by the test particle in hand, we next calculate the field equation by the standard technique of taking the divergence of the field strength and setting it equal to the local source density. Note, however, that it is the *four-divergence* of the *four-field strength* that is calculated. To keep the calculation simple, this computation is done in the instantaneous rest frame of the test particle so that Lorentz factors can be suppressed (as mentioned above). Since we will not be interested in situations where relativistic velocities are encountered, this simplification has no physical significance. The relativistic nature of this calculation turns out to be crucial, however, for all of the interesting behavior arises from the time-like part of the four-forces (and their corresponding field strengths). The four-divergence of Eq. A8 is:

$$-\frac{1}{c^2}\frac{\partial}{\partial t}\left(\frac{1}{\rho_0}\frac{\partial E_0}{\partial t}\right) - \nabla.f = 4\pi G\rho_0. \tag{A9}$$

Carrying out the differentiation with respect to time of the quotient in the brackets on the LHS of this equation yields:

$$-\frac{1}{\rho_0 c^2}\frac{\partial^2 E_0}{\partial t^2} + \frac{1}{\rho_0^2 c^2}\frac{\partial\rho_0}{\partial t}\frac{\partial E_0}{\partial t} - \nabla.f = 4\pi G\rho_0. \tag{A10}$$

Using $\rho_0 = E_0/c^2$ again:

$$-\frac{1}{\rho_0 c^2}\frac{\partial^2 E_0}{\partial t^2} + \left(\frac{1}{\rho_0 c^2}\right)^2\left(\frac{\partial E_0}{\partial t}\right)^2 - \nabla.f = 4\pi G\rho_0. \tag{A11}$$

We have written the source density as $G\rho_0$, the proper *active* gravitational matter density. *F* is irrotational in the case of our translationally accelerated test particle, so we may write $f = -\nabla\phi$ *in these particular circumstances, ϕ being the* scalar *potential of the gravitation* field. *Note that writing $f = -\nabla\phi$ employs the usual sign convention for the gravitational field where the direction of the force (being attractive) is in the opposite sense to the direction of the gradient of the scalar potential.* With this substitution for *f* Eq. A11 is:

$$\nabla^2\phi - \frac{1}{\rho_0 c^2}\frac{\partial^2 E_0}{\partial t^2} + \left(\frac{1}{\rho_0 c^2}\right)^2\left(\frac{\partial E_0}{\partial t}\right)^2 = 4\pi G\rho_0. \tag{A12}$$

This equation looks very much like a wave equation, save for the fact that the space-like (Laplacian) part involves a scalar potential, whereas the time-like (time-derivative) part involve the proper rest energy density. To get a wave equation that is consistent with local Lorentz invariance we must write E_0 in terms of ρ_0 and ϕ so as to recover the d'Alembertian of ϕ. Given the coefficient of $\partial^2 E_0/\partial t^2$, only one choice for E_0 is possible:

$$E_0 = \rho_0\phi. \tag{A13}$$

Other choices do not affect the separation of variables needed to recover a relativistically invariant wave equation. But this is just the condition that follows from Mach's principle (and SRT). [Note that another sign convention has been introduced here; namely that the gravitational potential energy of local objects due to their interaction with cosmic matter is positive. This differs from the usual convention for the potentials produced by local objects, which are negative. Unless the cosmic matter is dominated by substance with negative mass, this convention must be simply imposed to replicate the fact that by normal conventions the rest energies of local objects are positive. Note farther that "dark energy", with its "exoticity", fills this requirement very neatly, making the imposition of a special sign convention here unnecessary.]

Substituting $\rho_0\phi$ for E_0 in Eq. A12 makes it possible to, in effect, separate the variables ρ_0 and ϕ to the extent at least that the d"Alembertian of ϕ can be isolated. Consider the first term on the LHS of Eq. A12 involving time-derivatives. Substituting from Eq. A13 into Eq. A12 gives:

$$-\frac{1}{\rho_0 c^2}\frac{\partial^2 E_0}{\partial t^2} = -\frac{1}{\rho_0 c^2}\frac{\partial}{\partial t}\left(\rho_0\frac{\partial\phi}{\partial t} + \phi\frac{\partial\rho_0}{\partial t}\right)^2$$
$$= \frac{1}{c^2}\frac{\partial^2\phi}{\partial t^2} - \frac{2}{\rho_0 c^2}\frac{\partial\phi}{\partial t}\frac{\partial\rho_0}{\partial t} - \frac{\phi}{\rho_0 c^2}\frac{\partial^2\rho_0}{\partial t^2}.$$

(A14)

Making the same substitution into the second time-derivative term on the LHS of Eq. A12 and carrying through the derivatives produces:

$$\left(\frac{1}{\rho_0 c^2}\right)^2\left(\frac{\partial E_0}{\partial t}\right)^2 = \left(\frac{1}{\rho_0 c^2}\right)^2\left(\rho_0\frac{\partial\phi}{\partial t} + \phi\frac{\partial\rho_0}{\partial t}\right)^2$$
$$= \frac{1}{c^4}\left(\frac{\partial\phi_0}{\partial t}\right)^2 + \frac{2\phi}{\rho_0 c^4}\frac{\partial\phi}{\partial t}\frac{\partial\rho_0}{\partial t} + \left(\frac{\phi}{\rho_0 c^2}\right)^2\left(\frac{\partial\rho_0}{\partial t}\right)^2.$$

(A15)

Now, taking account of the fact that $\phi/c^2 = 1$, we see that the coefficient of the second term on the RHS of this equation is $2/\rho_0 c^2$, so when the two time-derivatives terms in Eq. A12 are added, the cross-product terms in Eqs. A14 and A15 will cancel. So the sum of these terms will be:

$$-\frac{1}{\rho_0 c^2}\frac{\partial^2 E_0}{\partial t^2} + \left(\frac{1}{\rho_0 c^2}\right)^2\left(\frac{\partial E_0}{\partial t}\right)^2 = -\frac{1}{c^2}\frac{\partial^2\phi}{\partial t^2} - \frac{\phi}{\rho_0 c^2}\frac{\partial^2\rho_0}{\partial t^2} +$$
$$+ \left(\frac{\phi}{\rho_0 c^2}\right)^2\left(\frac{\partial\rho_0}{\partial t}\right)^2 + \frac{1}{c^4}\left(\frac{\partial\phi}{\partial t}\right)^2.$$

(A16)

When the first term on the RHS of this equation is combined with the Laplacian of ϕ in Eq. A12 one gets the d'Alembertian of ϕ and the classical wave equations (A17) below is recovered.

$$\nabla^2\phi - \frac{1}{c^2}\frac{\partial^2\phi}{\partial t^2} = 4\pi G\rho_0 + \frac{\phi}{\rho_0 c^2}\frac{\partial^2\rho}{\partial t^2} - \left(\frac{\phi}{\rho_0 c^2}\right)^2\left(\frac{\partial\rho_0}{\partial t}\right)^2 - \frac{1}{c^4}\left(\frac{\partial\phi}{\partial t}\right)^2.$$

(A17)

The remaining terms that follow from the time-derivatives of E_0 in Eq. A16, when transferred to the RHS, then become transient sources of ϕ when its d'Alembertian is made the LHS of a standard classical wave equation. That is, we have recovered Eq. A6 above.

Reprinted from James F. Woodward, "Flux Capacitors and the Origin of Inertia," *Foundations of Physics*, vol. 34, pp. 1475–1514 (2004) with permission from Springer Verlag.

Part II

4

Getting in Touch with Reality

INTRODUCTION

Spacetime may be absolute, but neither space nor time have that property. The locally measured invariance of the speed of light, further, forces us to accept that spacetime is pseudo-Euclidean. So, while space is flat, and in a sense the same can be said of spacetime (because the Pythagorean theorem applies in four dimensions modified only by the space and time intervals having opposite signs), all of the well-known weird consequences of relativity theory follow. Time dilation. Length contraction. Mass increase with velocity. "Twins paradox" situations.

General relativity theory (GRT) extends SRT by assuming that SRT is correct as long as one considers only sufficiently small regions of spacetime – that is, regions so small that they can be considered flat. And GRT is predicated on the Einstein Equivalence Principle (EEP), which takes account of the fact that in each of the sufficiently small regions of spacetime the action of a gravitational field – as a consequence of the universality of the gravitational interaction – is indistinguishable from an accelerated frame of reference. The key feature of the EEP is the non-localizability of gravitational potential energy provision. This stipulation is required to enforce the indistinguishability of accelerated frames of reference from gravity fields, for otherwise one can always make such distinctions by observing, for example, the charge to mass ratios of elementary particles. Indistinguishability is the feature of GRT that is required to make the theory truly background independent – and wormholes real structures in spacetime.

Indistinguishability is also required for another reason. As we saw in Chap. 2, if the locally measured value of the total scalar gravitational potential is equal to the square of the speed of light, then GRT dictates that inertial reaction forces are produced by the gravitational interaction with chiefly distant matter in the cosmos. The condition in GRT that leads to the equality of the total scalar gravitational potential and the square of the speed of light is spatial flatness at the cosmic scale. This is now measured to be the case. Indistinguishability and non-localizabilty then guarantee that the equality be true everywhere and at every time. That is, they require that the total scalar gravitational potential be a locally measured invariant, just as the vacuum speed of light is. This, in turn, guarantees

J.F. Woodward, *Making Starships and Stargates: The Science of Interstellar Transport and Absurdly Benign Wormholes*, Springer Praxis Books, DOI 10.1007/978-1-4614-5623-0_4, © James F. Woodward 2013

that inertial reaction forces are everywhere and are at every event strictly gravitational in origin. That is, Mach's principle is universally true.

Since inertial reaction forces arise from the action of gravity, it makes sense to explore circumstances where interesting effects might arise that can be put to use for practical purposes. We will want large effects, so we will look for Newtonian order effects. We find that if we allow the internal energy of an object to change as the object is accelerated by an external force, a relativistically correct calculation shows that transient changes in the rest mass of the object take place during the acceleration. These changes – Mach effects – are sufficiently large that they should be detectable in laboratory circumstances using inexpensive apparatus. Although Mach's principle has a confusing and contentious history, the fact of the matter is that it is contained in GRT since the universe is spatially flat at the cosmic scale, and the calculation of Mach effects employs standard techniques of well-established physics. So, from the theoretical point of view, there is no "new physics" in the prediction of laboratory-scale Mach effects. Accordingly, we have every right to expect to see the predicted effects should we carry out the sorts of experiments suggested by theory.

Before turning to the experiments that have been conducted over the years to see if Mach effects really exist, we turn for a moment to some general considerations on the roles of theory and experiment, as they will help place those experiments in a context that is not widely appreciated.

EXPERIMENT AND THEORY

The roles of experiment and theory in the practice of science is a much discussed and sometimes contentious topic. The usual story about their roles follows the "hypothetico-deductive" method model, a naïve schema that purports to capture how science is done that was created in the wake of Karl Popper's devastating critique of the role of induction in the creation of knowledge in his *Logic of Scientific Discovery,* published in the early 1930s.

According to the hypothetico-deductive method, after identifying a topic or an area of interest for investigation, one first surveys what has been done and what is known about the subject. This, of course, is just common sense, and hardly constitutes a process that is especially scientific. Having mastered what is known, one formulates a "hypothesis" that, one way or another, purportedly delineates how features of the subject are causally related to each other. This is the tough part, for seeing all but the most obvious correlations, indeed, even figuring out what to measure in the first place, without theory as a guide is far more difficult than most imagine. In any event, having formulated an hypothesis, one then designs an experiment to test the hypothesis. Depending on the outcome of the experiment, the hypothesis is confirmed or refuted, and if refuted, may be modified to be brought into conformity with the experiment that led to the refutation.

The hypothetico-deductive method, devised by philosophers and pedagogues to deal with Popper's critique, has almost nothing to do with the actual practice of science. This was best captured by Thomas Kuhn in his *Structure of Scientific Revolutions,* which put the word "paradigm" into common (and often incorrect) use. A paradigm is a collection of theories, principles, and practices that the practitioners of a field take to be true and correct, and the experimental exemplars that confirm the correctness of the principles and theories. In modern physics allegedly two paradigms are in place – one based on the principle and theories of relativity, and the other constructed on the principles and theories of quantum mechanics. In the absence of a widely accepted theory of quantum gravity, these paradigms remain distinct. The expectation is widely shared that eventually a way will be found to merge these paradigms, and string theory and loop quantum gravity have been advanced as ways to affect such a merger.

For our purposes, however, such grand concerns are not important. What is important is how experiments fit into knowledge structures that operate with a paradigm. As a general rule, the fact of the matter is that experiments proposed and undertaken by those who are not leaders in the field the experiment falls in, especially experiments designed to test the foundations of the field, are not taken seriously by most practitioners of a paradigm. An example here is the work of John Clauser, designed to test the predictions of John Bell's work on the foundations of quantum mechanics in the early 1970s. Even when Clauser's results seemed to vindicate the standard interpretation of the paradigm, it was still regarded askance by most members of the quantum community.

You may ask, why? Well, according to Kuhn, scientists practicing "normal science" do not seek novelty and are uninterested in seriously questioning the foundations of the principles and theories that make up their paradigm. Bell and Clauser violated this informal rule of proper "normal" behavior. The lesson here is that doing experiments that put the foundations of the paradigm at risk, especially experiments that have no sanction from the conservators of the paradigm, no matter what results are obtained, are likely to be ignored. And the execution of such experimental work is unlikely to put one on a positive career path within the paradigm.

What about the remark one often hears that one solid experimental result that conflicts with a prediction of the paradigm theory is sufficient to warrant the rejection of the theory? Well, fact of the matter is that this is just nonsense. Two recent examples spring to mind. One is the claim that neutrinos traveling from the Large Hadron Collider (LHC) near Geneva to a facility in Italy do so ever so slightly superluminally and arrive 60 ns or so sooner than they would traveling at the speed of light. Those convinced that tachyons – particles that travel faster than light – should exist are heartened by this result. No one else is. Theoretically, neutrinos are not tachyonic particles. So should they really travel faster than light, their existence should lead to the rejection of SRT. Relativists, however, know that SRT is *massively* confirmed by the results of countless experiments. And the principle of relativity is so obviously true that seriously proposing that it is wrong in some fundamental way is simply off the table. So the likelihood that the community of physicists have rejected SRT (and GRT) on the basis of this neutrino flight time experiment by the time you read this is essentially zero.

Another current example is the search for the Higgs particle predicted by the Standard Model of relativistic quantum field theory. The Higgs process is an essential part of the Standard Model. It transforms zero rest mass particles that are constrained to travel at the speed of light into particles with rest mass that travel at speeds of less than the speed of light. Since we ourselves are made up of non-zero rest mass stuff, it is easy to believe that the Higgs process is crucial to the success of the Standard Model. The role of the Higgs process in the creation of non-zero rest mass stuff has led to the widespread misstatement that the Higgs field and its particle are the origin of mass. As we saw back in Chap. 1, all energy has inertia, and since the measure of inertia is mass, all energy has mass. As Frank Wilczek has pointed out, this is captured in Einstein's second law: $m = E/c^2$. And in Chap. 2 we saw that since $\phi = c^2$, Mach's principle has the same gravitational interaction as the origin of mass. Even if you ignore the Machian aspect of the origin of inertia and its measure, mass, the Higgs process does not have the same status as the principle of relativity and SRT. Indeed, it is widely acknowledged that the Standard Model, for all of its manifold successes, is not the last word on elementary particle physics. It has the aesthetic defect, in the eyes of some at least, that too many numbers have to be measured and put in by hand. More fundamentally, though gravity is widely thought to be unimportant for elementary particles, the Standard Model is plainly incomplete as it does not include gravity.

Recently, the search for the Higgs particle at the LHC has ruled out its existence over a wide range of possible mass-energies. Only a narrow window of energies remain to be explored. Those inclined to believe that the Higgs particle must exist are not yet deeply concerned, for several technical arguments can be made that the unexcluded window is the most likely place for the particle to be found. Those with a lesser level of commitment to the Standard Model are willing to publicly contemplate the possibility of the non-existence of the Higgs particle. But do not be deceived. Almost no one really believes now that the Higgs particle will not be found. The Standard Model is massively confirmed by experiment, and given the crucial role of the Higgs process in the model, the likelihood that it does not exist is vanishingly small. Indeed, rumor has it that an announcement is to be made in a week or so that a Higgs-like particle with a mass of 126 GeV has been observed at the LHC.

By the time you read this, you will likely know the resolution to both of the forgoing examples. However they have in fact turned out, the point here is that experiment and theory do not have the same status in the practice of science that they are often superficially accorded. And there are good reasons why this is so. In the seventeenth century, before the major fields of the physical sciences were explored and established, it was easy to assert the primacy of experiment over theory as the arbiter of what constitutes reality. And while in principle this remains true, anomalous experimental results no longer have the theory discrediting role that they once (if ever) had. That is because the theoretical edifice of modern science is not the tentative collection of proposals advanced several hundred years ago to organize the knowledge of the world extant at that time. So, even when the anomalous experimental result in question is produced by a team of world-class scientists with impeccable credentials, as in the examples used here to illustrate the issues involved in the roles of experiment and theory, the theory is not called into serious question because of the anomalous result.

When we consider instances where the participants are not members of the mainstream part of the scientific community in good standing, or the theory involved is not manifestly a widely accepted part of the paradigm, or the experimental procedures and techniques are not those of standard practice, the situation becomes even more complicated. John Clauser's investigation of the predictions of John Bell's analysis of the "measurement" problem in quantum mechanics is an example. Experimental investigation of Mach effects is another. In the case of Mach effects, we have theoretical predictions that purport to follow from the widely accepted principles and theories of physics – without the introduction of any "new physics" at all. But the predictions are based on the presumption that Mach's principle, as expressed in the Mach-Einstein-Sciama laws of inertia, is correct and in fact encompassed by GRT. Had the arguments for Mach's principle and the laws of inertia, and the predictions of Mach effects, been carried out by a recognized authority on GRT, the claim that no "new physics" is involved might have been accepted. But they weren't.

If the theory and predictions were made by someone unknown, others would find it easy to believe that anyone claiming that the theory and predictions are flawed, for whatever reason, may have a meritorious case. If one believes that the theory on which an experimental result is based is flawed, it is easy to dismiss any claimed experimental result as incorrect. The precise origin of the presumed incorrect result is unimportant, and the whole business can simply be ignored. After all, even the simplest of experiments are actually subtle and complicated, and the mistake or flaw could be due to any one of a number of things. Tracking down the presumed mistake or flaw, even with access to the experimental apparatus used, could easily prove to be a very difficult and thankless task. Why an incorrect result was obtained is not important; that the result is incorrect is important.

For onlookers to affect this sort of attitude toward work on Mach effects is not surprising. But what about those actually working on the problem? Those so engaged find themselves with a different problem: since the derivation of the effects involves no new physics, and as that derivation proceeds from Newton's third law and the EEP, getting an experimental result that deviates markedly from prediction – in principle – calls into question the validity of foundational aspects of all physics. It doesn't really matter that onlookers are uninterested and unimpressed by Mach's principle and Mach effects. For what they think of things has nothing really to do with what results should be obtained in any carefully executed experiment. If the derivation of the effects is not flawed, then an unexpected result in an experiment means that the execution of the experiment is flawed. It does not mean that Mach's principle is wrong, or that Mach effects do not exist. And theory trumps experiment, rather than vice versa. The problem one then faces is figuring out what was screwed up in the experiment that produced the unexpected results. This is usually much more difficult than one might guess, as experiments are always complicated by extraneous effects and the evolution of components of the apparatus.

MACH EFFECT EXPERIMENTS

As discussed in Chap. 3, experiments to test Mach effects follow almost obviously from the basic Mach effect equation (derived in Addendum #2 to Chap. 3):

$$\nabla^2 \phi - \frac{1}{c^2}\frac{\partial^2 \phi}{\partial t^2} = 4\pi G \rho_0 + \frac{\phi}{\rho_0 c^4}\frac{\partial^2 E_0}{\partial t^2} - \left(\frac{\phi}{\rho_0 c^4}\right)^2 \left(\frac{\partial E_0}{\partial t}\right)^2 - \frac{1}{c^4}\left(\frac{\partial \phi}{\partial t}\right)^2. \qquad (4.1)$$

ρ_0 is the proper matter density (where "matter" is understood as everything that gravitates) and E_0 is the proper energy density. "Mach effects" are the transient source terms involving the proper energy density on the right hand side. When are the transient source terms not zero? When the accelerating object considered in the derivation *absorbs "internal" energy as it is accelerated.* That is, if our accelerating body is not deformed by the acceleration, these terms are zero. As before, we can separate them out from the other terms in the field equation, getting for the time-dependent proper source density:

$$\delta\rho_0(t) \approx \frac{1}{4\pi G}\left[\frac{\phi}{\rho_0 c^4}\frac{\partial^2 E_o}{\partial t^2} - \left(\frac{\phi}{\rho_0 c^4}\right)^2\left(\frac{\partial E_0}{\partial t}\right)^2\right]. \tag{4.2}$$

A variety of different physical systems can be imagined that store internal energy during accelerations. So, in principle, several options for exploring Mach effects should be available to us. Simple macroscopic mechanical systems, however, aren't good candidates for the production of Mach effects, since the rate at which internal energy changes can be affected is quite limited. That means that the time-derivatives in Eq. 4.2 will be small at best. What we want are systems where the time-derivatives can be made quite large. In terms of practical apparatus, that means that we are going to be looking at systems involving electromagnetism. There are several electromagnetic energy storage devices – capacitors, inductors, and batteries. Since we are looking for fast transient effects, batteries are ruled out. Inductors store energy in the magnetic field, and to put some of this energy into an accelerating material medium, we would need to provide the inductor with some core material with, preferably, a high permeability.

Although a system using magnetic energy storage in some high permeability medium is plainly feasible, capacitors have several features that make them preferable to inductors. Compact devices capable of storing energy at very high energy density without breakdown are available. Shielding electric fields is much easier and cheaper than shielding magnetic fields. Ferroelectric materials can be used to make electromechanical actuators that operate at high frequencies. Such ferroelectric devices are themselves capacitors and can be integrated into the design of a device intended to display Mach effects. And capacitors are cheap and come in a wide variety of configurations. For these reasons, experiments designed to test for Mach effects have relied on capacitors of one sort or another.

If we integrate the contributions of this transient proper matter density over a capacitor being charged or discharged *as it is being accelerated,* we will get for the transient total proper mass fluctuation, written δm_0:

$$\delta m_0 = \frac{1}{4\pi G}\left[\frac{1}{\rho_0 c^2}\frac{\partial P}{\partial t} - \left(\frac{1}{\rho_0 c^2}\right)^2\frac{P^2}{V}\right], \tag{4.3}$$

where P is the instantaneous power delivered to the capacitor and V the volume of the dielectric. As noted in Chap. 3, if the power is applied sinusoidally, then the first term on the right hand side of Eq. 4.3, which depends on the derivative of the power, will scale linearly with the frequency since $\partial \sin \omega t/\partial t = \omega \cos \omega t$. Accordingly, operation at higher frequencies is to be preferred. It is important to keep in mind that the effects contained in Eqs. 4.1, 4.2, and 4.3 are only present in bulk objects subjected to accelerations, the explicit acceleration dependence being that given in Chap. 3. Simply varying the internal energy in an object will not by itself produce Mach effects. The second term on the right hand side of Eq. 4.3, of interest for making wormholes because it is always negative, though normally exceedingly small, shows no such explicit frequency dependence. Since the first term on the right hand side of Eq. 4.3 yields prediction of relatively large effects, predictions based on this term for systems using capacitors of one sort or another were those pursued almost without exception. After all, if evidence for the larger first term can be produced, it follows that the second term effect will necessarily be present.

Realization that the time-dependent terms in the Newtonian order relativistically invariant gravity field equations represented inertial effects dates from the fall of 1989. It took a year or two to work through both the details of the physics and formalism to understand the basic meaning of this insight. But experimental work designed to look for the expected effect started early on, and has been under way pretty much continuously since. That effort, for convenience, can be broken down into several periods, each with a distinctive cast to its activity owing to equipment and apparatus, and the participation of others. All periods, however, are characterized by seeking inexpensive means to carry out experiments. In the earliest period, where mass fluctuations were sought – the late 1980s – the two chief technical problems addressed were finding a weigh/force sensor with high sensitivity and fast response time, and securing an analog to digital data acquisition system appropriate to the needs of the work.

EARLY EQUIPMENT ISSUES

In the modern era, the days when electronic data acquisition was new seem part of a very remote past. The fact of the matter is, though, that this time was not so long ago. Analog to digital converters (ADCs) that could be integrated into experimental systems and computer controlled only became available in the mid-1980s, and at the outset getting them to work properly was a chancy business at best. One doing this sort of thing quickly learned that it was a good idea to purchase your ADC board from a local source – so you could go visit with them when you encountered problems. Trying to resolve issues by phone or mail was a poor substitute for a personal visit. As chance would have it, several recent Cal Tech graduates decided to go into the ADC and signal processing business about the time a data acquisition system became an obvious necessity. They formed Canetics, and started building ADC and signal processing boards. One of their founding owners went around drumming up business at local universities. The boards they built were really quite good. One of their boards – their PCDMA – is still in use. As it was designed for a computer using an 8088 processor, now long superseded by faster processors, but still works, this is very handy.[1]

---

[1] The board supports eight multiplexed channels with 12-bit resolution. The processing is a bit slow. But if you only need an acquisition rate of 100 Hz or less per channel, all is well. If fewer than the full eight channels are used, faster acquisition rates are possible. This is much slower than present systems, but more than adequate for the purposes of this work.

The complement to an ADC system is code for the computer that collects the data. In the 1980s, programs that would run on PCs and Apples to carry out data collection and analysis didn't exist. But Microsoft was in the process of turning their elementary version of Basic into a serious programming language, one that eventually provided for compilation of the code into executable files. They called it QuickBasic. The early versions of QuickBasic, the compilers anyway, were pretty clunky. But when Microsoft released version 4.0, pretty much all of the problems with the earlier versions were resolved. An especially nice feature of QuickBasic was its graphics support. Not only could you write simple acquisition, processing, and data-storage routines, writing graphics code to display the acquired data was straightforward. And changing the code to do new tasks as they presented themselves was easy, too. By the time the work on Mach effects got underway in late 1989, the Canetics data acquisition system and several QuickBasic programs to support it were in place. The code has been modified again and again over the years. But the basic system still in use was complete at that time.

Work on mass fluctuations started before the physics of their production was sorted out in the fall of 1989 and thereafter. As mentioned above, the chief apparatus problem was finding a way to measure very small changes in the mass of an object that take place very quickly. Several sensors were available that could make very precise measurements of the masses of things. But this precision was achieved by using long "integration" times. That is, the sensor would, in effect, make many determinations of the mass over some extended time and then effectively take the average of them. The settling time of such devices was (and still is) typically several seconds. The two types of weigh sensors that looked promising for fast, high sensitivity measurements were semi-conductor strain gauges attached to stainless steel foil springs, and Hall effect probes mounted on a sprung shaft in fixed magnetic fields with gradients. As the probe moves in the field, its resistance changes owing to changes in the conduction path produced by the changed magnetic field. Strain gauges and Hall probes of modest cost became available in the late 1980s.

Shortly after these sensors became available, Unimeasure Corporation developed a commercial device – their U-80 – using Hall probes in magneto resistive mode for a modest cost (less than $200). In addition to being a very high sensitivity device,[2] the U-80 was also designed to be unusually versatile. Unmodified, it could be used as a position sensor by attaching its shaft to the object whose position was to be measured. But if what was wanted was a force transducer, a diaphragm spring could be affixed to one end of the mobile shaft within the case. Springs of several force constants were available to tailor the response to one's particular needs. Pictures of the Unimeasure U-80 in various states of disassembly are shown in Fig. 4.1. Implementation of the U-80 is simple. It is mounted to measure the force in question, and wired as the active leg of a Wheatstone bridge. The bridge voltage then gives the force on the sensor. The chief drawback of the U-80 is that it is a very sensitive electronic device, so in harsh electromagnetic environments, considerable care must be taken with shielding. For the decade of the 1990s this was accomplished by placing the U-80 in a thick walled (>1 cm) aluminum enclosure lined inside with mu metal foil, and outside with a thick steel sleeve and more mu metal. The bridge electronics were

---

[2] The position sensitivity is 5 Ω per 0.001 in. of displacement.

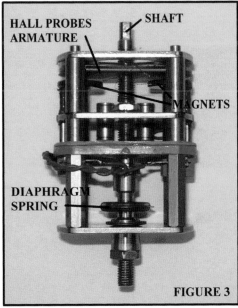

Fig. 4.1 The Unimeasure U-80 position sensor. On the *left* in its plastic casing and on the *right* with casing removed. The armature carries two Hall probes that move in a shaped magnetic field produced by the labeled magnets. The diaphragm spring transforms this position sensor into a force sensor

placed in a chamber in the aluminum case immediately below the U-80. Pictures of this arrangement are found in Fig. 4.2.

In the 1970s and early 1980s, electronics for use in research systems were fairly limited and generally sufficiently expensive so that their purchase from out of pocket resources wasn't to be realistically contemplated. Discrete semi-conductor components and a few integrated circuits were available for hobbyist use. But if you needed, say, a serious differential instrumentation amplifier, or a lock-in amplifier, or an automatic gain control, and so on, you were at the mercy of commercial outfits. You could expect to spend several thousand dollars for any one of these devices, along with oscilloscopes of practically any variety. Back then, several thousand dollars was real money.

All this changed drastically over the decade of the 1980s as chip manufacturers made increasingly sophisticated integrated circuits of constantly improving accuracy, sensitivity, and reliability. At the beginning of the decade the 741 operational amplifier was about as good as things got. By the end of the decade, Analog Devices for example, was making a series of single, dual, and quad op amp chips with markedly better gain-bandwidth. They were also making four-quadrant multipliers, instrumentation amplifiers, synchronous demodulators, and a variety of other integrated circuits for specialty use. Typically, these chips cost less than $100, often much less. So for a small investment and some soldering of a modest number of external discrete components, electronics that hitherto

Fig. 4.2 Views of the early weight/thrust sensor. In the *upper left* panel are the Wheatstone bridge electronics in their chamber mounted directly below the U-80 sensor chamber. In the *lower left* panel the U-80 sensor is shown mounted in its concentric aluminum and steel chamber. An aluminum ring that supports three rods that pass through the base plate of the main test chamber is shown attached to the end of the shaft of the U-80 sensor. In the *upper right* panel the main chamber base plate is added, along with the support ring tensioned by fine steel wires. In the *lower right* panel the entire test chamber is shown fully assembled

were hopelessly costly could be had. And world-class serious research systems no longer were the preserve of those with substantial funding.

Software and electronics are not the only things that put serious research projects out of the reach of those without serious funding. Research systems almost always involve the fabrication of specialized parts, especially in the core of the apparatus, without which experiments simply cannot be done. That is, you can't just go out and buy all of the stuff you need to do an experiment because some of the parts needed simply do not exist to be bought at any price. Fabrication of such parts by machine shops can be very expensive as machinists are highly skilled labor and the machines used in their fabrication are costly, too. Research universities maintain machine shops staffed with skilled machinists. But the shops do work that can be charged to some funding source first, and other unfunded projects second. And the machining facilities are only available to those who want to do their own machining if they demonstrably know what they are doing. Experimental grad students often pick up some very modest machining skills as part of their education.

The skill level required for most major projects, however, is a good deal more than student level. These skills are acquired by hanging out with machinists and being instructed by them. In effect, you have to do an apprenticeship if you want to do serious machining. But if you are lucky enough to get such an informal apprenticeship, and you have a machine shop available, apparatus that would otherwise simply be out of reach can be had for the cost of the materials.

One other matter that had an impact on the research program deserves mention. In the early 1990s EDO Ceramics had a manufacturing operation in Fullerton, about a mile down the road from the University. It was a period of tough economic times, and EDO decided to close down this operation and move it to Salt Lake City, where taxes were less burdensome. When the shutdown was almost complete, the staff found that they had some random loose crystals left. Rather than pack them up and send them to Utah, they decided to donate them to the Physics Department. The department staff put me in touch with the plant manager, as they knew I would be interested. I came into possession of a modest stash of PZT (lead-zirconium-titanate) crystals that were put to use over the next several years. More important, I got a short-course in PZT stack construction from Jim Sloan, the plant manager, that still serves me well. Talk about synchronicity and dumb luck. Years later Bruce Tuttle, a professional ceramics type, visited. He asked who had fabricated the devices in use and how the fabrications had been done. On learning the answers to his questions, he asked how we had acquired such specialized expertise. Turns out he knew Jim Sloan. Further explanation was not needed.

Early Experiments, 1989 to 1996

Through most of the 1990s the basic system in use consisted of the U-80 position sensor adapted as a force transducer mounted in the aluminum and steel case already described. Parts of the apparatus underwent changes during this period. For example, the method of electrically connecting the power feed to the test devices went from mercury contacts located beneath the sample chamber mounted on the U-80 to simple fine stranded wire, as it was found that the liquid metal contacts did not produce superior performance. And various mechanical mounting schemes and test device enclosures were built in an ongoing program to try to improve the performance of the system.

In addition to the ongoing refinement of the main system, completely different measuring systems were also built and tested. The most ambitious of these was a laser interferometer system with all of its parts mounted on a table that could be rotated between horizontal and vertical dispositions. The test devices for this system were small arrays of high voltage capacitors mounted on a 2.5 cm diameter aluminum rod about 3 cm long. The end of the rod opposite the capacitors was affixed to a brass reaction mass about 8 cm in diameter and of comparable length held in place on the table with a three-point suspension. The plan was to drive the capacitors with a voltage signal with a frequency equal to half of the mechanical resonance frequency of the aluminum rod, thus producing a mass fluctuation at the resonant frequency of the rod. While the excitation of the capacitors might cause the rod to oscillate, in the horizontal position the mass fluctuation would not have

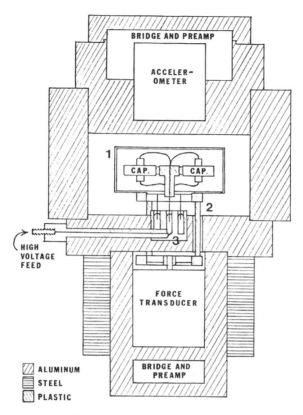

Fig. 4.3 Figure 4.1 from "A New Experimental Approach to Mach's Principle," a longitudinal section of the apparatus shown in the *lower right* panel of Fig. 4.2 here

much effect. In the vertical position, however, owing to the action of gravity, one might expect the predicted mass fluctuation to have an effect on the oscillation of the rod. A scheme using parametric amplification of the signals by adding a DC voltage to the power signal was devised to discriminate real from spurious signals. It worked nicely, but the results were not sufficiently compelling to warrant formal publication.

The first reported experimental results on Mach effects were included in the first paper on Mach effects, published in *Foundations of Physics Letters* (Volume 3, pp. 497–507, and corrigendum, Volume 4, p. 299) in the fall of 1990. Figure 4.1 from that paper is included here as Fig. 4.3. A pair of the test capacitors used and their enclosure (Faraday cage) are shown here in Fig. 4.4. The capacitors are Ceramite high voltage "door knob" capacitors with their insulation stripped to reduce their weight (and then painted with insulating acrylic). When all of the parts were mounted on the U-80, its mechanical resonance frequency turned out to be a bit over 100 Hz using the stiffest spring available in the U-80. This meant that the operating frequency had to be a bit over 50 Hz to take advantage of the mechanical resonance amplification of the predicted weight fluctuation signal. The Mach effect predicted for 50 Hz, even when the amplitude of the voltage signal applied was about 3 kV, is very small. A signal of roughly the expected magnitude was seen.

Fig. 4.4 Capacitors (*above*) of the sort used with the system shown in the previous figures, along with dummy capacitors (*below*) used in the system of the previous figures. The capacitor case used (aluminum with mu metal lining) is shown in the *bottom right*

Refinements to the system were carried out, and more types of high voltage capacitors were brought into the test program over the next year. Some of these capacitors are displayed in Fig. 4.5. The same basic system and techniques of the first reported results were employed in this work, published in *Foundations of Physics Letters* (Volume 4, pp. 407–423). The results were much like those in the first report. A year later another paper on Mach effects was produced (*Foundations of Physics Letters*, Volume 5, pp. 425–442), but it did not include specific experimental results. Rather, it was devoted to explaining the production of stationary forces in systems of the sort discussed above and in Chap. 3, and laying out the first version of the first principles derivation of the effects. Experimental work continued through the next several years, but the chief emphasis at that time was on getting issues of theory sorted out, a task made more difficult by the fact that there was then (and still is now) no clearly articulated version of Mach's principle.

Eventually, the capacitors settled upon as the best and most reliable were specially fabricated multi-plate ceramic "chip" capacitors made by KD Components (of Carson City, Nevada).[3] These were mounted in arrays of six capacitors glued between two small aluminum rings as shown in Fig. 4.6. The technique of "rectifying" the mass fluctuation using a second mechanical device discussed above and in Chap. 3 was employed in this work.

The second mechanical actuator was a stack of PZT crystals 19 mm in diameter and about 2.5 cm long. The PZT stack was affixed to a brass disk with glue and six machine screws, as shown in Fig. 4.7. The operating frequencies used were 7 and 11 kHz, orders of magnitude higher than the 50 Hz frequency employed in earlier work. Data were taken at

---

[3] These capacitors were provided by Jeff Day, President and CEO of the company, as he had taken an interest in the work some years earlier when his company was located near to CSUF in Santa Ana, California. Jeff visited the lab to see what was going on and invited me on a detailed tour of his manufacturing facility in Santa Ana, explaining in detail the fabrication procedures then in use.

Fig. 4.5 Some of the other high voltage capacitors use in early work on Mach effects. Note that no provision was made to generate "bulk" accelerations in these devices

0° and 180° of phase, and 90° and 270° of phase. A phase change of 180° should reverse any Mach effect present, so differencing the 0° and 180°, and 90° and 270° results should yield net signals, and the magnitude for the 0–180 difference should differ from the 90–270 difference. Results of this work were reported in a paper published in *Foundations of Physics Letters* (volume 9, pp. 247–293) in 1996. The main results were contained in Figs. 4.7 and 4.8 in that paper, reproduced here as Fig. 4.8. A net signal was found for both differenced phase pairs, and the magnitude for the two different phase pairs was also different, as expected.

The arresting feature of the results presented in Fig. 4.8 is the very large offset in the weigh sensor traces by comparison with the small differential signal produced by changing the phase by 180°. At the time the work was done, no attempt was made to discover the detailed cause of this large weigh signal present in traces for both phases. Since it was removed as a common mode signal by the subtraction process, it was not seen as relevant to the central result, the difference between the weigh signals for the two phases of each phase pair. This sort of signal was present in results for systems other than the one under consideration, and always ignored for much the same reasons.

Only recently (see the next chapter) has the likely cause of these signals been identified. Electrostriction. Whereas the piezoelectric response of the PZT crystals depends on the voltage, the electrostrictive response depends on the square of the voltage. As such, the electrostrictive response has twice the frequency of the exciting voltage signal – exactly as

Fig. 4.6 An array of 6 KD multi-plate high voltage capacitors wired in parallel clamped between two grooved aluminum rings used in the work of the mid-1990s

is the case for the first term Mach effect. The small differential effect claimed as the result is almost certainly due to the behavior of the array of KD capacitors, as discussed in the paper on this work. The much larger signal ignored at the time was likely produced in the actuator PZT stack independently of whatever might have been going on in the KD array.

What wasn't appreciated until recently is that the PZT crystals used in the actuator stacks have strong piezoelectric *and* electrostrictive responses. In this case, the KD Components capacitors have strongly suppressed piezoelectric and electrostrictive responses. After all, you don't want your fancy multi-plate high voltage chip capacitors to vibrate. It might screw up your expensive electronics. Suppression is achieved by contaminating the titanate base material with a proprietary mix of trace impurities that lock the lattice of the ceramic. In the actuator's PZT crystals you want exactly the opposite: a large mechanical piezoelectric response to produce large excursions and accelerations of the relevant parts of the system. If you have a large piezoelectric effect, you get a large electrostrictive effect, too, because you can't selectively turn off the second order electrostrictive effect. It has the right phase to produce a stationary thrust/weight effect. Had all this been properly appreciated at the time, the course of the experimental work would likely have been quite different.

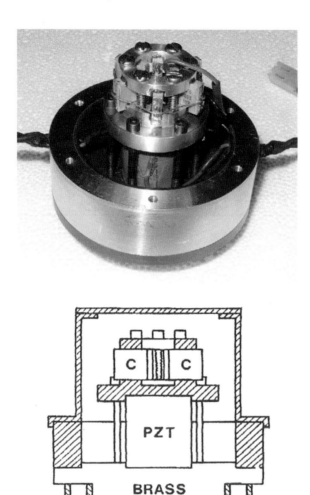

Fig. 4.7 A picture of a KD array mounted on a PZT actuator on its brass base plate with a schematic longitudinal section drawing of the device and carrying case below

EXPERIMENTS IN THE YEARS 1997 TO 2002

The year 1997 was a benchmark year for several reasons. NASA Administrator Daniel Goldin's efforts to create a program of research into "advanced" propulsion (see Chap. 6 for more details) bore fruit in the form of a workshop held at the Glenn Research Center in the summer. It was organized by Marc Millis and a group of scientists and engineers he had gathered to support the effort. That group was dominated by people who had convinced themselves that inertia could be explained as an electromagnetic quantum

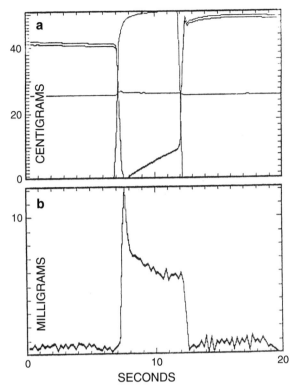

Fig. 4.8 The results obtained with the device of Figs. 4.6 and 4.7 running in the system shown in Figs. 4.1, 4.2, and 4.3 in 1996. The difference of the weight traces is shown full-scale in the *lower* panel B

vacuum effect, and that their quantum ideas held the seeds of the physics needed to do advanced propulsion. So, while a wide spectrum of enthusiasts were in attendance, the formal plenary and "breakout" sessions were dominated by "zero pointers." Others, if they were so inclined, had the opportunity to present their ideas. But no one was much interested in the ideas of others, so such presentations took the form of monologues delivered to a mostly uninterested audience. It was a very strange workshop.[4]

From the Mach effects perspective, the most important development of 1997 was the arrival of Tom Mahood in the then brand-new Masters program in physics. For several years before showing up at CSUF Tom had been a member of a small group of enthusiasts who styled themselves the "Groom Lake Interceptors." They had dedicated themselves to debunking a growing collection of colorful individuals who had coalesced around "Area

---

[4] No one was interested in Mach effects except for Greg Sobzak, a Harvard astrophysics grad student. But I got a chance to meet John Cramer and introduce him to Peter Milonni. Greg and I had the pleasure of sharing the banquet with John, Peter, and Ray Chaio and listen to a fascinating discussion of the measurement problem of quantum mechanics.

51," just north of Las Vegas, Nevada. Their highest visibility target was Bob Lazar, who claimed to have helped to reverse engineer flying saucers at "Papoose Lake" near the main complex at Area 51. By 1996 or so, they considered their mission pretty much accomplished. So Tom decided to learn some gravitational physics. CSUF was not known as a center of gravitational physics then.[5] But it was convenient for Tom as well as Stephen Goode and Ronald Crowley, both gravitation types, who were there and willing to be committee members for Tom's Master's thesis.

After getting a tour of the lab facilities and the work in progress, Tom's first project was to go through the computer code that had been written for data acquisition and analysis. That code had been written in the pre-color monitor days. Four traces were typically displayed, so different line styles were used to discriminate them. And if you had a lot of experience with the system, identifying which trace belonged to what was simple. For a neophyte, however, the traces were just a confusing mess. Tom took the code home and went through it, adding color coding so that the various traces could be immediately identified, even by one unfamiliar with the system. He also checked to make sure the code worked as claimed.

Tom also got some of his Groom Lake Interceptor friends involved. First he recruited Michael Dornheim, a senior technical editor at *Aviation Week and Space Technology*, to probe into the physics of Mach effects. Those conversations eventually resulted in a series of short papers with links to one another on various aspects of Mach effects that Tom put into web format and had attached to my Physics Department webpage.[6] After examining the type of devices – those mentioned at the end of the previous section – Tom suggested a new configuration. Noting that the stationary weight/thrust effect is produced by oscillating a component in which a mass fluctuation is driven at twice the frequency of the voltage applied to that component, he suggested that instead of just push/pulling on the fluctuating component from one end, that the component be placed between two actuators that would "shuttle" the component back and forth. This made sense because the devices had to be clamped with retaining bolts to keep them from tearing themselves apart. So, with an actuator at one end, only small excursions were possible.

Using the shuttling scheme would make larger excursions and accelerations possible. And implementation was simple; all you had to do was reverse the polarity of the crystals in the actuators at the ends of the devices. Then, when the same voltage signal was applied to both ends, one would expand and the other contract. After careful consideration, Tom recruited Jim Bakos (a Groom Lake Interceptor) to do the machining for a couple of shuttlers in his specialty machining shop. The resulting devices, one shown here in Fig. 4.9, were quite elegant. For a while they worked very nicely. But eventually they, like many other devices before them, died.

The degradation and eventual death of devices was and remains a serious technical issue in this business. The source of the problem seems to be the slow (and sometimes fast) degradation of the crystal structure of the components by mechanical and thermal

---

[5] Only now is this changing with the creation of an NSF Ligo center at the university, the Gravitational Wave Physics and Astronomy Center under the direction of Joshua Smith.

[6] These short articles are still there in the form they were first produced in 1997.

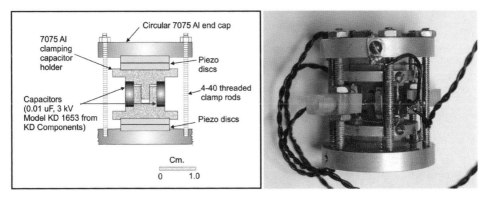

Fig. 4.9 On the *left* is Mahood's diagram of the first "shuttler" device. On the *right* is a picture of one of the two devices machined by Jim Bakos. A pair of KD capacitors are clamped in the center and driven at the base frequency to produce a mass fluctuation, while the PZT disks above and below produce an excursion at the second harmonic frequency to generate a stationary force

relaxation. In principle, the operating characteristics should be restorable by repolarization of the dielectric materials. That requires raising them above their ferroelectric Curie temperatures – typically more than 400° Fahrenheit – applying a strong electric field, and cooling the devices with the field applied. The problem is that the epoxy used in the assembly of the devices will only withstand temperatures more than 100° less than the Curie temperature. So repolarization is not a feasible option.

Tom was also suspicious of the U-80 based weight/thrust detection system. And being an independent sort, when the time came for him to do his formal work for his thesis, he decided to build a torsion pendulum to eliminate direct electrical detection of any thrust signal present in his system.[7] He decided to use the suspension fiber of the torsion pendulum as the power feed to the device to be tested, providing a ground return at the bottom with a contact dipped into a pool of mercury. Tungsten was his first choice of fiber material. It was a disaster. The current in the fiber heated it so quickly that as soon as the power was turned on, the fiber expanded thermally so much that the device sagged to the bottom of the vacuum chamber. Copper was substituted for the tungsten.

The devices Tom designed for his thesis work were based on a re-simplify analysis carried out in the spring of 1999. Several problems motivated this analysis, not the least of which was acoustic impedance mismatches in the devices. At every surface of such a mismatch, significant reflection of acoustic waves in the device could be expected, leading to complicated patterns of interfering waves in the devices. One way to approach this is with detailed mathematical modeling of the devices. For this to be useful, you need to have pretty precise information on a number of mechanical and electrical properties of the materials that wasn't readily available to us. Another way to deal with the problem is to design your devices so that there are few or no acoustic impedance mismatches of any

---

[7] Rather than use the system in use in the lab.

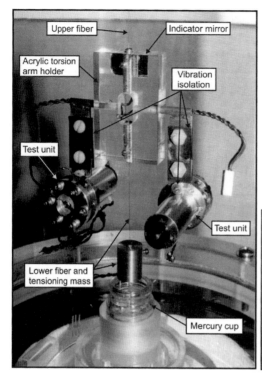

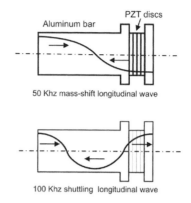

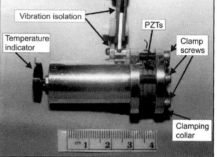

Torsion arm inside test chamber.

Test unit details. The top two cross sections show the positions of the resonant longitudinal waves within the units at 50 KHz and 100 KHz. The bottom image is a closeup view of a test unit.

Fig. 4.10 Pictures of Mahood's Master's thesis apparatus. On the *right* is one of the devices he built, along with drawings detailing their construction and operation. Two of the devices suspended in his vacuum chamber are shown on the *left*

importance in the devices. We both adopted the latter approach. Tom's response to this was to make short stacks of several PZT crystals and clamp them on the end of an aluminum rod. See Fig. 4.10.

Aside from the clamping parts, the aluminum rods were machined to the 19 mm diameter of the crystals, and their lengths tuned to resonate at some chosen operating frequency (typically about 100 kHz). To detect any thrusts that might be generated, Tom mounted a laser pointer outside his vacuum chamber (made of clear acrylic) and reflected the beam off of a mirror mounted on the torsion suspension. The reflected beam terminated on a piece of grid paper taped to a concentric plastic sheet about a half meter distant from the center of the chamber. Motion of the red spot during tests was videotaped for later analysis. Armed with his own experiment, Tom completed his investigation for his Master's over the summer of 1999. The results were equivocal (and can be found in his Master's thesis available on his website: theotherhand.org).

A different style of device was made for the main line of work in the laboratory. The acoustic impedance mismatch problem was addressed by making the entire part of the

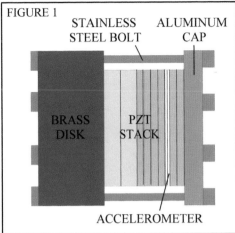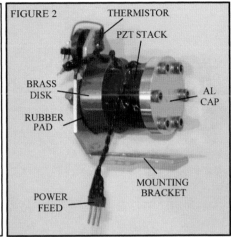

Fig. 4.11 A sectional diagram (*left*) and picture (*right*) of one of the PZT stacks designed and built for the ongoing work in the lab in 1999. The stacks had a thin pair of PZT crystals inserted where large effects were expected so as to monitor the electromechanical action there. Further details of construction and operation are found in Fig. 4.26

device intended to mechanically resonate out of PZT disks. The PZT stack was then clamped with a thin aluminum cap to a brass reaction mass with six small machine screws. The machine screws not only held the device together where free oscillation would tear it apart, they also acted as springs. Since the largest excursions of the stack were expected near the aluminum cap (opposite the brass reaction mass), the crystals at this end of the stack were chosen to be thin, maximizing the energy density produced by the exciting voltage here. The crystals near the reaction mass were chosen to have twice the thickness of the thinner crystals, as not much Mach effect was expected in this part of the stack. Pictures and a sectional diagram of these devices are displayed in Fig. 4.11. An aluminum bracket was attached to the back of the reaction mass and arranged so that the devices could be suspended at their center of mass along the axis of the devices. These PZT stacks were designed to be run with a single voltage waveform with both first and second harmonic components applied to the entire stack. The relative phase and amplitudes of the two harmonics could be adjusted to get optimum performance.

After Tom finished his thesis work in the lab, the decision was made to upgrade his torsion pendulum system so it could be used to test the new PZT stacks. The suspension fiber was replaced by a bundle of AWG #40 copper wires to minimize resistive heating when current was flowing. Both the delivery and return circuits were included in the bundle of wires so that the net current in the bundle was zero, eliminating the possibility of interactions with any ambient magnetic fields that might be present. This did not degrade the torsional sensitivity of the pendulum. A new device carrier was used that shortened the torsion arms to reduce the rotational inertia of the mobile part of the system. And a new oil pot damper that could be vertically positioned by remote control to make the amount of damping adjustable was added.

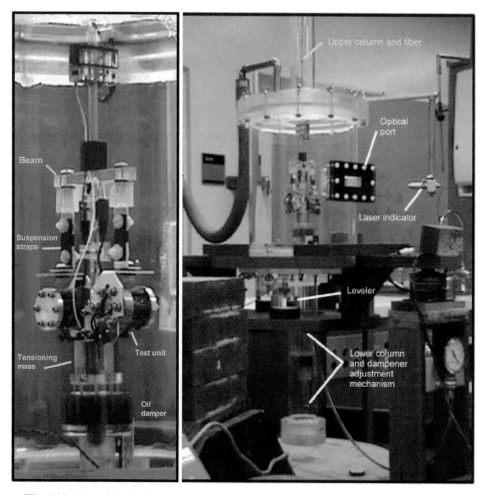

Fig. 4.12 Two of the devices suspended on the torsion fiber (a bundle of fine copper magnet wires used to feed power to the devices) – *left* panel – in the vacuum chamber adapted from Mahood's thesis work – *right* panel. During tests the oil damper was retracted so that motion of the beam was unimpeded

The simple paper scale used to measure the position of the reflected laser light was upgraded with an electronic detector. This consisted of a photovoltaic cell mounted behind an optical gradient density filter. As the light beam moved across the filter, the voltage produced by the cell changed. Tom designed and built an elegant positioning system for the detector that made calibration of the system trivial. A montage of pictures of these upgrades is found in Fig. 4.12. All of this worked remarkably well. Alas, the same could not be said of the new PZT stack devices when they were initially tested.

Everything was in place by October of 1999. Frequency sweeps were performed to locate the electromechanical resonances in the devices. Then voltage signals with first and second harmonics were applied at the chief resonance frequency. The power was a couple of hundred watts. The devices heated up quickly, so power could only be applied for a few seconds at a time.[8] A single pulse produced almost no detectable motion of the laser beam on the photovoltaic detector at all.

Since the oscillation period of the pendulum was about 15 s, the next step was to apply power pulses at the pendulum oscillation frequency in hopes of building up a detectable signal. This worked, but not very well. Tom, who stopped by from time-to-time, was not impressed. When things aren't going well, often the best you can do is to tweak things, even though you have no expectation that what you do will do any good. In this case, since vibration getting to the suspension was a background concern, thin rubber pads were added to the system between the brass reaction masses and aluminum mounting brackets. (See Fig. 4.13.) The consequence of doing this was spectacular. A single short power pulse (a second and a half or so) produced excursions of the laser spot on the detector of more than a centimeter. In order to do this repeatedly, it was necessary to establish stable thermal conditions. This was done by timing the power pulses and repeating them at intervals of 5 min. When Tom stopped by and saw this performance of the system, he was impressed.

Work over the next couple of months led to a presentation of these results at STAIF 2000, but they were too late to be included in the formal proceedings of that conference. STAIF had started a section on "advanced" propulsion the previous year, and Tom had presented Mach effects work. His presentation garnered him a half share in the best student presentation award. The results presented at STAIF 2000 were considerably more complete. And a variety of tests had been done to reassure interested parties that the thrusts observed were not just artifacts of spurious processes.

One of the procedures used was to vary the relative phase of the first and second harmonic voltage signals driving the stack in use. These results are displayed in Fig. 4.14. As with the earlier U-80 results for KD capacitors mounted on a PZT stack, the thrust varies with the variation of the phase. But it does not reverse for half of the phases as would be expected were no other effects present. As before, the cause of this behavior was attributed to the presence of a constant background signal of unknown origin, and one of no interest because it didn't respond to the variation of the relative phase. In retrospect, almost certainly, it was a Mach effect produced by electrostriction. Sigh. Missed opportunities.

Backtracking a bit, shortly after the BPP workshop, we were contacted by Paul March, then an engineer working for Lockheed-Martin in Houston on space shuttle electrical systems. He and his boss, Graham O'Neil, had been asked by Jim Peoples, manager of Lockheed-Martin's Millennium Projects in Fort Worth, to do a survey of all of the "advanced" propulsion schemes on offer and recommend those that passed muster to have people come out and give presentations on their work. Paul's first approach was just a request for information. Eventually, an invitation to do a presentation at Lightspeed was issued for the fall of 1999. In addition to the engineering staff of Millennium Projects,

---

[8] The temperature of the devices was monitored with bimetallic strip thermometers stripped from commercial freezer thermometers attached to the aluminum mounting bracket. Later, thermistors were glued to the mounting bracket to monitor the temperature electronically.

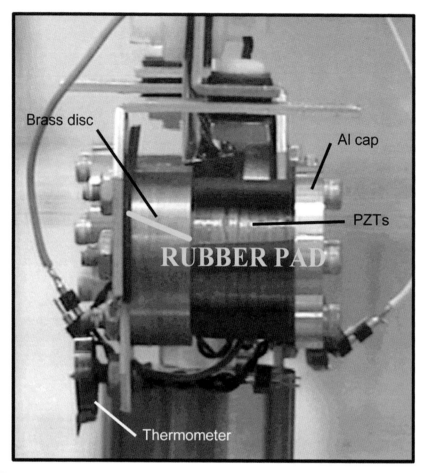

Fig. 4.13 A picture of one of the PZT stacks (from an unpublished paper circulated at STAIF2000) with the thin rubber pad that dramatically improved the performance of the device indicated

Graham and Paul, several consulting physicists were in attendance, and others teleconferenced in as well. The presentation on Mach effects took place in the morning and went well. In the afternoon, members of the local staff did presentations on matters they were investigating. Some of these topics were, to be frank, quite strange.

For example, one of the staff had decided to drop tennis balls from a balcony in the cafeteria with rare earth magnets enclosed and time their fall. The magnets were arranged into dipole and quadrupole configurations. And a check was done with a tennis ball enclosing weights to bring it up to the same mass as the magnet containing balls. Allegedly, the balls containing the magnets fell slower than the check ball – by some 25%. Outrageous! Tom, who had teleconferenced into the meeting, was so put off by this claim that he decided to check it by constructing a proper timing system and associated equipment and repeating the experiment with proper controls for obvious spurious effects.

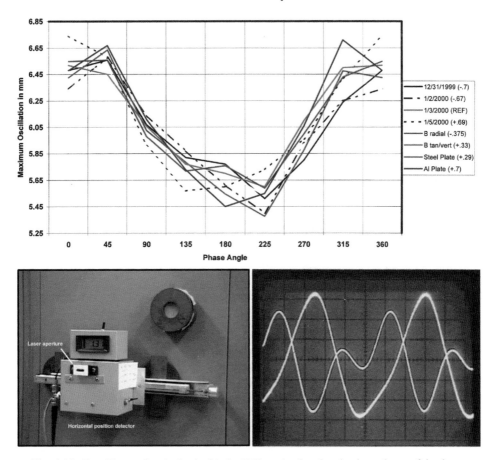

Fig. 4.14 *Top*: The results obtained with the PZT stacks showing the dependence of the thrust on the relative phase of the first and second harmonic components of the power signal applied to the devices. *Bottom*: On the *left* is a picture of the laser beam position detector based on a gradient density filter and photovoltaic sensor. On the *right* are scope traces of the signal generator waveform (with obvious first and second harmonic components) and the voltage waveform of the signal as delivered to the device for a particular relative phase of the harmonics

Needless to say, when done a few weeks later, no difference was found in the fall times for the various magnet and check configurations.

Follow-up trips to Lightspeed occurred in the falls of 2000 and 2001. The mix of participants changed a bit from trip to trip, but for the most part, the same people were in attendance. The visits often provided helpful suggestions for new experimental directions that were implemented following the meetings. Internal Lockheed-Martin politics after 2001 led to changes in the tasking of Millennium Projects, and the annual visits came to an end.

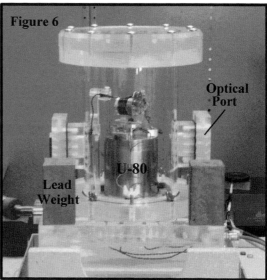

Fig. 4.15 Color versions of the black and white figures from "The Technical End of Mach's Principle" showing one of the PZT stacks mounted on the U-80 thrust sensor stage and that assembly mounted in the small vacuum chamber used in this attempt to produce the second term, always negative mass fluctuation of the Mach effect predictions

At the last meeting, one of the physicists suggested that instead of looking for thrusts, since the wormhole term effect should allegedly be detectable in the laboratory (see Chap. 3), that a search for the predicted weight reduction be sought. So, over the winter of 2001 and 2002, the apparatus was reconfigured to do this with the U-80 sensor. Rather than use the massive aluminum and steel case and chamber used in earlier work shown in Fig. 4.2, the U-80 was fitted in a 1-cm thick steel case and mounted in a small plastic vacuum chamber as shown in Fig. 4.15. This was done so that the entire vacuum chamber and contents could be inverted to see if the alleged thrust/weight signal changed as expected under inversion. Spurious signals could be eliminated by subtracting the upright from inverted signals to get a net signal. One of the PZT stacks used in the STAIF 2000 work was attached directly to the base mounting stage of the U-80.

About the time that this work program was set in motion, I was invited by Andre K. T. Assis, the program organizer, to give a presentation and write a paper for the 50th anniversary of the founding of the Kharagpur campus of the Indian Institute of Technology. The celebration was to include a scientific program, and the director of the campus, Amitabha Ghosh, being an aficionado of Mach's principle, had decided to focus the program on the topic of his interest. Funds for travel, alas, were not to be found. But the organizers agreed to accept a paper for their proceedings notwithstanding that I would not be able to attend. The proceedings are published as: *Mach's Principle and the Origin of Inertia* (Apeiron, Montreal, 2003) and the paper, "The Technical End of Mach's Principle," is found therein.

The apparatus used was intended for the follow-up to the 2001 Lightspeed meeting: one of the EC-65 dielectric PZT stacks mounted on the U-80 sensor in the small plastic vacuum chamber. A variety of vibration damping mounts had been tried to isolate the U-80 from the high frequency mechanical effects of the stack. They were all found to degrade the performance of the system in one way or another, so in the end, the stack was attached directly to the aluminum stage that protruded from the steel case enclosing the U-80.[9] This, of course, laid open the possibility that vibration might be the cause, at least in part, of any weight changes detected. But notwithstanding that vibration was an obvious problem, it was decided to go ahead and find ways to check for the presence of spurious signals caused by vibration.

Early work on this project consisted of finding the conditions of electromechanical resonance in the PZT stack used by sweeping a range of frequencies where resonances could be expected. A prominent resonance was found at about 66 kHz. Then, tuning to the resonance, power pulses of a second or two at constant frequency were delivered to the device. The device heated rapidly, but stayed on resonance long enough to produce weight changes that seemed to be contaminated by thermal drift. Since thermal drift was in the same direction as far as the U-80 readings were concerned, and inverted results were to be subtracted from upright results, the drift would be largely eliminated in the final results as it would cancelled as a "common mode" signal. So thermal drift was ignored.

The other obvious potential source of problems was that the PZT stack was mounted on the U-80 stage unshielded. In operation, the device was a source of non-negligible electromagnetic radiation, and that radiation could either contaminate the U-80 signal or it might couple the device to the local environment, causing a spurious apparent weight change. Since, however, the spring in the U-80 was very stiff, the position of the device on the U-80 and in the vacuum chamber changed only imperceptibly when the system was inverted, and such spurious effects would be eliminated along with the thermal drift as common mode signals by the inversion/subtraction process.

The data collected in the early runs with this arrangement after everything was working fairly well looked much like the results obtained with the KD capacitors and PZT stack for relative phase reversal shown earlier in Fig. 4.8. Only the large offset now understood as the effect of electrostriction in the PZT stack of the earlier configuration was not present, as no PZT stack acting along the axis of the U-80 sensor was present. These are shown in Fig. 4.16. When the upright and inverted results were differenced, a net weight shift was found present, as shown in Fig. 4.17. This weight shift is not the same as that found with the earlier configuration. The earlier "weight" shift was presumably produced by a thrust generated by the first transient term in the Mach effect equation ("rectified" by the action of the PZT stack). In this case, the weight shift was presumably generated by the production of an effect due to the second term in the Mach effect equation – that is, an effect produced by a real stationary weight change in the mass of the PZT stack.

The roles of productive mistakes and dumb luck in experimental work is not nearly well enough appreciated by those who do not do this sort of work. When they happen, you've got to be ready to take advantage of them. In this case, dumb luck struck in the form of getting essentially all of the "just so" conditions needed to get almost perfect operating conditions right. As recounted in "The Technical End of Mach's Principle," one set of runs

---

[9] The stage was hardened against lateral motion with an array of three tensioned fine steel wires to insure that only motion along the axis of the sensor would be detected.

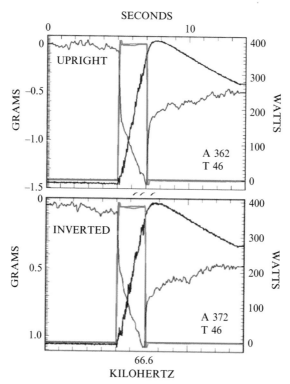

Fig. 4.16 Early results obtained with the system shown in Fig. 4.15. The weight is the *red* trace; the power is the *green* trace; and the *blue* trace tracks the temperature of the device. The full-scale amplitudes of the accelerometer and temperature traces are designated in the *lower right* of each panel by the "A" and "T"

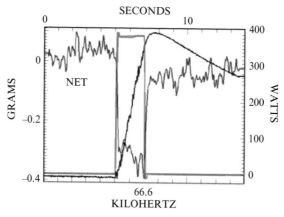

Fig. 4.17 The weight trace (*red*) obtained by subtracting the inverted trace of Fig. 4.16 from the upright trace. Note that the net is about a half gram, and that the subtraction process all but eliminates the obvious secular thermal drift in the weight traces in Fig. 4.16

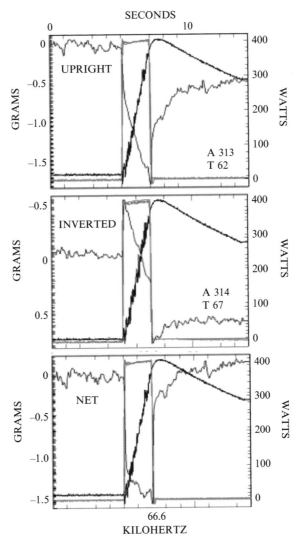

Fig. 4.18 For one run everything was "just so," and the inverted weight trace went positive rather than negative. Those "just so" conditions also produced the largest observed net weight change – nearly a gram and a half

with the system produced first the expected weight shift when a power pulse of constant frequency was delivered at the resonance frequency. When the system was inverted and run again, instead of getting a weight shift in the same direction as the initial results, only of different magnitude, the weight shift actually reversed, too – as it should in ideal circumstances. It took several minutes for me to understand the significance of what had happened. To say I was amazed would be to slightly understate the case. These results are displayed in Fig. 4.18.

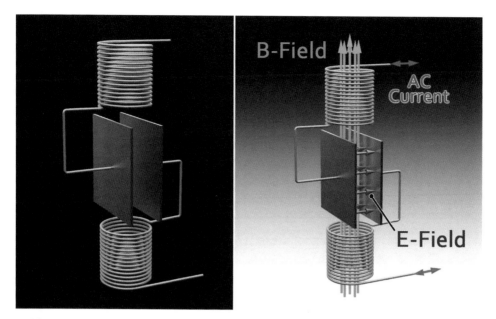

Fig. 4.19 The Slepian circuit that produces a periodic, alternating thrust that time averages to zero. Current flows through the inductor produce a magnetic field between the plates of the capacitor perpendicular to the electric field between the plates. Through the Lorentz force, the combined action of the electric and magnetic fields on electrically charged matter between the plates is a thrust perpendicular to both the electric and magnetic fields

A series of tests, recounted in "The Technical End of Mach's Principle," followed. And though those tests seemed indicate that the weight shifts observed in this experiment were genuine, I couldn't convince myself with absolute certainty that vibration in the system wasn't contaminating the results. So, when another way of going at producing Mach effects presented itself, it wasn't hard to abandon this line of work and move on to another.

Mach-Lorentz Thrusters

Hector Brito, an Argentine aerospace engineer, started investigating advanced propulsion schemes in the 1980s. About the same time, James Corum also became involved in the same sort of investigation. They both had hit upon the idea that it might be possible to produce thrust in a simple electrical circuit that had been the subject of an article published in the late 1940s by a fellow named Joseph Slepian examining the forces on the dielectric in a capacitor immersed in the magnetic field of a coil, part of the same circuit as the capacitor, shown schematically here in Fig. 4.19. An alternating current (voltage) is applied to the circuit. Since it is a simple series inductive-capacitive circuit, the current and voltage have a relative phase of roughly 90° (except at or near the resonant frequency of the circuit).

The "theory" of operation is that the electric field in the capacitor causes the ions in the dielectric to oscillate back and forth, and in consequence they become an oscillating current that is then acted upon by the oscillating magnetic field produced by the coil through the magnetic part of the Lorentz force of classical electrodynamics:[10]

$$\mathbf{F} = q\left[\mathbf{E} + \left(\frac{1}{c}\right)(\mathbf{v} \times \mathbf{B})\right], \tag{4.4}$$

where $\mathbf{F}$, q, $\mathbf{E}$, $\mathbf{v}$, and $\mathbf{B}$ are the force, electric charge on an ion in the oscillating current in the dielectric, electric field, ion velocity, and magnetic flux respectively. Slepian, in his analysis, pointed out that although forces could be expected on the dielectric due to the action of the Lorentz force, averaged over a cycle, the net force would be zero. Corum and Brito had convinced themselves that notwithstanding that the laws of classical electrodynamics prohibit the production of a net force on such a system, as it would be a violation of momentum conservation – and it is well-known that such violations simply do not occur in classical electrodynamics. Nonetheless, a net force might be produced. Corum argued that this might occur because of the existence of "non-linear boundary conditions," whereas Brito created a complicated scheme involving "solidification points" and suchlike to justify his position. These machinations were, at best, misguided, for classical electrodynamics does prohibit violations of momentum conservation.

The reason why the Slepian circuit is of interest from the point of view of Mach effects is that should Mach effects really occur, they change the situation regarding momentum conservation. In the purely classical electrodynamic case, there is no way to transfer momentum from the circuit to anything else,[11] but if the mass of the dielectric fluctuates (at twice the frequency of the AC voltage applied to the circuit), this situation changes. Where the electromagnetic effects in the Slepian circuit cannot convey momentum out of the local system, in the case of Mach effects such momentum transfer is possible via the gravitational interaction in suitably designed devices.

When Brito became interested in Slepian devices, he had had special toroidal capacitors made with high dielectric constant ferroelectric material, which he then wound with toroidal coils, as shown in Fig. 4.20. The appeal of devices of this sort is that the electric and magnetic fields that act on the dielectric and polarization currents therein propagate through the device at light speed. In PZT stacks, while the applied voltage signals appear throughout the stack essentially instantaneously, the mechanical effects involving the reaction mass are acoustic and involve propagation velocities much less than that of light. In the same vein, PZT stacks are complicated by acoustic reflections at impedance mismatches, whereas the Slepian devices are free of such issues.

When the decision was made to explore Slepian type devices – that eventually acquired the name "Mach-Lorentz thrusters" – funding for specialty toroidal capacitors was not to

---

[10] What appears to be an echo of Slepian's scheme can be found on pages 182 and 183 of Wolfgang Panofsky and Melba Phillips' classic text: *Classical Electricity and Magnetism* (2nd ed., Addison-Wesley, Reading, MA, 1962).

[11] Aside from a minute flux of electromagnetic radiation that can be expected from such a circuit.

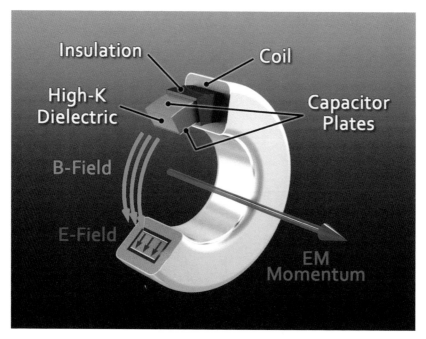

Fig. 4.20 One of the devices built by Hector Brito in an attempt to produce stationary thrusts with the Slepian circuit. For a while, Brito claimed to have seen such thrusts, but they could not be consistently reproduced in systems other than his original setup

be had. So, Ceramite "door knob" capacitors were adapted to a configuration shown in Fig. 4.21, where a pair of capacitors is mounted between the pole pieces of sectioned toroidal inductors. The inductor and capacitor circuits were wired separately so that the relative phase of the magnetic and electric fields in the capacitor dielectric could be varied.

Several issues have to be addressed in order to show that Mach-Lorentz thrusters will actually work, for example, that the phasing of the fields and the velocities they produce in the dielectric have the phase needed to act on the mass fluctuation in such a way as to produce a net force. These issues are considered in some detail in "Flux Capacitors and the Origin of Inertia." Results of experimental tests of the device, shown in Fig. 4.21, mounted in a steel Faraday cage on the U-80 sensor are also reported there. The main result is presented here as Fig. 4.22, which shows the differenced thrust signals for 0 and 180, and 90 and 270° of phase between the inductor and capacitor circuits. A net thrust signal was seen for 270 minus 90° of phase, but none worth mentioning for 180 minus 0° of phase. Note that the inductor and capacitor applied powers were staggered to show that a thrust effect was only present when both circuits were powered.

After early work with devices like that in Fig. 4.21 was pretty much wrapped up, new generations of devices were developed. The first was a slight modification of the original design. The pole faces of the powdered iron inductor rings were shaped to the exterior shape of the capacitors used and glued to the capacitors. The resulting toruses were then

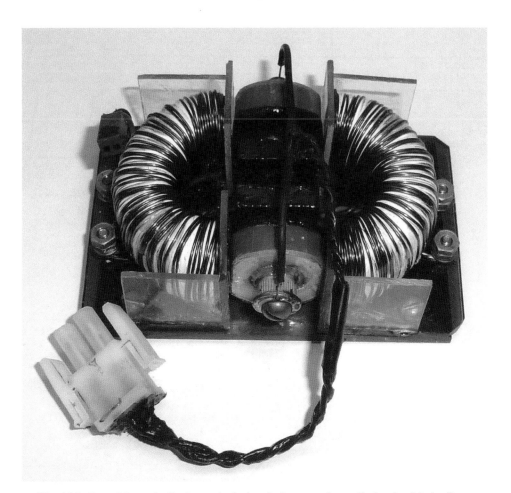

Fig. 4.21 One of the early Slepian-style devices built to test the prediction that Mach effects might make it possible to produce stationary thrusts as the mass fluctuations change their operating dynamics. The coils and capacitors were separately wired so that the relative phase of the voltage signals delivered could be adjusted

wound with magnet wire and the circuit hard wired. No provision for phase variation was provided. This design evolved into rings of smaller capacitors glued into rings with small segments of sintered iron toruses between the capacitors, mimicking Britos' custom-fabricated capacitors at a small fraction of their cost.

As these developments in device design were implemented, the U-80 sensor was moved to a larger vacuum case, and a system of plastic rings and rods were introduced so that the whole system could be oriented horizontally as well as vertically. In horizontal operation, only thrusts should be detected, eliminating all effects that might mimic a weight change. One of the later designs is displayed in Fig. 4.23; the modified U-80 system is shown in Fig. 4.24.

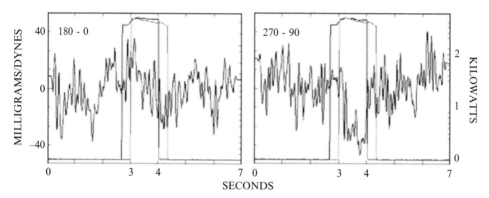

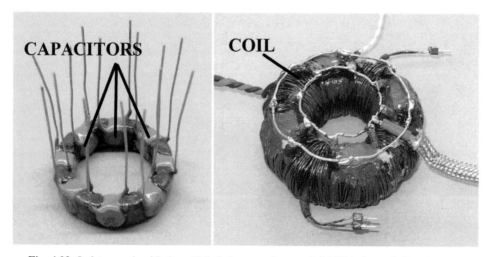

Fig. 4.22 Thrust results (noisy traces) obtained with a device like that shown in Fig. 4.21 after some signal averaging and subtraction of the phases of the capacitor and inductor power waveforms indicated in the panels of the figure. The inductor and capacitor power traces are the noise-free traces

Fig. 4.23 In later work with these "Mach-Lorentz thrusters" (MLTs) rings of 500 pf 15 kV capacitors were glued to segments of powdered iron toroidal inductor rings, as shown, and then wound with magnetic wire to recreate Brito-type devices (Fig. 4.20) cheaper than the cost of custom-made ring capacitors

Shortly after the publication of "Flux Capacitors and the Origin of Inertia," Martin Tajmar, an aerospace engineer at the Austrian Research Center (now Austrian Institute of Technology), head of their space propulsion program with a scheme of his own for rapid spacetime transport, put up a paper on the arXiv server claiming that mass or inertia

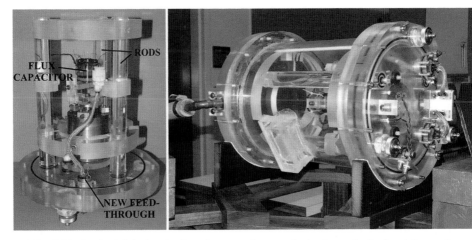

Fig. 4.24 On the *left* is the plastic rod system that was devised to make horizontal operation of the U-80 thrust sensor-based system possible. The vacuum chamber with the enclosed sensor and test device is shown on the *right*

modification could not be used for practical improvements in spacecraft behavior.[12] After allowing that the Equivalence Principle and GRT were correct, Tajmar went on to ignore the Equivalence Principle and argue that mass fluctuations were without practical value for propulsion. The reason why the Equivalence Principle is important in this case is that it asserts that the active gravitational, passive gravitational, and inertial masses of an object are the same. So, if you vary one of the masses, the other masses change, too. If this aspect of the Equivalence Principle is correct (and it is), then it is almost trivial to show that mass variation has serious propulsive advantages. Tajmar, of course, knew this. To make the argument he wanted to make – that Mach effects, if real, are useless – he had to violate the Equivalence Principle, which he went on to do. His paper was a polemic, not serious science.

Notwithstanding that the flaw in his argument was pointed out in public, Tajmar's paper was eventually published unmodified in the *Journal of Propulsion and Power*. This was a failure of the peer review system. Aerospace engineers are rarely experts in GRT. The referees of Tajmar's paper were unable to detect a defect that no competent general relativist would have missed. It pays to be cautious of even the peer-reviewed professional literature.

Tajmar's attack on Mach effects figures into this narrative because a year after "Flux Capacitors" was published, he volunteered to let one of the engineers working in his

---

[12] Tajmar's paper was called to my attention by Phillip Ball, a writer for the journal *Nature*. Ball had correctly surmised that Tajmar's arguments were an attack on Mach effects and their use for rapid spacetime transport. Ball wanted my response. I pointed out that Martin had set aside the Equivalence Principle, but that the Equivalence Principle was true, so while all Martin's conclusions were correct if you ignored the Equivalence Principle, they were all completely bogus because the Equivalence Principle as a matter of fact is true.

division – Nembo Buldrini, the illustrator of this book – have access to a very sensitive thrust balance and associated apparatus to test one of the devices made at CSUF. The work on Mach-Lorentz thrusters seemed to be proceeding nicely, with large effects that seemed genuine being produced without much difficulty. The signals, as before, were contaminated by other than Mach effect sources. But using the subtraction of differing phases protocol, these spurious signals were thought to be eliminated from the net results. So sending Nembo several of the devices and some supporting circuitry to test on the ARC thrust balance seemed a reasonable thing to do, despite Martin's negativity toward Mach effects. Nembo had been a very smart, interested participant in email discussions of the Mach effect work by that time for years. His capability and competence were beyond question.

Martin gave Nembo access to the thrust balance facility at ARC for several months in the fall and winter of 2005–2006. Some special parts – a Faraday cage to mount on the balance – were made at ARC for the tests, Carvin power amplifiers of the type used at CSUF were procured, and the power circuitry used with the loaned device at CSUF was sent to ARC to assure proper operation of the device.

In retrospect, the one questionable feature of the ARC test was the delivery of power to the device. This was done with simple wires running through the bearings in the central column of the balance. But this issue notwithstanding, Nembo did a careful investigation. His results did not conform to the sort of results obtained at CSUF with the U-80 thrust sensor.[13] Some of those results are presented here in Fig. 4.25.

Where tens of micro-Newtons were routinely measured with the U-80, the thrust of a micro-Netwon or two were the largest that could have been produced with the ARC thrust balance. Either the small signal response of the U-80 was not that extrapolated with the assumption of linearity from the calibrations with a one gram mass, or something very fishy was going on. In addition to this issue, there were the problems that the effect would die out with long operation of the devices – presumably due to dielectric aging – and the variability of the effects produced from device to device.

Tom, who had kept contact with the project since finishing his Master's degree, decided that the best way to address these issues was to build an ARC-style thrust balance for use at CSUF. Since he had equipped himself with a small machine shop in his garage, this was a feasible option. We agreed to split the costs of the project, the chief cost being the purchase of a Philtec optical positioning sensor. The thrust balance that evolved from this project is still in use, and described in the next chapter. Before turning to that work, we first address various attempts to replicate Mach effects other than the work at ARC in 2005–2006.

---

[13] This work is found in N. Buldrini and M. Tajmar, "Experimental Results of the [Mach] Effect on a Micro-Newton Thrust Balance," in: *Frontiers of Propulsion Science* (AIAA, Reston, VA, 2009), pp. 373–389, eds. M. Millis and E. Davis.

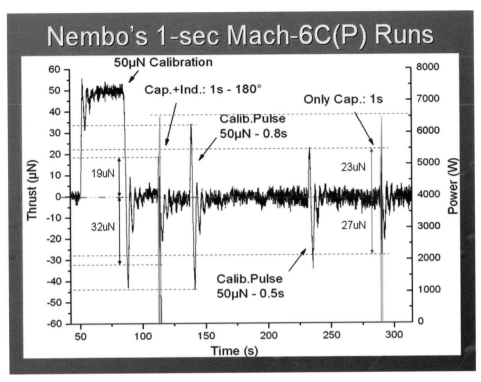

Fig. 4.25 Typical results obtained by Nembo Bldrini with the ARC thrust balance at what is now the Austrian Institute of Technology in 2005. Signals on the order of a few tens of micro-Newtons produced with the MLT being tested are *not* present. The short-term noise in the system, however, is too large for a signal on the order of a few micro-Newtons to have been seen in this work

REPLICATORS

Any serious work that suggests it might be possible to make a revolutionary advance in propulsion technology is likely, at least eventually, to attract enough attention that others will want to know if there is really anything to the claimed results. In the case of Mach effects, the first seemingly serious replication effort was proposed not long after the workshop at NASA Glenn in 1997. The interested party was David Hamilton, an electrical engineer at the Department of Energy headquarters in Washington. Hamilton headed a group at Oak Ridge National Laboratory that was developing "soft switching" techniques for high power electrical systems intended for automotive applications. Hamilton also was given access to the ceramics operations at Sandia National Laboratories. A large amount of money was committed to fund his replication project. In January of 1999 Hamilton and several members of his team came out to CSUF to see what we were doing, and to discuss

November 04, 2004   Paul March - Friendswood, TX   44

Fig. 4.26 Some members of Hamilton's team at the January 1999 meeting at CSUF. They were recruited to carry out a replication of the work done with Mahood in the CSUF lab in the late 1990s. Photo by Paul March, who also attended the meeting

their plans for the work they intended to do.[14] (See Fig. 4.26.) After presentations of the work in progress and its underlying physics, the members of Hamilton's team told us what they had in mind.

At the time we were using stacks of PZTs glued together following Jim Sloan's techniques. The stacks that evolved over the spring and with the addition of rubber pads the following fall produced spectacular results. Rather than buy commercial crystals, they planned to use one of the ferroelectric specialty fabrication lines at Sandia and create their devices by growing single crystal elements, vacuum depositing electrodes as appropriate in the process. They planned to make their devices several inches in diameter and about an inch thick (hence, hockey pucks). The planned crystal thickness was to be small, less than the 1.5 mm thick crystals we were using. Their obvious intent was to scale up the parameters we had been using by an order of magnitude and more. This struck both Tom and me as an unwise course of action, especially since the devices we

---

[14] The team members were: David Hamilton (DOE Headquarters), Bruce Tuttle (ferroelectric ceramics, Sandia National Laboratories), Don King (Sandia Directors Office liason), John McKeever (ORNL). Also in attendance were JFW, Tom Mahood, and Paul March (Lockheed liason to Lightspeed).

had running were crudely characterized at best. We suggested that they do a more modest scaling so that they might expect their devices to work at least more or less as ours did. They declined to reconsider their planned course of action. A tour of the lab followed, and they went away.

Next September two of Hamilton's team were back, Don King and a technical fellow who had not been at the January meeting. We surmised that the return visit was occasioned by a failure to get the "hockey pucks" to work. Given the power available through the ORNL end of the project, we assumed that the hockey pucks had been smoked. The examination of our lab equipment this time was much more careful than the inspection of the first meeting. Tom had prepared moderately detailed drawings of our recommended device configuration (see Fig. 4.27).

The drawing found a more receptive audience than our suggestions had at the first meeting. Where silence had followed the first meeting, some email traffic ensued after the return visit. Mid-fall we learned that they planned to use a two meter pendulum mounted in a vacuum case with an optical position sensor as the thrust detection system. The actual devices to be tested were to be constructed according to the drawings we had supplied. But instead of commercial grade PZT, they planned to have their devices fabricated using a special ferroelectric mix with a much higher dielectric constant.

Shortly thereafter, a member of the ORNL group had the idea that this was all a waste of time because we had failed to take into account the vdm/dt term in Newton's second law. For a while, everyone was attracted to this proposal. Then Paul March allowed as how he didn't see that the vdm/dt argument made sense. The fact of the matter is that vdm/dt "forces" are not like m**a** forces. If you can show that the system on which the vdm/dt force allegedly acts can be taken to be in some reasonable instantaneous frame of rest, the "force" is zero as v vanishes. When a "force" or acceleration vanishes in any inertial frame of reference, it vanishes in all such reference frames. This zeroing of the vdm/dt "force" takes place in Mach effect systems, for it must act at the active mass through the actuator on the reaction mass. That point of action does not involve an invariant velocity and can always be taken as in an instantaneous frame of rest where the vdm/dt "force" vanishes. Paul was right. Getting people to understand this, however, turned out to be a lot more challenging than one might expect. (This issue is also mentioned in Chap. 3.)

While all of this business about Newton's second law can be diverting, the real question is: What happened at the experimental end of things? Well, in early January, Tom and I learned through a reliable back channel that the ORNL team had gotten the system described to us a couple of months earlier running – the system with PMN (lead-manganese-niobate) stacks of our design running on a 2-m pendulum in a vacuum case with optical position sensing. Their devices produced "interesting" results. We were told that the apparatus and all associated materials had been collected and removed from ORNL. A couple of months later, we tried to confirm this story with people who would have known whether it was true. Silence. But eventually the ORNL team put up the results of an experiment they had done on the ORNL website. It makes for interesting reading as it bears almost no relationship to the experiment we were told was constructed and executed the previous fall.

A much less well funded replication effort was carried out by John Cramer and a sequence of his Master's degree students at the University of Washington at about this

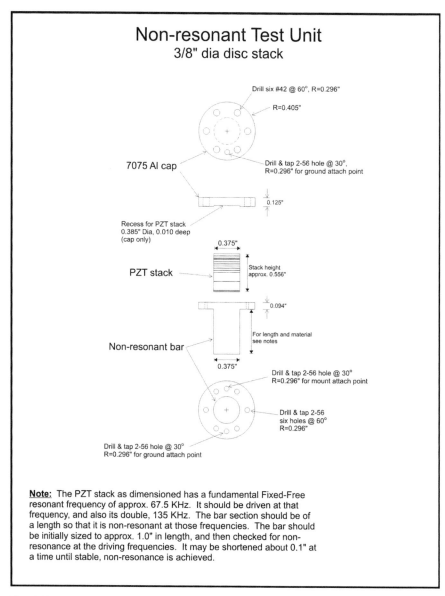

Non-resonant Test Unit
3/8" dia disc stack

Drill six #42 @ 60°, R=0.296"

R=0.405"

7075 Al cap

Drill & tap 2-56 hole @ 30°,
R=0.296" for ground attach point

0.125"

Recess for PZT stack
0.385" Dia, 0.010 deep
(cap only)

0.375"

PZT stack

Stack height
approx. 0.556"

0.094"

Non-resonant bar

For length and material
see notes

0.375"

Drill & tap 2-56 hole @ 30°
R=0.296" for mount attach point

Drill & tap 2-56
six holes @ 60°
R=0.296"

Drill & tap 2-56 hole @ 30°
R=0.296" for ground attach point

Note: The PZT stack as dimensioned has a fundamental Fixed-Free
resonant frequency of approx. 67.5 KHz. It should be driven at that
frequency, and also its double, 135 KHz. The bar section should be of
a length so that it is non-resonant at those frequencies. The bar should
be initially sized to approx. 1.0" in length, and then checked for non-
resonance at the driving frequencies. It may be shortened about 0.1" at
a time until stable, non-resonance is achieved.

Fig. 4.27 An earlier rendition of this figure, essentially the same as that here, was given to members of Hamilton's team as part of our advice in September 1999. We were told later that devices of this sort were in fact built for testing, which started in December 1999

time. John had applied for and received funding through NASA's Breakthrough Propulsion Physics program, though it took quite a while for the funding to come through. His first plan was to put a Mach effect device on a rotating platform that was to float on an air table. By adjusting the phase of the effect with the rotation of the platform, he hoped to make the platform move in whatever desired direction. The problem with this approach is that the rotation frequencies realistically achievable for devices that stably float on an air table are so low that only minuscule Mach effects are expected.

John abandoned this approach for "Mach's guitar." High voltage capacitors were to be suspended by tensioned steel guitar strings that would also act as the power feeds. A mass fluctuation was to be induced at the resonant frequency of the strings. The resulting expected oscillation of the capacitors induced by their fluctuating weight was to be detected with an optical position sensor. Eventually, the guitar strings were abandoned for a tuning fork, as higher frequencies could be achieved that way. The final results of this effort were deemed inconclusive,

Two other replication efforts merit mention. The first was carried out by Paul March, the Lockheed Lightspeed liason around the turn of the millennium. Since Mach effects scale with the operating frequency, he decided to try to get results in the 1–10 MHz range. As an inducement, I built a small device (shown in Fig. 4.28) designed to be run at either 1 or 2 MHz, depending on the wiring of the components.

Paul also built devices of his own using more than two capacitors and wound with fancier coils. This was a spare time project for him, and his lab was located in a bedroom vacated by a daughter who had left home. He used a commercial force sensor and fashioned a vacuum chamber from plastic parts. All this is shown in a photo montage in Fig. 4.29. Early work with both his own device and that sent him seemed to show the sort of behavior predicted by Mach effects. His results are displayed in Fig. 4.30. Note that the effects he saw were *larger* than the Mach effect predictions. He has interpreted this anomaly as indicating a "vacuum fluctuation" scheme advocated by a friend of his, which may be true. Eventually, though, the effects seen at the outset disappeared. The reason for the disappearance of the early observed effects was never identified. Equipment failures and other obligations led to discontinuation of the work. Only in 2011 has Paul returned to work on this problem. At the time of this writing, no results have emerged from the new work.

After Paul suspended work on Mach effects, Duncan Cumming became interested in doing a replication. Paul sent Duncan the 1 or 2 MHz devices in his possession. Duncan, a Ph.D. electrical engineer from Cambridge University and owner of a southern California aerospace consulting company, was well equipped both by training and support facilities. Rather than do the sort of experiments that had been done by Paul and those that were in progress at CSUF, Duncan decided to do a "self-contained" experiment. He secured and refurbished a Mettler H-20 precision balance. This balance has a load limit of 180 g, so he set about designing a system that could be put, literally, into a Coke can. Everything. Batteries, signal generator, power amplifier and the Mach Lorentz thruster. He provided an optical switch so that the device could be turned on and off with a simple beam of light.

No provision was made to operate the device in a vacuum, but since the Coke can was carefully sealed, no thrust effect was expected if Mach effects do not exist. Heating of the device during operation could be expected to cause the Coke can to expand, and that would

Fig. 4.28 One of two MLTs based on two Ceramite 500 pf, 15 kV capacitors built by JFW for Paul March to test in the laboratory he assembled in a bedroom at home (see next figure)

produce a change in the buoyancy of the can. But this would be a slow, cumulative effect, easily distinguished from a prompt thrust effect.

Duncan reported a null result on concluding his work with the system he built. Actually, the scale on the Mettler balance moved a bit when the device was activated with the light beam. But the movement of the scale was less than one of the finest scale divisions, and much less than the expected thrust from a simple analysis of the MLT in operation. Others found details of Duncan's work questionable. Nonetheless, his null result report was a perfectly reasonable, defensible conclusion.

Fig. 4.29 Paul March's home lab. On the *left* is his vacuum chamber with a test device he built mounted and ready for testing. Some of the power electronics and instrumentation are visible on the *right* of this panel

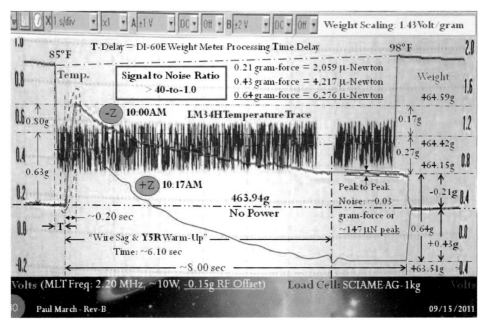

Fig. 4.30 A recent rendition of results that Paul March obtained with a MLT of his own construction in 2005. Observe the two traces, labeled +Z (*red*) and –Z (*green*), that were produced by the weight/thrust sensor for two relative phases 180° apart

WHAT'S GOING ON?

Sooner or later we have to deal with the fact that the results of the various experimental efforts over the years were, to put it circumspectly, variable at best. In part, this can be attributed to things like variation in construction details, the quality of components, and the aging of materials operated toward their electrical and mechanical limits. But something more fundamental seems to have been going on.

The person who put his finger on that more fundamental issue was Nembo Buldrini. What he pointed out was that given the way the transient terms of the Mach effect equation are written – in terms of the time-derivatives of the proper energy density – it is easy to lose sight of the *requirement* in the derivation that the object in which the mass fluctuations occur *must* be accelerating at the same time. In some of the experimental cases, no provision for such a "bulk" acceleration was made.[15] As an example, the capacitors affixed to the tines of the tuning fork in the Cramer and students' experiment made no provision for such an acceleration. Had the tuning fork been separately excited and the electric field applied to the capacitor(s) been properly phased, an effect might have been seen. But to simply apply a voltage to the capacitors and then look for a response in the tuning fork should not have been expected to produce a compelling result.

Other examples could be cited and discussed. Suffice it to say, though, that after Nembo focused attention on the issue of bulk accelerations in the production of Mach effects, the design and execution of experiments changed. The transition to that work, and recent results of experiments presently in progress, are addressed in the next chapter.

---

[15] By "bulk" acceleration we are referring to the fact that the conditions of the derivation include that the object be both accelerated *and* experience internal energy changes. The acceleration of ions in the material of a capacitor, for example, does not meet this condition. The capacitor as a whole must be accelerated in bulk while it is being polarized.

5

In Reality's Grip

A meeting took place in March of 2007 at CSUF, instigated by Gary Hudson.[1] Gary had attended the STAIF 2007 meeting a month or so earlier and had urged those of us working on Mach effects to get serious about getting organized. Having then recently done well with some Air Force and DARPA contracts, he, in his words, was looking to reinvest the then recently available government money in some promising projects that might not attract government funding.

Gary's effort in this direction was not the first such. A couple of years earlier, after retiring from Lockheed-Martin Aerospace, Jim Peoples had tried to do the same sort of thing. Both Jim and Gary, and later David Mathes, had sized up the situation and come to the conclusion that some material and organizational support was what the project needed.

They all were, of course, right – save for one consideration: I was determined *not* to let anyone invest serious resources in the project until I was convinced that such an investment involved only reasonable risk. In particular, I wanted a device in hand that would produce a thrust effect so large that it could be seen in a single cycle so clearly that there could be no question whatsoever of whether an effect was present. And a simple set of protocols should accompany such a "demonstrator" device to show beyond reasonable question that the origin of the easily observed thrust was the predicted Mach effect. In 2005 it seemed that such devices were in hand. Then Nembo Buldrini did the replication effort with the ARC [Austrian Research Center] thrust balance. His results called into question the validity of the results being produced at CSUF with the U-80 thrust sensor. So, when Jim and then Gary tried to get everyone organized, I deliberately slowed the

---

[1] For those unfamiliar with the aerospace community, Gary had made a reputation for himself by designing and building a prototype of the "Roton." The Roton was intended as a single stage to orbital system where the vehicle took off vertically with a pair of rocket powered counter-rotating rotors on its nose, and conventional rockets took over when the craft reached an altitude where the rotors were folded up as they were no longer efficient. The rotors could be redeployed for descent. Seriously proposing this project in the era of the space shuttle took real courage, and to stick with it, uncommon persistence.

J.F. Woodward, *Making Starships and Stargates: The Science of Interstellar Transport and Absurdly Benign Wormholes*, Springer Praxis Books, DOI 10.1007/978-1-4614-5623-0_5, © James F. Woodward 2013

process – as it turned out, right into the financial crisis of 2008. That's a story we'll come back to. If you are not too much interested in technical details, you may want to scan or skip the next two sections.

THE THRUST BALANCE

Tom Mahood's response to Nembo's STAIF 2006 presentation, as mentioned in Chap. 4, was to go build an ARC-style thrust balance. Tom got the specifications for the balance from the folks in Austria and set to work building one of his own design based on theirs. The heart of these thrust balances are a pair of C-Flex flexural bearings. One of these bearings is shown from two perspectives in Fig. 5.1.

Crossed steel strips attached to concentric steel tube segments keep the parts of the bearing aligned and provide a restoring torque when the two sections are rotated with respect to each other. The steel strips also produce their axial load-bearing capability. They are quite elegant, and inexpensive. They come in a variety of sizes and stiffnesses, so you can select those best suited to your application. In the ARC balance, "G" sized bearings were used, as they would carry a load of several kilograms while providing adequate sensitivity to make better than micro-Newton thrust determinations possible.

Alas, when Tom read the ARC documents, he misread "C" for the "G" in the drawings. "C" bearings are much smaller and more sensitive than "G" bearings – and they will not tolerate loads much in excess of a several hundred grams. Since the parts of the balance themselves were a significant fraction of a kilogram, this left almost no working load capacity. It took a while to figure out that the bearings that Tom had used in his first rendition of his thrust balance were not the right ones.

Fig. 5.1 Two views of a C-flex flexural bearing of the type used in the USC/ARC style thrust balance. The crossed steel leaves attached to the two halves of the tubular shell provide a restoring torque when the upper and lower halves are rotated with respect to each other

The bearing problem aside, another difficulty that became apparent immediately was that handling the power circuit to a device on the beam, especially when voltages of several kilovolts were involved, without introducing unacceptable spurious torques, was not going to be a simple matter of adjusting some wires. In the ARC balance, with its larger "G" bearings, Nembo had fed the power leads to the beam through the bearings. Since he was looking for thrusts in the range of 50 micro-Newtons, he determined that this arrangement would not affect his results.

With the "C" bearings in Tom's initial rendition, this was simply out of the question, as the bearings were far too small to allow power leads through. So a variety schemes using wires were tried. None worked very well. As this was proceeding, to avoid vertically loading the obviously delicate bearings, the balance was rearranged to operate in horizontal "teeter-totter" mode. In this configuration, dealing with the power feed problem had a simple solution: use liquid metal contacts to eliminate any hard mechanical connection between the power feeds on the beam and those leading to the beam. Instead of mercury, "galinstan" – galinium, indium, and tin – was used, as it is much less toxic. Wires on the beam were dipped into cups holding the galinstan. Since the motion of the beam was teeter-totter, when the beam moved, the wires were raised or lowered very slightly with respect to the cups. The success of the liquid metal contacts indicated that whatever the final arrangement of the thrust balance beam, such contacts would have to be used if small thrusts were to be measured.

Eventually, it became clear that the "C" bearings were simply too delicate to be used in any configuration for the work at hand. The bullet was bitten, and the balance was redesigned to work with "E" or "F" class bearings carrying a vertical load so that the beam would swing in a horizontal plane. The galinstan contacts of the teeter-totter arrangement were no longer suitable, so the contacts were redesigned, too. The contacts were relocated above the bearings that supported the beam. To eliminate torques, the contacts were relocated coaxially above the bearings, as shown in Fig. 5.2. Tom had fashioned the arms of the beam out of a ¾ in. square aluminum channel and designed their attachments so that the arms could be interchanged. He made several arms of various lengths so that a wide range of arm lengths was available. He also made several different mounting attachments for the ends of the beam, and several attachments that could be clamped onto the beam at varying distances from the bearing column in the center.

With the parts machined by Tom in hand, several modifications and additions to the basic thrust balance were made in the lab. In addition to the coaxial galinstan contacts, one of the beam end mounting plates was provided with a threaded fixture so that the power feeds could be run inside the aluminum channel to whatever device was being tested, and by simply loosening a nut, the device mount on the end of the beam could be rotated in the plane perpendicular to the beam, enabling the reversal of the direction of the device and allowing the device to be oriented in any direction of choice in the vertical plane. This was done to make a direction reversal test possible to insure that only thrusts that reversed direction when the device was rotated by 180° on the end of the beam were considered as candidate genuine signals.

A variety of mounts were made that held devices, often in Faraday cages, to be tested. See Fig. 5.3. The balance was calibrated using three 10 turn coils, wound on 35-mm film canister caps, wired in series. Two of the coils were fixed on the platform of the balance,

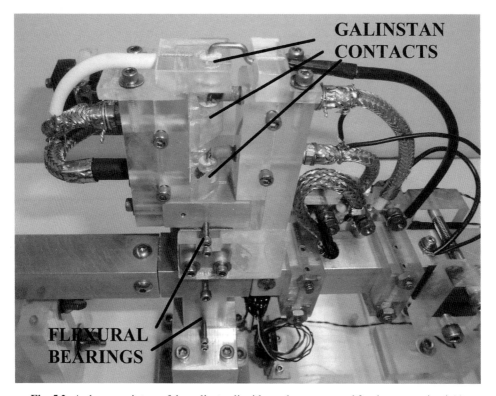

GALINSTAN CONTACTS

FLEXURAL BEARINGS

Fig. 5.2   A close-up picture of the galinstan liquid metal contacts used for the power circuit(s)

while the third was attached to the beam and located midway between the fixed coils mounted on the platform.

Provision was made for current direction reversal in the mobile coil so that the direction of the calibrating force could be reversed. The calibration coils are shown in Fig. 5.4. Close to the central column of the balance fine wires (# 40 AWG stranded copper) from a plug on the platform to one on the beam were provided so that thermistors and accelerometers on the test devices and beam parts could be monitored, as well as providing leads to the calibration coil attached to the balance beam.

The opposite side of the balance beam was also put to use. A platform was mounted on the end of the beam so that counterpoise brass masses could be stably set on the beam. Just inboard from those masses a reflective surface was attached to the beam for use with the Philtech optical position sensor. The fiber optic probe for the sensor was clamped in a mount secured to the stage of a stepper motor so that the distance between the end of the probe and reflective surface attached to the beam could be adjusted. Next to the position sensor a damper was mounted. See Fig. 5.5. This consisted of several small neodymium-boron magnets mounted in a plastic block affixed to the balance platform and a pair of aluminum sheets positioned on either side of the magnets attached to the beam.

When the beam and aluminum sheets moved, the magnets drove eddy currents in the sheets that dissipated the motion of the beam. Just inboard from the damper an adjustable

Fig. 5.3 The end of the thrust balance beam where the test devices are mounted, located in a Faraday cage – a small aluminum box lined with mu metal – suspended on a plastic mount with machine screws and attached to the end of the beam with a large threaded fitting that allows the power feeds to pass through the beam. The nut on this fitting can be loosened so that the mount, Faraday cage, and test device can be rotated on the end of the beam in any direction (perpendicular to the beam)

plastic block was placed so that the beam could be locked during work on the various parts of the balance, for example, changing the test device. The entire balance is shown in Fig. 5.6.

ELECTRONICS

The Philtech optical sensor comes with some basic electronics. It is equipped with a light source and photodetector. Light from the source is conveyed to the probe with a fiber optic cable, and the reflection from the surface whose position is to be measured is conveyed with a bundled, separate cable to the photodetector. The detector returns a voltage that

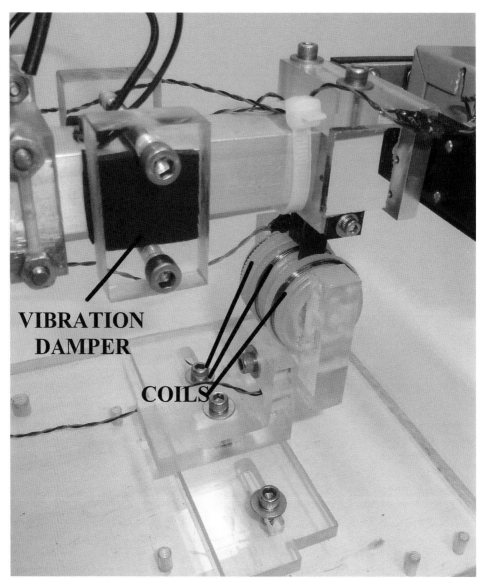

Fig. 5.4 Just inboard from the test device mounting on the end of the beam where the calibration coils are located. Each coil is ten turns of magnet wire wound on a 35 mm film canister cap

depends on the intensity of light. For high-sensitivity measurements of small displacements, the probe must be positioned near the reflective surface, and accordingly the reflected beam is quite intense. This produces a voltage near to the peak voltage of the sensor: 5 V. The signal sought, however, is orders of magnitude down from this 5 V signal for very small displacements of the reflecting surface.

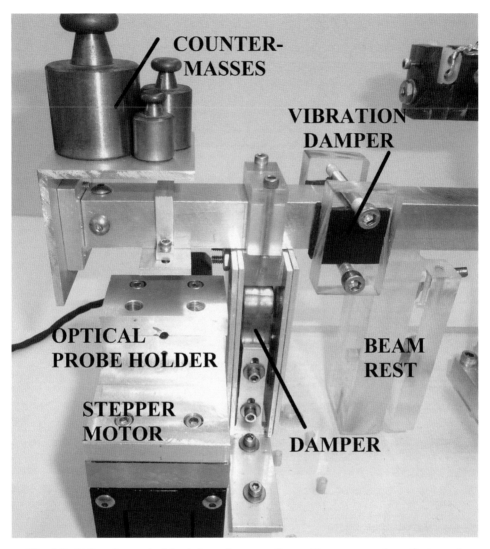

COUNTER-
MASSES

VIBRATION
DAMPER

OPTICAL
PROBE HOLDER

STEPPER
MOTOR

BEAM
REST

DAMPER

Fig. 5.5 At the other end of the balance beam are the position sensor probe, reflector and magnetic damper

In order to detect small changes in this 5-V signal a very stable offset voltage of the same magnitude must be introduced, and the resulting differential voltage amplified. Integral offset and amplification sub-circuits are options that must be separately specified and paid for from Philtech with the original purchase. Tom didn't order them. So a separate module was built to perform these functions. In addition to the basic offset and amplification sections, filters were included in this module both to address aliasing and to suppress pickup from ambient electromagnetic noise in the environment. Since this

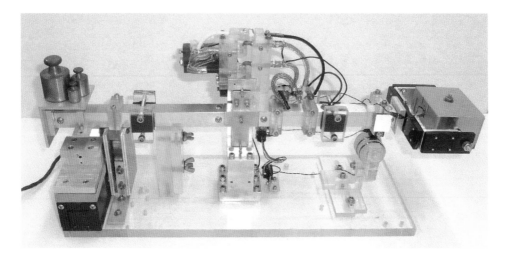

Fig. 5.6 The complete USC/ARC-style thrust balance in its present configuration. Note especially the vibration damper on the side (plastic and Sorbothane pads clamped on the beam) and a static beam rest used to support the beam when it is being serviced

section of the electronics is particularly important, it was enclosed in a cast aluminum box, and wiring of the sections inside the box were connected using coaxial cable with grounded shielding. The output of the position sensor electronics went to an oscilloscope and meter for direct monitoring, and to the ADC for data storage during runs.

Several other parameters of the system were monitored with data acquisition and storage during runs. In all cases, the voltage across the test device was read. Since the amplitude of the AC voltages across the devices ranged from 100 V and more to as much as 5 KV, a divider circuit had to be employed. A resistive divider was used. Since a resistive divider can distort the signal that it is used to monitor it owing to frequency dependence, it must be capacitatively compensated to make the RC time constants of the legs of the divider equal. Such compensation was done with this and all dividers employed in the experiment. While AC signals can be directly displayed on an oscilloscope, or fed into a spectrum analyzer, they cannot be displayed on a meter or recorded with an ADC as they time-average to zero unless they include a DC offset. To get quasi-DC signals from AC signals, they must be rectified and filtered. The Analog Device's AD630 synchronous demodulator chip can be configured as a high frequency rectifier by self-referencing the signal to be rectified. In order to measure only the signal of interest with respect to its local ground, and to protect expensive circuits from voltage transients, all circuits measuring sources with voltages in excess of a few volts were provided with buffers. A block diagram of a typical sensor circuit is displayed in Fig. 5.7.

In all cases the voltage across the test device produced by the power amplifier was monitored. The test devices, being electromechanical, were routinely provided with one or more accelerometers to record their mechanical activity, and the output of at least one of these accelerometers was also monitored. The temperature of the test devices was

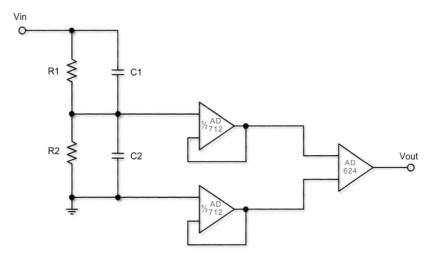

Fig. 5.7 A schematic diagram of a typical sensor buffer circuit. An AD 712 dual op amp chip is used in gain mode as a first stage (since they are cheap should they need to be replaced) followed by a high precision instrumentation amplifier (the AD 624 chip) with adjustable gain to give the instrumentation ground-referenced differential signal produced by the source

monitored with a thermistor (and associated circuitry) glued to the mounting bracket carrying the device.[2]

In the work with these devices around 2000, the thermistor had been supplemented with a bimetallic strip freezer thermometer so the temperature of the device could be read in real time by simple visual inspection. When run in the torsion balance of that time, this was also required, as running leads to the thermistor on the torsion balance proved impractical. The strip thermometers were abandoned in the fall of 2011. And in the spring of 2012, the thermistor glued to the mount was moved to a small hole drilled in the aluminum cap that retains the PZT stack. This location has the advantage of the closest possible proximity to the active region of the stack without actually drilling into the stack. While a temporal lag of a few milliseconds can be expected, the response to changes in temperature of the stack is quite prompt. When thermal effects were subjected to detailed examination, a second thermistor was added to the system – located in a hole in the reaction mass. Silver epoxy, with excellent thermal conductivity, was used to affix both thermistors so as to minimize the response time.

The electronics associated with the thermistors is elementary. A fixed voltage is applied to a fixed resistor (normally 40 K ohms) with the thermistor (nominally 10 K ohms at room temperature) wired in series to ground. One then simply monitors the voltage drop across the thermistor. Owing to the negative coefficient for the thermistors used, the voltage was fed into the inverting side of an instrumentation amplifier so that an increasing output

---

[2] IR sensing of the device temperature with a suitable detector in close proximity to the test devices is a planned upgrade of the system.

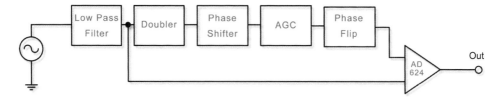

Fig. 5.8 A block diagram of the signal conditioning circuit. The low pass filter is a standard VCVS four-pole Butterworth filter (described in, for example, Horowitz and Hill, *The Art of Electronics*, 2nd ed., Cambridge University Press, 1989) based on the AD 712 chip. The frequency doubler is based on the AD 633 four quadrant multiplier chip that squares the first harmonic signal to get the second harmonic

voltage from the amplifier corresponded to increasing temperature. The amplifier also acted as a signal buffer. The signal(s) went to both the ADC and digital panel meters.

The only other non-off-the-shelf electronics in the system is the signal generator. This was built early in the work program, that is, in the late 1980s. Commercial signal generators were then moderately expensive, so several Elenco "Function Blox" signal generator boards were purchased at modest cost and adapted into a versatile signal source.[3] Only the sine waveform of the several available is used. The frequency modulation by means of a small DC voltage made generating frequency sweeps by applying a varying voltage through one of the D to A channels of the Canetics ADC board trivial.

Since the production of Mach effects generally depends on the use of voltage signals of two frequencies, one twice the other, that are phase locked, an additional signal processing module was constructed. The first stage is a low pass filter to insure that the original waveform is a pure sine wave of single frequency. The filtered waveform is then directed to a "bounceless doubler" to generate the second harmonic signal required. The doubler is actually a four-quadrant multiplier (Analog Devices AD633 chip) that squares the input signal, as squaring produces a signal of double frequency. The external components are chosen to minimize the offset that the squaring process would otherwise produce. The double frequency signal is then run into an AC coupled buffer to remove any residual offset, and then fed into phase shifters so that the relative phase of the second and first harmonic signals can be adjusted to any desired value. The phase shifters are "unity gain" design, but the reality of operation is that the amplitude of their output varies perceptibly as the phase is adjusted. To deal with this problem, an automatic gain control circuit based on the Analog Devices AD-734 chip was added after the phase shifters.

To make it possible to flip the phase of the second harmonic signal with respect to the first harmonic signal, the second harmonic signal was made a switchable input to a differential input instrumentation amplifier (Analog Devices AD-624). Separate offset and volume controls for the first and second harmonic signals is provided before the two signals are added by subtraction as the inputs to an instrumentation amplifier (Analog Devices AD-624). The signal conditioning circuit just described is shown in block diagram in Fig. 5.8. This mixed signal, switched by a computer controlled relay, is the

---

[3] These nifty boards have long since gone the way of all good things. No longer available.

source that drives a 1 KW Carvin power amplifier. Since the output voltage swing of the Carvin amplifier is about plus or minus 70 V, a custom wound broad band step-up transformer is used to bring the voltage amplitude up to several hundred volts.

The rest of the electronics was/is commercial off-the-shelf equipment. In addition to the Canetics PCDMA data acquisition system, it consists of a Hewlett-Packard frequency counter, several oscilloscopes, and several Picoscope PC-based oscilloscopes, one of which is used as a power spectrum analyzer. Another displays in real time the evolution of the test device voltage and thrust as measured by the position sensor. At any rate, that is how two of the Picoscopes are presently used. All of the apparatus and equipment, of course, has evolved over the years, both in its design and use. The details of this evolutionary process, for the most part, are not important. But a few of those details are important.

GETTING ORGANIZED

As mentioned earlier, Tom's reaction to Nembo's presentation at STAIF 2006 was to start designing and building a thrust balance in the style of the one used at ARC. The design, construction, and debugging of his balance took many months. In the interim, the source of the problem with the U-80 sensor seemed to be that its small signal sensitivity was greater than that expected on the basis of calibration with masses much larger than the signals seen.

One way of dealing with this problem was to see if the thrust sensitivity changed when the sensor was not loaded with the weight of the device and all of the mounting hardware. This could be achieved by adding support structures that would make possible operation with the sensor and vacuum case oriented horizontally. While the ARC-style thrust balance was being built, this was done. The result is shown in Figs. 4.24 and 5.9, taken from the presentation on Mach effects made at STAIF 2007.

Preliminary results obtained with Tom's ARC style thrust balance were also reported at that meeting. The test device used in this work was a Mach-Lorentz thruster of the sort described in Chap. 4. The net thrusts seen with the U-80 sensor were on the order of a micro-Newton, and those measured with the ARC style balance ranged from a few tenths of a micro-Newton to a micro-Newton or so. Nembo, who had been looking for thrusts on the order of 50 micro-Newtons, could not have seen thrusts of this magnitude in his work with the ARC balance a year earlier. But small thrusts were evidently present, and they passed the suite of tests that had been developed to identify and/or eliminate spurious thrusts (Fig. 5.10).

As mentioned at the outset of this chapter, Gary Hudson had been sufficiently impressed to deem it worthwhile to try to get things moving more smartly than the loosely coordinated effort that seemed to be going on. So the March meeting at CSUF was planned. Paul March, and after some arm twisting, Sonny White, came in from Houston. Andrew Palfreyman came down from San Francisco, and Gary and his friend Eric Laursen were there. George Hathaway came in from Toronto. Jim Peoples came in from Fort Worth. Frank Mead came in from the Edwards Air Force Base. Len Danczyk came down

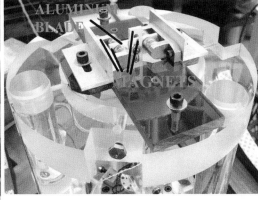

Fig. 5.9 The rod system (also shown in Fig. 4.24) that enables horizontal operation with the U–80 thrust sensor. The upper stabilizing ring and magnetic damper, later additions to this system, are displayed here

from Santa Barbara. The locals in attendance were Tom Mahood, Duncan Cumming, Carole Harrison, and me.

It was a large group, and a mixture of management and tech types. When Duncan learned who planned to be there, he took it upon himself to drive over to Fullerton to warn me that a large group for a meeting of this sort was a bad idea. Sigh. He was right, as it turned out. Tech types and managers coexist uneasily at best. And when you add to that mix managers from different backgrounds who don't know each other very well, it's a recipe for disaster. It took quite a while to repair the damage.

At the meeting it was agreed that the target for the work in the lab should be to increase the size of the thrust signal from a few tenths of a micro-Newton extracted from much larger signals that did not reverse when the direction of the device on the beam was reversed, to a few micro-Newtons, preferably with a direction reversing thrust signal. This was to be accomplished in 6 months to a year, and when achieved, would be the basis for reconsideration of serious investments. The devices in use at that time were Mach-Lorentz thrusters, and George Hathaway volunteered to have his ceramicist make some special capacitors to be used in devices of a new design. Aside from George's contribution, no other significant investment was to be made until benchmark results were achieved.

Not long after the meeting in March, Sonny White and Paul March approached Gary with a proposal that he fund an experimental effort they had planned. At that time Paul had convinced himself that a fellow in England named Shawyer, who claimed to be able to produce thrust with a tapered cavity resonator, might be onto something. Several of us had tried to convince Paul that the claims must be bogus, for if the device worked as claimed, it

Fig. 5.10 A montage of photos of the shuttler devices tested in early 2010. The device itself is in the *upper left*. The Mach effect is driven in the *center* section consisting of thin EC–65 crystals. The two pairs of thick crystals on either end produce the shuttling action. In the lower *left* is the device mounted on the end of the balance beam. And to the *right* is the Faraday cage, with a plastic and copper foil wrap

must be violating conservation laws in ways that would have been obvious in other situations decades earlier. Paul was not dissuaded. He was determined to build a tapered resonator and try out Shawyer's scheme.

Sonny had been paying attention to the Mach effect work and the experimental results, many of which suggested that a real effect might be present. But Sonny wasn't interested in forwarding the Mach effect agenda per se. Rather than focus on the transient mass terms in the Mach effect equation, he focused on the minuscule term that is routinely discarded in Mach effect work. This, he argued, suggested that a quantum mechanical "zero point" effect might be put to use for propulsion.

Rather than argue for the electromagnetic field zero point effect, Sonny claimed that the important process was the transient fluctuation, where an electron and positron are fleetingly created from nothing, and almost instantly annihilate each other. His idea was that the combined electric and magnetic fields present in a Mach-Lorentz thruster should act on these "virtual" electron-positron pairs in the vacuum, and the reaction on the device would produce thrust. Since its inception, Sonny has replaced virtual electron-positron pairs by the real sigma particle pairs, so-called "Q" particles, that the Standard Model predicts to be present in the vacuum. That is, according to Sonny's conjecture, you can push off the vacuum with suitable electric and magnetic fields. Mach-Lorentz thrusters, according to Sonny's conjecture, are really just Lorentz plasma thrusters, the plasma being the virtual electron-positron or "Q" pairs of the vacuum. Some pretty obvious arguments can be marshaled against this scheme. But exploring those arguments would lead us far astray from our path. Suffice it to say that Sonny and Paul, who have always had an affinity for quantum vacuum fluctuations, spent some of Gary's money and built some equipment to test their schemes. The results of their efforts were the production of no effects to support their ideas. Work was suspended. But times have changed, and they are now pursuing their dreams in their "Eagleworks" lab at NASA Johnson with at least modest support from management.

The next investigation was suggested by Duncan Cumming after Andrew Palfreyman urged that the effort should go "back to basics." He suggested a rotary system. Paul March and I had briefly considered rotary systems in the early 2000s, but abandoned those efforts when the prospect of Mach-Lorentz thrusters became appealing. Duncan's proposal was different, and sufficiently interesting that some effort was expended in this direction. If you're interested, the results were reported at Space, Propulsion, and Energy Sciences International Forum in Huntsville, Alabama in 2009.[4] Suffice it to say that they were interesting, and not null. No one much paid attention. Those interested in propulsion saw no propulsion application. And those willing to look at this as a scientific test of Mach effects could always find something to criticize, especially as an electrostrictive signal complicated the results.

The main lesson of this exercise was that the only types of experiments that get any attention are those that show direct thrust that if scaled can be put to practical use. And the activity contributed nothing to getting organized. Jim Peoples had given up on us and moved on to other ventures. He kept in touch, but abandoned efforts to create a commercial venture. Eventually, he started advocating the writing of a book about Mach effects, for there was no single place where someone interested could go and find everything pulled together in digestible form. He was convinced that no one would pay much attention until a book was written. Yes, this is the book.

Gary, dealing with the impact of the financial crisis of 2008, had scaled back after Sonny and Paul's experiments had produced no interesting results. In the later part of 2008, David Mathes, with a background in defense systems, nanotechnology, and private sector aerospace technology, found his way to the project. It did not take him and his associates long to figure out that any organized effort wasn't going to happen, at least for a while. The core of the project was a tabletop experiment producing interesting, but not compelling, results that were far from the point of transition to a serious development effort. If that would change, and such change were in the cards, it was not apparent.

---

[4] STAIF, organized chiefly by Mohamed El-Genk, Regents Professor of Engineering at the University of New Mexico, had gone the way of all good things after the 2008 meeting and El-Genk's retirement. SPESIF, organized chiefly by Tony Robertson, was a heroic effort to keep the meeting alive for the advanced propulsion community.

TIME OF TRANSITION

In retrospect, 2008 and 2009 were a time of transition. Marc Millis's and Eric Davis's edited book on advanced propulsion was published by the AIAA. And, as recounted in the next chapter, Greg Meholic put together a presentation on advanced propulsion that he gave first to the Los Angeles AIAA section, and then to a number of interested groups.

For me, Greg's presentation had a different impact. As a survivor since 2005 of metastatic lung cancer, and metastatic lymphomas since 2007, it was clear that unlimited time to pursue my interests was not to be had. After pointing out the extreme implications of Mach effects for exotic "propulsion" in the mid-1990s, the main focus of the Mach effects project had been directed to the first transient Mach effect term implications for propellant-less propulsion. In part, this choice of focus had been made because the first term effects held out the promise of practical devices in a foreseeable future. And in part the choice had been motivated by concerns that talking about starships and stargates would get the project labeled as hopelessly romantic at best, outright flakey at worst, and in any event without serious practical consequence. The prospect of limited time left made re-evaluation of means and objectives inevitable. And Greg's willingness to stick his neck out was an inspiration. Starships and stargates will never be built unless those interested in building them are willing to risk the ridicule of reactionaries.

Invitations to John Cramer's 75th birthday symposium (and formal retirement) went out in the early summer of 2009. On learning that I would be coming, John told me to get in touch with the organizers to give a presentation. After some consideration, I decided to take this opportunity to speak out on exotic spacetime transport – and to support his "transactional interpretation" of quantum mechanics that I still believe to be the only interpretation of quantum mechanics that is consistent with relativity theory. The presentation was cast as an explanation of why aficionados of science fiction have little to worry about regarding scientists taking up the challenge of exotic transport any time soon.[5]

After the symposium, the presentations were put up on Mike Lisa's website at Ohio State.[6] Several months later, Daniel Sheehan, organizer of the Society for Scientific Exploration's 2010 meeting in Boulder, Colorado, issued an invitation to talk about Mach effects as part of a session on advanced propulsion they had planned. Since Boulder is not far from our summer home in the Colorado mountains, I accepted. The talk was a revised version of the presentation at John's symposium. It left the audience, some of whom were expecting something different, mostly nonplused. However, it prodded me to take up a problem – the issue of spin in the ADM model of electrons, discussed in Chap. 8 – that I had walked away from in the mid-1990s. Finding an error in the earlier work on spin and ADM electrons, the issue was solved. Then this book followed.

The experimental program changed at this time, too. After the work on rotary systems was abandoned, shuttlers were made with new crystals secured from EDO Corporation,

---

[5] That you are reading this book is evidence that the reactionary views of mainstreamers of my generation are passing. We are all retired, replaced by those with different experiences and views. Retirees, though, get to write books. ☺

[6] Available at: http://www.physics.ohio-state.edu/~lisa/CramerSymposium/talks/Woodward.pdf

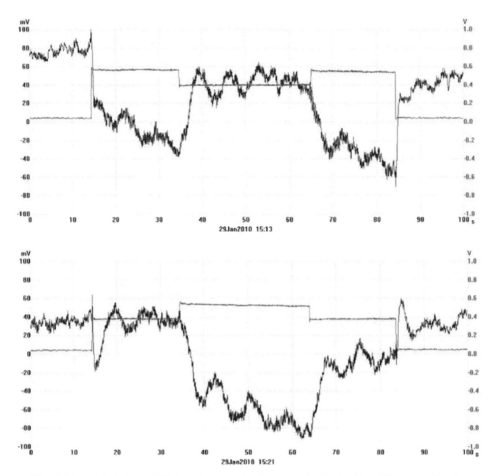

Fig. 5.11 Typical thrust (*blue*) and power (*red*) traces obtained with a Picoscope for the device shown in Fig. 5.10. The relative phase of the first and second harmonic signals is flipped for the central part of the powered portion of the traces. The ordering of the phases is reversed for the lower traces relative to that used for the upper traces

shown in Fig. 5.10. Interesting results were obtained. The device being tested could be turned on, producing a thrust shift. After a few seconds, the phase of the two harmonic signals could be flipped, changing the thrust, and then flipped back before powering down.

Pretty, single-cycle traces resulted, as shown in Fig. 5.11. But as before, the test devices eventually died. Death traces are displayed in Fig. 5.12.

Deciding to ignore the suggestions of others regarding what the means and objectives of the experimental program should be, a re-evaluation was done. The PZT stacks that had produced the spectacular behavior back in 2000 were selected as the devices to be tested. Why? Not because of the spectacular behavior, but because they were simple. They only required one driving signal. And because they manifestly satisfied the condition that bulk acceleration was present. The intent was to get these devices to produce thrusts that could be easily seen in a single cycle, so that one did not have to look at averaged results that could not

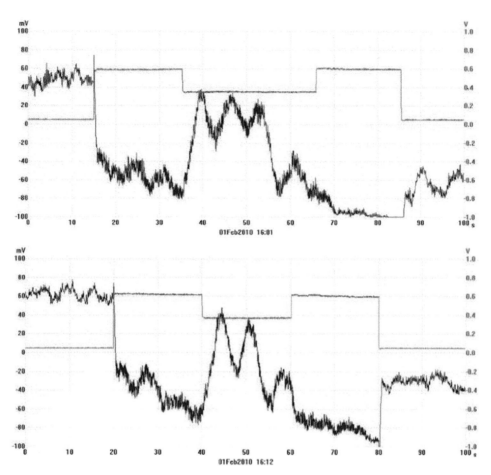

Fig. 5.12 Thermal and mechanical evolution eventually cause most Mach effect devices to die. These are two of the last traces showing an effect and its decay mid-cycle, for the device shown in the previous two figures

be simply demonstrated in real time to a casual bystander. Demonstration quality performance required the ARC style thrust balance and new devices. So a large supply of crystals were ordered from Steiner-Martins, and the ARC style balance was made operational.

DEMONSTRATION QUALITY RESULTS

Since the Steiner-Martins crystals were special order, it took several months for them to show up. In the interim the old PZT stacks were used with the thrust balance to character-ize the balance's behavior with these devices and make a series of minor upgrades to improve the performance – especially the signal to noise ratio. For example, a small aluminum project box was adapted for use as a Faraday cage enclosing the test device, and eventually pieces of mu metal were silver epoxied to the surfaces of the box to improve its attenuation of magnetic fields. The details of and shielding of the sensor circuits were explored, improving noise suppression.

The PZT stacks of 1999–2000, with these upgrades, produced a direction reversing signal of about a micro-Newton, that is, a signal that could be seen in a single cycle as the noise in the system had been suppressed to a few tenths of a micro-Newton. A variety of tests for the genuineness of the signal were carried out. In addition to the basic test of thrust reversal when the direction of the test device on the balance beam was reversed, other tests were done:

- Variation of the residual air inside the vacuum chamber (no change in the signals when the pressure changed from ~10 μm to >200 μm).
- Variation of the relative phase of the first and second harmonic signals. (Phase shift changed the thrust signals as expected, but the presence of electrostriction made the variation much less than would be expected if the applied second harmonic signal were solely responsible for the change).
- Variation of the power applied to the test device. (Power scaling was what was expected).
- Response when the test device was oriented vertically. (When the "up" and "down" results were differenced, no net thrust effect was seen).
- Changes when exposed parts of the power circuit on the beam were shielded (none observed).

All of this notwithstanding, the system was not yet a "demonstrator." The signals at a micro-Newton were too small. While the thrust signals reversed with device direction reversal, competing, non-reversing effects were present that made the direct and reversed signals markedly different. Since there wasn't much more that could be done to suppress noise and non-reversing spurious thrusts in the system, it was clear that a demonstrator would have to produce a thrust of at least 5 micro-Newtons, and preferably 10 or more micro-Newtons. That wasn't in the cards for the old EC-65 PZT stacks.

The first stacks using Steiner-Martins crystals were put to the test in the fall of 2011. The material used in the crystals –SM-111 – was chosen for its high mechanical Q and low thermal dissipation (almost an order of magnitude smaller than that for the EC-65 crystals). The price paid for these improved properties was a lower dielectric constant. The lower dielectric constant was compensated for by having the crystals made thinner than their Edo counterparts.

The usual sorts of problems were encountered. But in January of 2012, while preparations to move the lab to new quarters were in progress, a series of runs were done with one of the new devices after the system had been tweaked to especially nice performance levels. Part of the unusually good performance resulted from taking cycles at half-hour and more intervals. This provided adequate time for the test device to cool from heating incurred in the previous cycle. The reason why this is important is that the thermal expansion of PZT is *negative* in the direction of the axis of the stacks, whereas the expansion of everything else in the device is positive.[7] So, heating degrades the preload of the stacks – and this turns out to degrade the performance of the devices.

---

[7] The thermal expansion of the crystals in the radial direction is positive, but that does not affect preload.

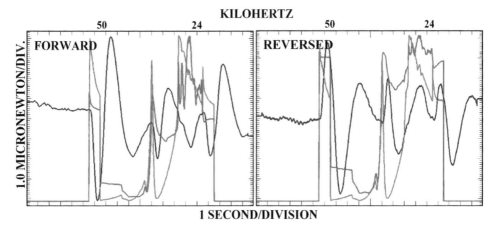

Fig. 5.13 The averages of about a dozen cycles of data in each of the forward and reversed directions of one of the Steiner-Martins crystals devices obtained under optimal conditions in January of 2012. The thrust traces are *red*, the power traces *light blue*, and the stack accelerometer traces *green*. After an unpowered interval of several seconds (to establish behavior in quiescent conditions), a 1 s pulse of power at the center frequency of a swept range (chosen to be on the electromechanical resonance of the device) is delivered. This is followed by a sweep from 50 to 24 KHz, followed by another 1-s pulse at the center frequency

The runs done in January during the lab move, in addition to the long cooling inter-cycle interval, profited from another unappreciated just-so circumstance only later understood. It turns out that these devices are more complicated than one might hope. In general, they display *two* principal resonances. One is the electrical resonance of the power circuit. The other is the electromechanical resonance of the PZT stack in the device. These two resonances, in general, do not coincide. When they do coincide, unusually good performance results.

Quite accidentally, these two resonances happened to coincide in the device of the January tests. In consequence, the thrusts produced in this set of cycles averaged at 10 micro-Newtons, much larger than any performance achieve before. The data protocol used was 20 s cycles. After a few seconds of quiescent data, a 1-s power pulse at the center frequency of the sweep to be done was acquired, followed by the frequency sweep (over a range of 25 kHz) centered on the resonance, followed by another 1-s power pulse at the center (resonant) frequency and several seconds of outgoing quiescent data.

A number of cycles with the device pointing in both the "forward" and "reversed" directions were collected and averaged. Then the reversed direction average was subtracted from the forward direction average (to cancel all non-direction reversing effects), yielding a net direction reversing signal. Plots of these thrust traces, along with the applied power, are displayed in Figs. 5.13 and 5.14.

The January performance, by any reasonable standard, was "demonstration" quality. Indeed, the behavior was so interesting that videos of several cycles were recorded so that the experience of the signal acquisition could be relived virtually. However, when the move to the new quarters was completed, the outstanding performance of several weeks

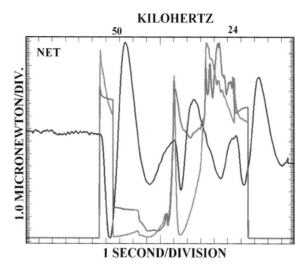

Fig. 5.14 The "net" thrust trace recovered by differencing the forward and reversed thrust results displayed in the previous figure. This suppresses all thrusts, of whatever origin, that do not reverse with the direction of the device

earlier had disappeared. Since almost everything had been affected by the move, the reason for the degraded performance was not immediately obvious. Tracking down the cause took several weeks, for everything had to be tested.

This was not the waste of time it might seem, for in all those tests, a number of upgrades were identified and made. And, in the course of those tests and upgrades, it became clear that doing presentation quality tests for possible spurious signals was more important than restoring the performance achieved in January. After all, if the thrusts produced are not demonstrably due to the first Mach effect, then there is no reason to believe with confidence that Mach effects exist. And if they don't exist, making wormholes is going to be a lot harder than we would like. So the following sections of this chapter are going to be a bit more formal and quantitative than the material covered so far.

THE NEW PZT STACKS

In general dimensions the PZT stacks are almost identical to the old stacks. They are 19 mm in diameter and when assembled 19 mm in length. Whereas the old stacks were made with one pair of thick crystals at the end closest to the reaction mass and three pairs of crystals at the active end next to the aluminum cap, the new stacks have two pairs of 2 mm thick crystals near the reaction mass and four pairs of crystals in their active ends. Crystals 0.3 mm thick are used as accelerometers. One accelerometer is located in the middle of the active end; two more are included, one at each end of the stacks. The new stacks have a capacitance of 39 nf, whereas the old stacks had 19 nf capacitance. The old stacks had a dissipation constant of about 3%, and the new stacks have 0.4%.

Since the new stacks have the same dimensions as the old, all of the mounting hardware made in 1999 was used for the present work. Two important changes were made to the mounting parts. First, the thermistors used to monitor the temperature of the devices were relocated. If you want to track thermal effects, the detectors must be as near to the region of interest as possible. Embedding a thermistor in a stack is not feasible. But thermal imaging with a far infrared camera loaned by Lockheed-Martin's Millennium Projects a decade ago confirmed the obvious: heating chiefly takes place in the active part of the stacks. The best place to monitor this heating is in the aluminum cap. So small holes were drilled in the aluminum caps, and thermistors were attached with silver epoxy.

Thermal effects are the result of motion induced by the expansion/contraction of parts of the device during and after operation, and the largest effects are to be expected in the largest parts of the device. This dictates that a second thermistor be embedded in the brass reaction mass, as it is 9.56 mm thick and has a larger coefficient of thermal expansion than the other parts of the devices – 18 ppm per unit length. PZT, with a coefficient of 5 ppm and a length of 19 mm is the other part chiefly affected by heating. The stainless steel retaining bolts are inconsequential, partly because of their low coefficient of expansion (10 ppm per unit length), but chiefly because of their low thermal conductivity. The aluminum mounting bracket has a high coefficient of expansion (22 ppm). But it heats slowly and acts to reduce any Mach effect thrust, so it need not be considered further.

SYSTEM UPGRADES

The devices mounted in their Faraday cages vibrate. An obvious concern is that vibration might be communicated from the device in operation to other parts of the thrust balance, and those vibrations might produce a signal that could be mistaken for a thrust in the device. This is referred to as a "Dean drive" effect, named for Norman L. Dean, who claimed to be able to produce thrust by slinging masses around on eccentric orbits at varying speeds.

Dean's claims were carefully investigated in the early 1960s and thoroughly discredited. The apparent thrust produced was traced to the fact that the coefficients of static and kinetic friction are not generally the same. So if you can produce a periodic mechanical effect where the static coefficient operates in part of the cycle, and the kinetic coefficient in the rest of the cycle, a vibrating object can be made to move relative to its environment.

Dean drive effects have been addressed all the way along in the course of the Mach effect experimental effort. The use of rubber, or sometimes Sorbothane vibration attenuators, has been a part of all designs of the apparatus. For the sequence of runs reported here, further efforts were made. The fork mount attached to the end of the balance beam, initially plastic with rubber washers on two bolts holding the "tines" of the fork to the cross-member (this mounting fork is visible in the lower left panel of Fig. 5.10) was replaced by a more elaborate fork made from plastic and aluminum held together with 4–40 brass screws, washers, and small O-rings. (See Fig. 5.15.) The aim here was to reduce the amount of contact surface in the mount to minimize vibration transfer while providing a reasonably rigid structure to transfer chiefly stationary thrusts.

In order to get a quantitative measure of the vibration communicated to the flexural bearings supporting the balance beam – the only place where a Dean drive effect could act

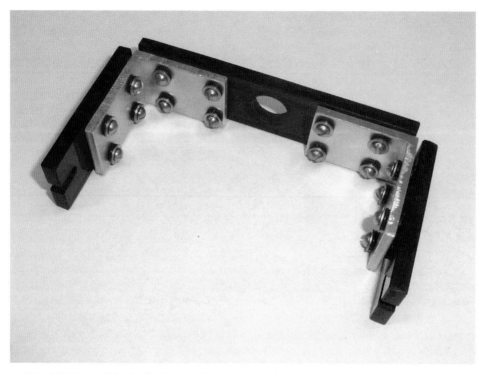

Fig. 5.15 One of the (*red*) plastic and aluminum fork mounts that replaced the fork mount shown in the picture in Fig. 5.10. The parts of the mount are held together with multiple 4–40 machine screws damped with small Buna-N O – rings and brass washers

to produce a spurious thrust-like effect – accelerometers were attached to the central part of the beam in proximity to the lower flexural bearing. The accelerometers were fabricated from 2-mm square pieces of thin PZT material, a brass electrode and a 2-mm square brass "anvil" mass (1 mm thick). See Fig. 5.16.

The electrode is placed between the pieces of PZT (oriented with appropriate polarizations), and the anvil mass is part of the ground circuit, locally grounded. In order to suppress pickup that might appear on the accelerometer leads, the signals from the accelerometer were detected with a differential amplifier so that pickup would be rejected as common mode noise. The usual precautions were taken with the electronics. These accelerometers made it possible to show that reversal of the direction of the device on the end of the beam had no effect on the vibration reaching the flexural bearings in the central column of the balance – but the observed thrust signal changed direction with the device. See Fig. 5.17.

A number of other system upgrades were done in the months after the relocation of the lab to new quarters. Some were routine matters, such as grounding and shielding, and the bringing up of new equipment and circuits to improve the performance of the system.

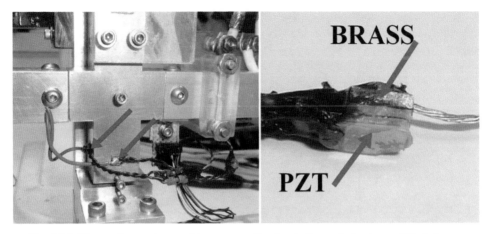

Fig. 5.16 The accelerometers used to measure vibration in the central column of the balance. On the *right*, one of the accelerometers is shown before attachment. On the *left*, the *red arrows* indicate where they were attached. The brass anvils of the accelerometers are grounded with the *green* leads

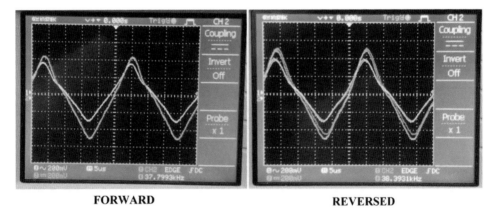

FORWARD   REVERSED

Fig. 5.17 The waveforms produced by the accelerometers during operation of the device for both the "forward" and "reversed" orientations on the end of the beam. Since the waveforms for the two orientations are the same, one can be confident that the vibration reaching the support column does not depend on the orientation of the device

There was, however, a more fundamental guiding plan put in place that determined what was to be done. When the spectacular performance of January disappeared with the move to new quarters, the natural course of action was to try to discover the cause of the degraded performance and restore the system to its previous operating characteristics. When this proved a bit more challenging than expected – notwithstanding the help of a group of world-class electrical engineers virtually kibitzing – another course of action presented itself.

If you look back at the traces in Figs. 5.13 and 5.14, you will see that there is a problem, from the point of view of systematic investigation of the system, with its operation. This is especially evident in the light blue traces, which are the square of the voltage across the device, a quantity proportional to the power delivered to the device.[8] During the constant frequency power pulses at the beginning and end of the frequency sweeps, the power does not remain constant, as one might hope. Evidently, the device heats up sufficiently quickly so that the resonance condition only exists for a small fraction of the 1-s interval of constant frequency operation. This means that the thrust produced during these intervals is not a response to steady conditions. Rather, it is a response to transient conditions that only persist for less than a few hundred milliseconds.

The obvious way to deal with rapidly changing conditions is to try to stabilize them. In this case, we're talking about moderately sophisticated feedback circuits designed to keep the power signal on resonance. This is not as simple as it sounds. And when the cause of the disappearance of the spectacular behavior was eventually found – the issue of the coincidence of the electrical and mechanical resonances – it became more complicated still. Now, these issues will have to be dealt with. And indeed, they are being dealt with. But before investing serious resources and effort in such a program, it is worth knowing whether the thrust effect you are looking at is a real Mach effect, or just some spurious junk that will not lead to the interesting physics and technology that is your goal.

It turns out that running at lower power, the devices don't heat up so quickly as to drive themselves off resonance, even when operated at constant frequency for as much as 10 s (or more). That means that you can do the tests needed to make sure you're not engaged in serious self-deception at lower power without having to first get a lot of fairly complicated circuitry built and debugged. Accordingly, the chief focus of work shifted to carrying out the tests designed to make sure the thrust effect seen was real. Work on the hardware needed for higher power operation didn't stop. But it assumed lesser priority than carrying out the tests of the system that would justify continuing work.

ARE THE THRUST SIGNALS REAL?

Running at lower power obviated the need for resonance tracking electronics; but it was not without a cost. Since the thrust effect scales with the power, running at lower power means that some signal averaging must be done to suppress noise in the thrust traces. If the thrusts you are looking at are on the order of a few microNewtons at most, and the noise in a single cycle is as much as a half microNewton, though you can see the signal in a single cycle, reliable estimates of behaviors of interest are not possible. Typically, the average of one to two dozen cycles of data was required to get the noise down to an

---

[8] The power – typically with a peak value of 100 to 200 W – is equal to the square of the voltage divided by the impedance of the device (at the operating frequency). Note that most of the power in this circuit is "reactive." That is, most of the power is not dissipated in the device as it is operated.

acceptable level, that is, plus or minus a tenth or two of a microNewton. Since the aim of this work was to get reliable, but not necessarily "pretty," results, signal averaging was kept to a minimum.

MECHANICAL BEHAVIOR OF THE SYSTEM

The mechanical behavior of the system can be inferred from the work done with the devices in operation. Nonetheless, if this is the only means used to assess the mechanical behavior of the system, you leave yourself open to the criticism that the inferred mechanical behavior is not the actual mechanical behavior, and as a result, the inferences made about the behavior of the devices cannot be assumed valid, as some mechanical quirk of the system might be responsible for the behavior observed. (Yes, that's the level of objection/paranoia that this sort of work entails.) It turns out that this sort of concern is easily dealt with by using the calibration coils to deliver a thrust to the balance beam so that the mechanical response can be assessed. And in the course of the tests done with the devices and a dummy capacitor, this test was carried out.

Actually, two types of responses were examined. One was the turning on of a stationary force on the beam that persisted for 8 s. Since the mechanical settling time for the balance had been found to be on the order of 5 s, an 8-s sustained thrust was deemed sufficient to explore this behavior. The second type of thrust examined was transient. In the course of work with the system, it became obvious that switching transients, both electrical and mechanical, were present in the system, at least some of the time. The correlated thrust transients could be as large as a few microNewtons, so emulation with the calibration coils was carried out by putting a current pulse of 0.3 s through the coils. The thrust traces recovered in these tests are displayed in Fig. 5.18. Note that a much larger thrust than a few microNewtons was required to produce the transient pulses of a few microNewtons.

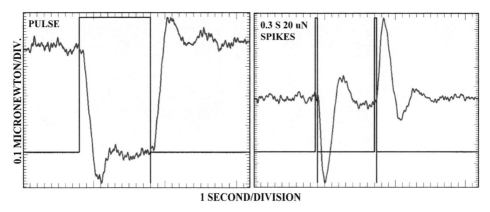

Fig. 5.18 Emulation of the thrust traces generated using the calibration coils. On the *left*, a force of several μN was switched on at 8 s and *left* on for 10 s. In the plot at the *right*, 300 ms force pulses of about 20 μN are applied at the beginning and end of an 8 s interval – in opposite directions (to qualitatively emulate thermal transient behavior)

RESIDUAL AIR IN THE VACUUM CHAMBER

Another obvious issue of concern is the possible action of the electromagnetic fields present during operation on the residual air in the vacuum chamber. If the air is ionized by the presence of the fields, it can be acted upon by them as well, producing an "ion wind" effect. And even if ionization is unimportant, given that vibration is driven in the PZT stack, directed acoustic effects are possible.

Testing for these effects is quite simple. You simply do a full set of cycles, forward and reversed, and compute the net reversing thrust signal at one vacuum chamber pressure. And then repeat the process at another vacuum chamber pressure markedly different from the first sequence of cycles.

Since the vacuum system runs with only a (Welch 1402) rotary vane vacuum pump, typical chamber pressures with reasonable pump-down times are a few milli-Torr. The vacuum system is quite "tight," so it can be bled up to, say, 10 Torr, sealed, and the chamber pressure does not change appreciably for an hour or more. Chamber pressures of less than 10 m-Torr and 10 Torr – a difference of three orders of magnitude – were chosen for this test. The results are displayed in Figs. 5.19 and 5.20. Evidently, the residual air in the vacuum chamber does not contribute to the thrusts seen in this system.

STRAY ELECTROMAGNETIC FIELDS

Another concern when electromagnetic fields of moderate to high field strength may be present is that those fields may couple to nearby fixed objects and cause the balance beam to be deflected. Extensive shielding, improved over the course of the work with the

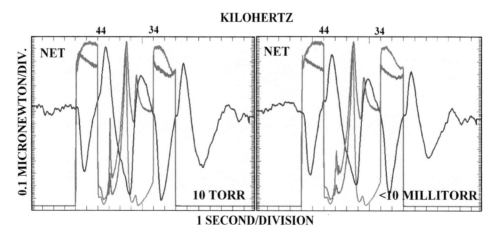

Fig. 5.19 The test for the dependence of the thrusts observed on the residual air in the vacuum chamber consisted of producing net thrust traces (forward minus reversed) at two very different chamber pressures: 10 Torr and less than 10 mTorr. Those thrust traces are shown here side-by-side

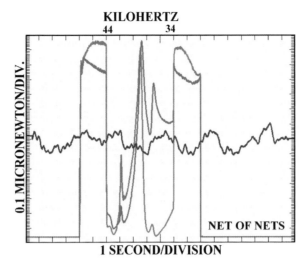

Fig. 5.20 The net of nets thrust trace for the two vacuum chamber pressures used in the residual air test. No signal at all is apparent at the noise level of a few tenths of a µN

balance, was implemented to prevent this sort of behavior. And such effects are removed as a common mode signal eliminated by subtraction of the forward and reversed direction results, as mentioned above.

The Faraday cage was designed to suppress the sorts of stray fields that might cause such spurious signals. Not only is the cage made of solid aluminum, plates of mu metal 0.5-mm thick were attached to the surfaces of the cage with silver epoxy to further suppress magnetic fields. This test was conducted by first acquiring enough cycles of data in normal operating circumstances to produce an averaged signal with small noise. The traces for this average are shown in Fig. 5.21. Next, the top half of the Faraday cage was removed, leaving the device exposed in the direction of the nearest insulating surfaces (see Fig. 5.22), and another sequence of cycles were acquired and averaged. That average was then subtracted from the average for normal operation.

The average obtained with the cage partially removed and the net thrust trace recovered when subtracted from the thrust trace for normal operation are presented in Fig. 5.23. Exposing the device during operation in this fashion produces no change in the observed thrust.

To rule out the possibility that coupling to nearby conductors might contribute to the observed thrusts, the coils in the calibration system attached to the balance platform were detached and moved out of their normal position, as shown in Fig. 5.24. Again, a sequence of cycles were acquired and averaged; and this average was subtracted from the average for normal operation. The average and the net thrust are displayed in Fig. 5.25. As in the case of the insulator test, no thrust effect is seen that might contribute to the thrusts routinely produced with these devices. These tests give one confidence that the thrust effects seen in normal operation are not caused by the presence of stray electromagnetic fields.

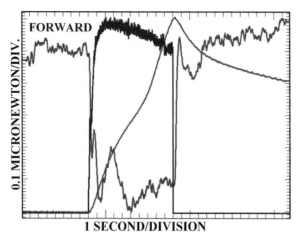

Fig. 5.21 In preparation for tests of the effects of stray electromagnetic fields coupling to the environment, a thrust trace for the forward orientation of the device with full shielding was acquired – the average of a dozen and a half cycles. This is the trace used for comparison purposes with traces generated for conditions where coupling, if significant, should alter the recorded thrust trace

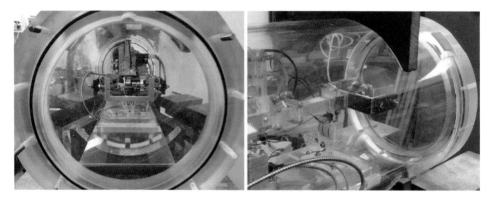

Fig. 5.22 The first of these tests was for coupling to nearby insulators. It was carried out by removing the top part of the Faraday cage, thus exposing the device and power circuit normally enclosed. The nearby plastic of the vacuum chamber should be affected if this potential source of spurious effects is actually present

WHAT IS THE ELECTRICAL RESPONSE OF THE SYSTEM?

A way to test for the effect of electromagnetic effects in this system, other than the tests just described, is to disconnect the device from the power supply and replace it with a simple commercial polyfilm capacitor of about the same capacitance. The capacitor actually used in this test is shown in Fig. 5.26, along with it placed in the Faraday cage (Fig. 5.27).

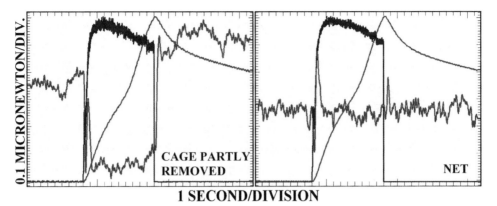

Fig. 5.23 On left is the thrust trace (an average of a dozen cycles) recovered with the top of the Faraday cage removed. On right is the net thrust trace obtained when the reference thrust trace is subtracted from the one on the left. Evidently, coupling to nearby insulators does not contribute to the thrust effects seen in this experiment

Fig. 5.24 To test for coupling to nearby conductors, the nearest conductors – the calibration coils – were moved, and data were acquired with the repositioned coils (shown here). Normal position is that on the *left*, and the moved coils are shown on the *right*

Since polyfilm capacitors have no electromechanical properties, no thrust is expected from the excitation of the capacitor per se. Nevertheless, a full set of cycles, both forward and reversed were acquired. The forward and reversed averages are displayed in Fig. 5.30.

As expected, there is no thrust associated with the powered interval. As one might expect, the power traces are quite different from those recovered with the test devices. Where the test devices, because of their electromechanical response, show a fairly wide and structured resonance, the only resonance present with the dummy capacitor is one with very high Q. That is, it is very narrow in the frequency sweeps. The system was tuned so

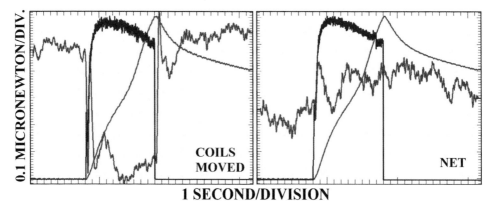

Fig. 5.25 The thrust trace produced with the moved coils is shown at the *left*; and the net generated by subtracting the reference thrust trace is displayed at the *right*. As for insulators, no thrust effect attributable to coupling to nearby conductors is present

Fig. 5.26 The capacitor used in the "dummy" test is shown with its wiring and connector on *left*. At *right* is a picture of the dummy capacitor installed in the Faraday cage (top half removed)

that the initial center frequency pulse would be on resonance, and the power traces in Fig. 5.30 show that this part of the powered interval was indeed on resonance. But the center frequency drifted a bit, smearing the resonance recorded during the sweeps and in the outgoing center frequency power pulse.

The net of the forward and reversed averages for the dummy capacitor is presented in Fig. 5.28. Note that the subtraction process eliminates the small ground plain offset seen in the thrust traces for the individual forward and reversed orientations. And, of course, there is no reversing thrust recorded, as expected. It may seem that this test was redundant and unnecessary. But it was this test that made plain that the electrical and electromechanical resonance conditions in these devices need not be coincident, explaining the change in behavior of the first Steiner-Martins device.

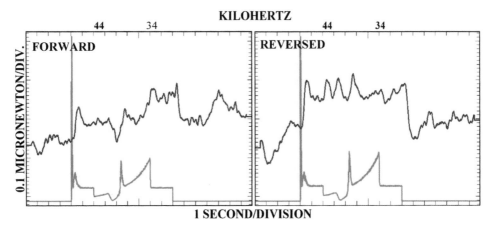

Fig. 5.27 The forward and reversed orientation thrust traces (*red*) for the dummy capacitor. Only the power traces (*light blue*) are plotted as the stack accelerometer is irrelevant in this test. Note the small ground plain shift when the power is applied (caused by ground return currents in the system)

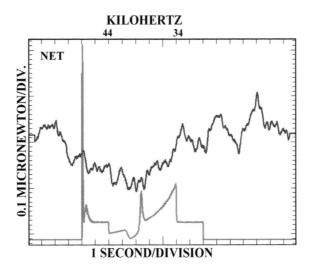

Fig. 5.28 The net of the forward and reversed thrust traces for the dummy capacitor. Note that the ground plain offsets present in the individual traces in the previous figure are canceled as a common mode signal, as intended

ORIENTATION OF THE DEVICE ON THE BEAM

The use of the protocol of subtracting reversed from forward device orientation has already been described. For smaller signals where non-reversing signals are also present, this protocol is essential to getting reasonable results. The reason for checking the behavior of the system with the devices oriented in the vertical, as well as horizontal, directions is that mode coupling may be present. The C-Flex bearings allow horizontal motion of the balance beam, their crossed steel blades providing the restoring torque required of a thrust balance.

Ideally, these bearings would be completely impervious to vertical forces on the beam. But the bearings are not perfect in this regard, and vertical forces on the beam can produce horizontal displacements. Indeed, the ARC version of this thrust balance was "zeroed" by moving a small mass along one of the arms of the beam. So, the question is: Does a steady force in the vertical direction produce a stationary horizontal displacement of the beam? When the largest signal being produced was on the order of a microNewton, this test was conducted, and no evidence that vertical thrusts produced horizontal displacements was found. Now that larger thrusts are routinely produced, the test was repeated.

A montage of pictures of the device in the various orientations tested is displayed in Fig. 5.29. The device used in this test produced large electrical and electromechanical transients, so the duration of the power pulse was set to 10 s, since in earlier work it had been found that 5–6 s were required for the thrust balance to settle from large transients.

About a dozen and a half cycles were averaged for each of the orientations, producing the thrust traces shown in Fig. 5.30. The reversed result was subtracted from the forward result, and the down from the up. A simple, linear secular drift correction was applied so that steady – especially prominent in the up trace – drift did not complicate the analysis. The two net traces so generated are presented in Fig. 5.31.

The subtraction order was chosen to make the thrust traces as similar as possible. Nonetheless, as highlighted by the lines and ovals provided to assist interpretation, while a large transient is present in both cases, the forward minus reversed thrust trace shows a steady thrust during the powered interval – predicted by Mach effects – whereas the up minus down thrust trace does not – as would be expected if mode coupling at a steady thrust level of a few micro-Newtons is not important. These results show that net horizontal thrusts are not artifacts of vertical forces acting on the balance, either at the device itself, or elsewhere in the system.

DEAN DRIVE EFFECTS

Back at the beginning of the system upgrades section we briefly considered Dean drive effects, those that are produced by vibration in systems where components have both static and kinetic friction effects that operate during different parts of a cyclic process. Of all possible sources of spurious effects, Dean drive effects are far and away the most likely.

We've already indicated that these effects were repeatedly checked for, and provided the results of one of those tests with the accelerometers attached to the center column of the thrust balance, the only place where, in a system like this, a Dean drive effect would have to operate. That test was not enough. Some time before it was done, all of the vibration isolation materials were removed and runs done to check for differences from runs done with all isolation materials present. At the time, the vibration suppression materials were sheets of Sorbothane applied to the interior of the Faraday cage, Buna-N O-rings and washers where the machine screws that support the Faraday cage attach to the cage, O-rings and washers where the machine screws attach to the fork mount, O-rings and

Fig. 5.29 A montage of photos of the device in the Faraday cage (with the top half removed) showing the four orientations for which data was acquired. The forward and reversed, and up and down orientations are differenced. The end plate of the vacuum chamber is removed, and the thrust balance platform slid partly out of the chamber for these photos

washers where the multiple small machine screws attach the plastic parts of the fork mount to the aluminum parts, and rubber and brass washers at the attachment point of the fork mount to the balance beam. In addition, vibration dampers consisting of plastic blocks and sheet Sorbothane were clamped to both arms of the balance beam. In the test, the damper of this sort on the device side of the balance beam was removed.

In the test just mentioned, no appreciable difference in the thrust signals was noted. This can be seen in the thrust traces recorded in Fig. 5.32. A thrust effect of a couple of microNewtons is present in the resonance center frequency pulses as well as at the swept resonance, and they are essentially the same for both with and without the isolating materials present.

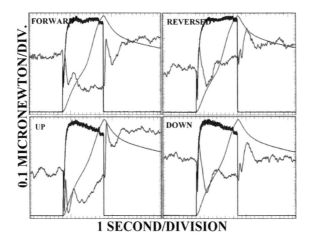

Fig. 5.30 The thrust (red), power (blue), and temperature (lavender) traces for the four orientations of the device shown in the previous figure. The power is the same in all cases, and the temperature rise is about 10°C

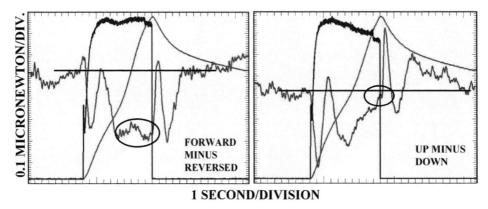

Fig. 5.31 The net thrust traces for horizontal (forward minus reversed) and vertical (up minus down) operation. The secular drift noted for the individual traces remaining has been corrected by a simple linear correction. Although transient thrust fluctuations are present for both orientations, only the horizontal orientation shows a sustained thrust in the powered interval – as it should if an impulse Mach effect is present. This behavior is highlighted by the circled parts of the traces

The two results are not exactly the same, as the duration of the sweep was changed in the week separating the two runs. But the first part of the sweeps can be subtracted to quantitatively compare the thrusts produced. That is shown in Fig. 5.33. No detectable difference in the thrusts at the first center frequency pulse is present.

KILOHERTZ

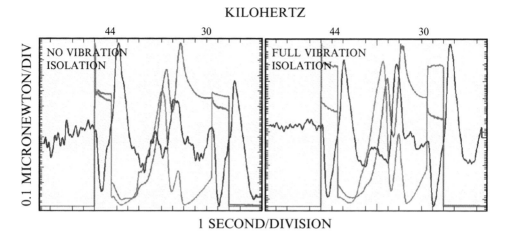

Fig. 5.32 The net thrust traces for the first of the Steiner-Martins devices in the initial test for Dean drive effects. On the *left*, this is the thrust produced when all of the vibration isolation was removed from the system. On the *right*, the thrust for full vibration isolation is shown. In spite of the apparent differences in the power (*blue*) and accelerometer (*green*) traces (*plotted full-scale*), aside from switching transients, these quantities are the same for both isolation conditions

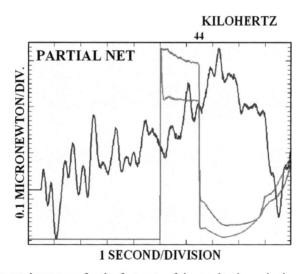

Fig. 5.33 The net thrust trace for the first parts of the results shown in the previous figure. Evidently, there is no evidence that the presence or absence of vibration isolation materials makes any difference, especially in the 1.5 s center frequency (on resonance) power pulse that starts at 5 s

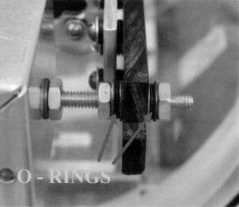

Fig. 5.34 After accelerometers were attached to the center column of the balance to measure the vibration present at the lower flexural bearing, quantitative Dean drive tests became possible. In order to minimally disturb the system when changing the vibration state, only the O-rings shown in the *right panel* were removed, producing the configuration of the mounting hardware shown in the *left panel*. This produced a 25% change in the vibration reaching the center column of the balance

When the vibration isolation materials were restored, a few were left out. In particular, the Sorbothane lining of the Faraday cage and the damper on the beam were not restored, as they added significant mass to the beam, increasing its moment of inertia and slowing the response of the beam to the application of thrusts. The O-rings where the machine screws attach to the Faraday cage were likewise not restored because it was found that vibration tended to loosen these attachments when the O-rings were present. This was not a problem with the attachment of the screws to the fork mount.

Work on other tests followed this Dean drive test. But eventually it was decided that a more formal, quantitative Dean drive test should be undertaken. The accelerometers had not been affixed to the balance column during the first Dean drive test, so a quantitative measure of the vibration present was not available.

In a similar vein, removal of all of the vibration isolation materials had noticeably changed the mass of the beam, affecting the thrust response. So it was decided to only remove the O-rings on the screws holding the Faraday cage where they attach to the plastic part of the mounting fork, as shown in Fig. 5.34. As measured by the column accelerometer(s), removal of these O-rings changed the amplitude of the vibration from 490 ADC counts to 370 ADC counts. That is, a change on the order of 25% in the vibration was attributable to these O-rings.

The usual forward/reversed protocol was used for this test. This led to the "net" thrust traces for the two O-ring configurations shown in Fig. 5.35. Simple visual inspection

KILOHERTZ

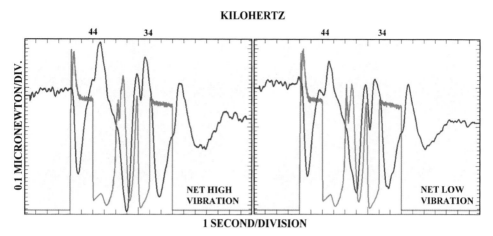

Fig. 5.35 The net (*forward minus reversed*) thrust traces (*red*) for low vibration (*left*) and high vibration (*right*) at the central column of the balance. Two second initial and final center frequency on resonance power pulses are used in this test. The traces for each are nearly identical to those of the other

KILOHERTZ

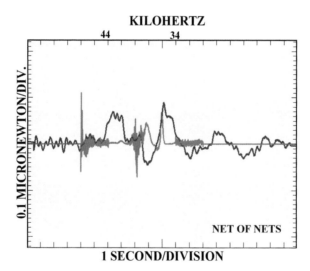

Fig. 5.36 The net of the nets for high and low vibration displayed in the previous figure. Note that no thrust signal like those in the previous figure, with reduced amplitude, is present for the initial and final constant frequency (on resonance) power pulses, notwithstanding that the vibration levels are different by about 25%

reveals that no obvious differences in the thrust traces are present. Differencing the two net thrust traces shows this to be so, as can be seen in Fig. 5.36. No signal on the order of a microNewton as might be expected from a Dean drive effect is present during the resonance center frequency pulses. This is easy to see when the net of nets thrust trace

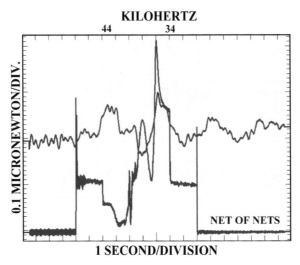

Fig. 5.37 The same plot as in the previous figure, but with the net power trace replaced with the net column accelerometer trace (plotted full-scale). No correlated behavior is present in the initial and final constant center frequency power pulses, where large thrust effects are present in both the high and low vibration data

is plotted against the center column accelerometer trace, as in Fig. 5.37. The conclusion that follows from all of these tests is that the observed thrusts in this experiment simply cannot be attributed to Dean drive effects.[9]

THERMAL EFFECTS

In addition to vibrating, these devices heat up as they are run. Since the recorded thrusts are small, on the order of several microNewtons, it is easy to believe that either vibration or heating could be responsible for the observations. Back in 2000, thermal effects were the spurious cause of choice for the Oak Ridge Boys (ORBs), folks already referred to in Chap. 4 of this book. They even built a crude apparatus to demonstrate their claim. Yes, crude is the correct adjective. Tom Mahood, whose Master's thesis they had attacked, wrote a detailed critique of their claims, showing them to be misguided and false.[10]

The physics of thermal effects is straightforward. As power is delivered to a device, part of the power is thermalized, and the PZT stack heats up. As heat increases in the stack, it is

---

[9] When I was a starting assistant professor, I shared an office with a Chinese historian, Sam Kupper. One evening, after a conversation on pedagogy, as Sam was leaving the office, he commented, "Remember! If you can't dazzle 'em with brilliance, baffle 'em with bullshit!" I hope that this doesn't strike you as a baffling exercise. The issue is too important, however, to be left to some casual comments.

[10] Tom's recounting of all this is available on his website: OtherHand.org.

conducted to the reaction mass, and eventually to the aluminum mounting bracket and beyond. The lengths of the stack and reaction mass along the symmetry axis of the device depend on their temperature, the dominant dependence being linear in the temperature. Defining the length of either the stack or reaction mass as l, the temperature of the part as T, and the coefficient of thermal expansion as k, we have:

$$l = l_0(1 + kT).\tag{5.1}$$

The rate at which the length changes with change in temperature is:

$$\frac{dl}{dt} = l_0 k \frac{dT}{dt},\tag{5.2}$$

and if the rate of change of temperature is changing:

$$\frac{d^2l}{dt^2} = l_0 k \frac{d^2T}{dt^2}.\tag{5.3}$$

Now, Eq. 5.2 describes a velocity, and Eq. 5.3 an acceleration. If a mass is associated with the acceleration, a force is present. Note, however, steady heating produces a "velocity" but no "acceleration" and thus no force.

We can apply this analysis to the case of a device mounted on the beam (actually, in the Faraday cage) by considering each element separately in the system and then summing their effects to get the total force on the balance beam. We take one end of the part in question to be "fixed" by the action of the balance beam and assume that the heating in the part is uniform throughout the part. These are idealizations, especially during rapid changes in circumstances. But since we are looking for a stationary force, we ignore the fact that we are using an idealization approximation. Under the assumed conditions, it is easy to show that the force exerted on the fixed end of the part in question is equal to:

$$F = \frac{1}{2} ma = \frac{1}{2} m l_0 k \frac{d^2T}{dt^2},\tag{5.4}$$

where m is the mass of the part and a the acceleration of the free end of the part.

Next, we consider how the parts of the system affect each other, ignoring the fact that the PZT stack is clamped on the reaction mass. (Clamping will reduce the effect of the stack, regardless of whether the stack expands or contracts when heated, so we are considering a worst case scenario in this regard.) In fact, PZT in these circumstances has the interesting property that it contracts in the direction of the axis of the stack and expands radially. Given the arrangement of the parts in the system, this behavior produces a force that acts in the direction opposite to that observed. That means that F computed with Eq. 5.4 for the stack alone is negative. The reaction mass, made of brass, however, has a positive coefficient of expansion, so its action on the PZT stack and itself is positive.

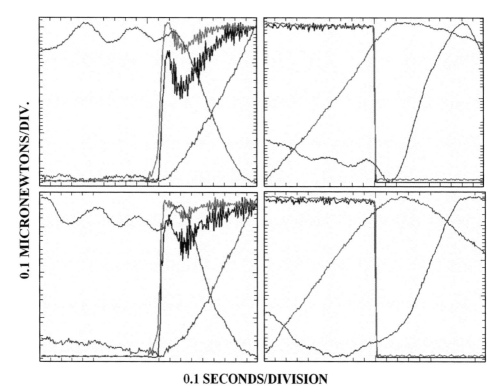

0.1 MICRONEWTONS/DIV.

0.1 SECONDS/DIVISION

Fig. 5.38 Traces for switch-on (*left*) and switch-off (*right*) for two individual cycles. The lavender trace records the temperature in the aluminum cap of the device. As these traces indicate, the swing from no heating to steady heating, or vice versa, takes place in about 200 ms. These are the intervals in which a thermally induced thrust effect should take place if detectable

Taking account of all this, we can write the force on the aluminum bracket as:

$$F = -\frac{1}{2}m_{PZT}a_{PZT} + m_{PZT}a_{BRASS} + \frac{1}{2}m_{BRASS}a_{BRASS}. \qquad (5.5)$$

The expansion of the aluminum bracket is in the direction opposite to the expansion of the device (notwithstanding that the expansion of the stack is negative), and so will reduce the force in Eq. 5.5 that acts on the beam. We ignore this, at least for the time being. To compute the values that go into Eq. 5.5 we need the physical parameters of the parts. They are:

Part	Length (CM)	Mass (GM)	k ($\Delta l/l/$ °C)
Reaction Mass	9.6	43	18×10^{-6}
PZT stack	19	46	-5×10^{-6}

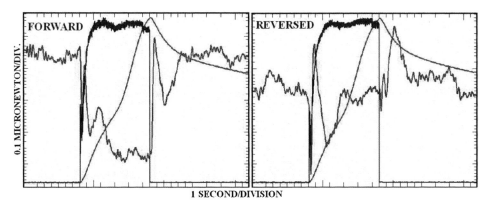

Fig. 5.39 Ten second duration forward and reversed thrust traces. Were there no ground plain offset, the traces for the two orientations would be near mirror images of each other, as in the spectacular results obtained with a different Steiner-Martins device, displayed in Fig.

All we need now is the second time derivative of the temperature for the parts. That we can estimate from the temperature traces for individual cycles recorded by the thermistor in the aluminum cap. Such traces are shown in Fig. 5.38.

The time taken for steady heating/cooling after the power is switched on/off is roughly 200–300 ms. The heating rate during the powered interval is about 1°C per second. So, the magnitude of d^2T/dt^2 is roughly 5. The value of d^2T/dt^2 in the stack may momentarily be as much as an order of magnitude larger than this at switch-on, but since the stack produces a thermal effect with the opposite sign to that of the detected thrust, we simply ignore this. We also ignore the fact that d^2T/dt^2 in the reaction mass will be smaller than that measured in the aluminum cap.

Substitution of the various values into the above equations produces forces for the terms in Eq. 5.5 of -1.1×10^{-8}, 4.0×10^{-8}, and 1.9×10^{-8} N, respectively. Summed, they come to 0.065 microNewtons, nearly two orders of magnitude smaller than the observed thrusts. And that's with all of the approximations taken to maximize the predicted thermal effect.

Calculating the effect of the aluminum mount, which will reduce the force, seems pointless given the predicted force. Note, too, that the predicted thermal effect is a transient spike at switch-on and switch-off, not the sustained stationary thrust predicted by Mach effects and that observed. That is, the thrust trace should look like the "spikes" trace in Fig. 5.18, whereas in fact it looks like the thrust trace in Fig. 5.40, the net of the traces in Fig. 5.39. Evidently, the thrusts observed with these devices cannot be written off as thermal effects.

GETTING PREDICTIONS RIGHT

Where do we stand at this point? Well, we have a thrust of the sort predicted by the first term Mach effect. And the tests done show that the thrust generated cannot simply be ascribed to some conventional, spurious source. But the magnitude of the thrust is very

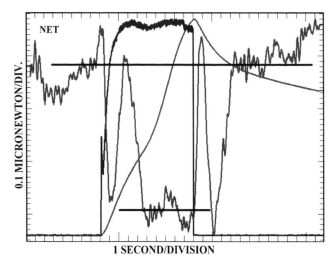

Fig. 5.40 The net thrust trace computed from the forward and reversed traces shown in the previous figure. Comparing this with those in the mechanical emulation test – Fig. 5.21 – it appears that in addition to the stationary thrust emphasized with the *black lines*, a larger transient, reversing thrust occurs when the device is switched

much smaller than that predicted employing the formal procedures in use before Nembo focused attention on the issue of bulk acceleration. By several orders of magnitude – a few microNewtons versus a few tenths of a Newton. This has been a cause for serious concern. The problem seems to be that the customary procedure doesn't properly account for acceleration dependence.

Indeed, if you calculate the predicted Mach effect in an electrical component like a capacitor based on the assumption that $P = iV$ using the leading term in Eq. 3.5 and then multiply the δm_0 so obtained by the presumed acceleration produced by a mechanical actuator, as in Eqs. 3.9 and 3.10, you get ridiculously large predictions. You may recall that in Chap. 3 we wrote out the Mach effect equation with explicit acceleration dependence – in Eqs. 3.6, 3.7, and 3.8. That is the formalism that must be used to get reasonable predictions. That calculation can be adapted to the case of the PZT stacks as follows. As in Chap. 3, we note that when a (three) force **F** acts on an object, the change in energy dE of the object is given by:

$$dE = \mathbf{F} \bullet d\mathbf{s} \tag{5.6}$$

So,

$$\frac{dE}{dt} = \mathbf{F} \bullet \mathbf{v} = P \tag{5.7}$$

And,

$$\frac{dP}{dt} = \mathbf{F} \bullet \mathbf{a} + \mathbf{v} \bullet \dot{\mathbf{F}} \tag{5.8}$$

where P is the power communicated to the object by the applied force and the over dot indicates differentiation with respect to time. Care must be taken with this analysis, for $\mathbf{v}$, which has a simple interpretation in the case of a rigid body acted upon by an external force, is more complicated when the object acted upon can absorb internal energy, as noted in Chap. 3. We also note that $\dot{\mathbf{F}}$ involves a third time derivative of position and the time derivative of the mass. So, for a lowest order calculation, we set the second term on the RHS of Eq. 5.8 aside. Substituting from Eq. 5.8 into Eq. 3.5, we get:

$$\delta m_0 \approx \frac{1}{4\pi G}\left[\frac{1}{\rho_0 c^2}\frac{\partial P}{\partial t}\right] \approx \frac{1}{4\pi G\rho_0 c^2}\,\mathbf{F}\bullet\mathbf{a} = \frac{1}{4\pi G\rho_0 c^2}\,m_0\,a^2. \tag{5.9}$$

Evidently, the simplest Mach effect depends on the square of the acceleration of the body in which it is produced. Note, by the way, that the units of these equations are Gaussian, in keeping with the traditions of field theory of yesteryear. The SI counterpart of this equation does not have the 4π in the denominator on the RHS, though G (in appropriate units) remains the coupling coefficient between sources and fields for gravity.

According to Eq. 5.9, the mass fluctuation induced by a sinusoidal voltage signal of angular frequency ω will be proportional to the square of the induced acceleration. Since the chief response is piezoelectric (that is, linear in the voltage), the displacement, velocity, and acceleration induced will occur at the frequency of the applied signal. And the square of the acceleration will produce a mass fluctuation that has twice the frequency of the applied voltage signal.

In order to transform the mass fluctuation into a stationary force, a second (electro) mechanical force must be supplied at twice the frequency of the force that produces the mass fluctuation. In the work with the Edo-based PZT stacks, this second harmonic signal was produced by squaring the first harmonic signal and providing offset and phase adjustment capability. While this capability remained a part of the signal conditioning electronics, it was found unnecessary in the work with the Steiner-Martins-based PZT stacks, for the electrostrictive response of this material was markedly stronger than that for the EC-65 material.

Calculation of the interaction of the electrostrictive electromechanical effect with the predicted mass fluctuation is straight-forward. We assume that a periodic voltage V, given by:

$$V = V_0 \sin \omega t \tag{5.10}$$

is applied to the PZT stack. We take the power circuit to be approximately a simple series LRC circuit. Operation of these devices on resonance is required to produce observable thrusts. When such a signal is applied at a frequency far from a resonance of the power circuit, the corresponding current will be roughly 90 degrees out of phase with the current. At resonance, where the largest Mach effects can be expected, the relative phase of the current and voltage in the power circuit drops to zero, and:

$$i = i_0 \sin \omega t \tag{5.11}$$

However, the relative phase of the voltage and current in the PZT stack remains in quadrature. The length of the PZT stack, x, including the piezoelectric displacement is:

$$x = x_0(1 + K_p V), \tag{5.12}$$

where x_0 is the length for zero voltage and K_p is the piezoelectric constant of the material. The velocity and acceleration are just the first and second time-derivatives (respectively) of Eq. 5.12. The mass fluctuation, accordingly, is:

$$\delta m_0 \approx \frac{1}{4\pi G \rho_0 c^2} m_0 a^2 = \frac{1}{4\pi G \rho_0 c^2} m_0 \left(-\omega^2 K_p x_0 V_0 \sin \omega t\right)^2$$

$$= \frac{\omega^4 m_0 K_p^2 x_0^2 V_0^2}{8\pi G \rho_0 c^2} (1 - \cos 2\omega t) \tag{5.13}$$

Keep in mind that in SI units, a factor of 4π must be removed from the denominators in this equation. Electrostriction produces a displacement proportional to the square of the applied voltage, or:

$$x = x_0\left(1 + K_e V^2\right) = x_0\left(1 + K_e V_o^2 \sin^2 \omega t\right), \tag{5.14}$$

where K_e is the electrostrictive proportionality constant. Using the customary trigonometric identity and differentiating twice with respect to time to get the acceleration produced by electrostriction,

$$\ddot{x} = 2\omega^2 K_e x_0 V_0^2 \cos 2\omega t \tag{5.15}$$

The force on the reaction mass (the brass disk in this case) is just the product of Eqs. 5.13 and 5.15:

$$F = \delta m_0 \ddot{x} \approx \frac{\omega^6 m_0 K_p^2 K_e x_0^3 V_0^4}{4\pi G \rho_0 c^2} (\cos 2\omega t)(1 - \cos 2\omega t). \tag{5.16}$$

Carrying out multiplication of the trigonometric factors and simplifying:

$$F = \delta m_0 \ddot{x} \approx \frac{\omega^6 m_0 K_p^2 K_e x_0^3 V_0^4}{8\pi G \rho_0 c^2} (1 - 2\cos 2\omega t - \cos 4\omega t). \tag{5.17}$$

The trigonometric terms time-average to zero. But the first term on the RHS of Eq. 5.17 does not. The time-average of F is:

$$\langle F \rangle \approx \frac{\omega^6 m_0 K_p^2 K_e x_0^3 V_0^4}{8\pi G \rho_0 c^2}. \tag{5.18}$$

Remember that a factor of 4π must be removed from the denominator of Equation (5.18) if SI units are used. This is the thrust sought in this experiment.

To get a sense for the type of thrusts predicted by the larger Mach effect, we substitute values for a PZT stack device used in the experimental work reported here into Eq. 5.18. The resonant frequencies for this type of device falls in the range of 35–40 KHz, and we take 38 KHz as a nominal resonant frequency. Expressing this as an angular frequency and raising it to the sixth power, we get 1.9×10^{32}. The mass of the PZT stack is 46 g. But only a part of the stack is active, say, 25 g, or 0.025 kg. The length of the stack is 19 mm; but a typical value for the active part of the stack is roughly 15 mm, or 0.015 m. In most circumstances the voltage at resonance is between 100 and 200 V. We assume 200 V here. And the density of SM-111 material is 7.9 g/cm^3, or $7.9 - 10^3$ kg/m^3. We use the SI values of G and c, and set aside the values of the piezoelectric and electrostrictive constants for the moment. Inserting all of these values and executing the arithmetic:

$$\langle F \rangle \approx 3.4 \times 10^{23} K_p^2 K_e. \tag{5.19}$$

Steiner-Martins gives 320×10^{-12} m/V for the "d$_{33}$" piezoelectric constant of the SM-111 material. That is, K_p has this value. Note, however, that the dimensions given are not correct. The piezoelectric constant does not have the dimension of meters because the spatial dependence is "fractional." That is, the constant contains spatial dependence as change in meters per meter, which is dimensionless.

Steiner-Martins list no value for the electrostrictive constant. But electrostrictive effects are generally smaller than piezoelectric effects. When the stated value for K_p is inserted into Eq. 5.19 we get:

$$\langle F \rangle \approx 3.5 \times 10^4 K_e. \tag{5.20}$$

If we take the electrostrictive constant to be roughly the same as the piezoelectric constant, we find that a thrust of about ten microNewtons is predicted. As we have seen, the observed thrust for these circumstances is at most a few microNewtons. That is, prediction and observation agree to order of magnitude. Two issues, however, should be kept in mind. One is that the electrostrictive constant is likely smaller than the piezoelectric constant. Indeed, examination of the first and second harmonic amplitudes of the stack accelerometer spectra show that the second harmonic is down by a factor of 0.3 to 0.1 from the first harmonic. The other is that mechanical resonance amplification doubtless takes place at the mechanical resonance frequency.

Resonance was taken into account in the relative phase of the current and voltage in the power circuit. But no allowance for mechanical resonance amplification was made. Since the mass fluctuation goes as the square of the acceleration, even modest resonance amplification can have pronounced effects. That said, the piezoelectric constant is measured with "free" crystals; and the PZT stacks use here are run with considerable preloading.[11] So the effect of mechanical resonance amplification will be less than one might otherwise expect.

---

[11] The retaining bolts that hold the stack together were torqued to 8–10 in. pounds during assembly.

All of these considerations are competing effects, and as such, taken together, are unlikely to radically change the prediction we have found here. So a match of prediction and observation to order of magnitude can be claimed. It seems fair to say that prediction and observation are consistent with each other, though an exact quantitative match is not to be had at present.

There is another reason to take the results reported here as evidence for Mach effects. They are predictions of standard theory. There is no "new" physics involved. As discussed in Chap. 2, inertial forces, given what we now know about general relativity and cosmology, are gravitational in origin. Since that is the case, it is reasonable to ask what the local sources of inertial (gravitational) effects are when one allows internal energies of objects to be time-dependent. Standard techniques produce the Mach effect equations. No oddball unified field theory is needed. No additional parallel universes. No special pleading. No, "and then a miracle occurs." Standard physics. From this perspective, it would be surprising were Mach effects *not* found when sought.

ORGANIZATION IN THE OFFING?

Early on January 8, 2011, I got on a Southwest flight out of Orange County (John Wayne Airport) to San Francisco. David Mathes met me at the airport. We drove to Gary Hudson's home in the Palo Alto area. Gary had arranged for Jim Peoples, Mark Anderson, and Eric Laursen to video conference in. The mistake of a large group mixing tech types and managers wasn't going to be repeated.

The purpose of the meeting, as before, was to get organized. The conversation was fairly general, as a demonstrator was still not in hand. But it was also forthright. Jim explained his reasoning that getting a book done was central to the effort. Gary brought up the realities relating to intellectual property, and David brought up the problems of finding support, and technical issues of qualifying any practical devices that might be developed. Everyone agreed to participate. And Gary volunteered to help put together the formal documents that were to be developed the following spring and summer to create a legal entity to encompass our activities.

Unbeknown to us, other events, some already in progress, would deflect the organizational effort. NASA, at the instigation of Simon "Pete" Worden, former Air Force general and Director of NASA's Ames research center, and DARPA had put in motion the "100 Year Starship" initiative. The premise of this operation was that it would likely take 100 years to bring about the technology needed to build a "starship," and that given the vagaries of politics, government would likely not be able to sustain properly an effort of that duration. The views of the organizers of the initiative were that the best way to sustain a project of this duration would be to create a "private," that is, non-governmental, organization dedicated to the goal of building starships.

The 100 Year Starship initiative had its first meeting within a week or two of David and I traveling to Gary's place and videoconferencing with Jim, Mark and Eric. In the spring, the initiative decided to have a meeting open to the public in Orlando at the end of September and invite anyone interested to attend, and those so inclined to propose presentations of various sorts on the issue of starships – including science fiction aficionados. It promised to be a circus in which any serious work would be lost in the noise. I am told by those attendees I know that the event, given its provenance, was enjoyable.

The other development that could not have been anticipated in January of 2011 was this book. As spring progressed and experimental work continued, leading eventually to the developments you've just read of, not too much work on the legal organization got done. Jim became increasingly insistent that a book on Mach effects be written. Given health issues, his concern was easy to understand. I might die before the book got written. Not too long after arriving at our summer place in the mountains of Colorado, I received an email from Springer soliciting a book proposal –one of those automatically generated solicitations when you trigger a web scanning program for some reason or other. Fully expecting to be blown off, I responded with a proposal for this book. I was amazed when the proposal passed preliminary editorial review. More information was requested, and a package was submitted to the Editorial Board and approved. Thoughts of creating legal entities were set aside, at least for the duration of book writing. Jim, Gary, and David remain committed to furthering the building of starships, and by the time you are reading this, perhaps they will be pursuing such activities. And the 100 Year Starship program? They are just getting organized as I write this. They seem not to have figured out what's going on in the starship building business yet. No doubt, that will change.

Part III

6

Advanced Propulsion in the Era of Wormhole Physics

If you've watched any of a number of TV shows dealing with either "advanced propulsion" or time travel in the past several years, you've doubtless encountered at least the mention of wormholes. Were the topic time travel, a fairly extensive discussion of wormholes might have been involved. Maybe not a very technical discussion, but usually one accompanied by artistic renditions of what a wormhole generator might look like.

When the topic of discussion is time travel, wormholes are pretty much unavoidable, as they are the only scheme that holds out the prospect of enabling time travel, especially to the past. When advanced propulsion is the topic, wormholes usually only get mentioned after lengthier discussions of several much less "out there" options. We're not going to consider time travel further here.

Popular discussions of advanced propulsion usually start with technologies that are decades old. That's because even in the 1950s and 1960s, it was obvious, really obvious, that seriously long-distance space travel could not be realistically accomplished using chemical rockets. So alternatives were explored, such as ion engines, nuclear rockets, both thermal and electric, and bombs. The Orion project examined the feasibility of ejecting nuclear bombs from the rear of a spacecraft and detonating the bombs when at a fairly small distance from the craft. The craft, of course, was equipped with a large "shield" with shock absorbers at the rear which the blast would act on – designed to withstand repeated blasts without destroying the shield or the ship. A conventional explosive version of this scheme was actually tested as a demonstration of principle. It worked. The project was shut down, however, before the nuclear variant could be put to the test.

The notion behind ion engines, and electric propulsion in general, is that much higher "specific impulses"[1] can be achieved by accelerating small amounts of propellant to very

---

[1] Specific impulse (I_{sp}) is defined as the ratio of the thrust produced by a rocket to the rate of the mass, or weight at the Earth's surface, of the propellant ejected per second. That is, thrust $= I_{sp}* dm/dt*$ g. When weight is used, I_{sp} is expressed in seconds. The higher the I_{sp}, the more efficient the rocket motor. Very high I_{sp} is usually achieved by ejecting a small amount of propellant at very high velocity producing only modest thrust. The most efficient rocket motor by this measure is a "photon rocket" which uses light as propellant, producing minuscule thrust.

J.F. Woodward, *Making Starships and Stargates: The Science of Interstellar Transport and Absurdly Benign Wormholes*, Springer Praxis Books, DOI 10.1007/978-1-4614-5623-0_6,
© James F. Woodward 2013

high exhaust velocities, than by explosively blowing a large amount of propellant out of a tail pipe, as in conventional chemical rockets. The gargantuan thrusts produced by chemical rockets cannot be produced in this way. But electric propulsion engines can operate steadily at low thrust for very long times, producing very large changes in velocity.

A number of technical issues attend electric propulsion schemes; but these are engineering problems, not issues of fundamental physics. The physics of electric propulsion is strictly conventional. If the engineering issues of a particular approach can be successfully addressed, there is no question that the resulting device will function as expected.

Setting aside the bomb approach of blasting your way through space, nuclear energy, still new and trendy 50 years ago, seemed a plausible energy source to power spacecraft. It still does. The advantage of a nuclear system is the very much higher energy and power density in a nuclear reactor system compared to a chemically based system. Two approaches were investigated: thermal and electric. In the thermal scenario, a reactor would create high temperatures to which some suitable propellant would be exposed before entering a reaction chamber.[2] Thermal expansion of the propellant in the reaction chamber would produce the desired thrust. The electric scenario would have a nuclear reactor produce electricity. And in turn the electricity would be used with an electric propulsion system.[3] One of the better known of the electric propulsion schemes was advanced many years ago by former astronaut Franklin Chang Diaz. So-called VASIMIR, a variable specific impulse magnetohydrodynamic system. Only in the last few years have any serious resources been put into Diaz's scheme.

Other non-chemical schemes have been proposed over the years. Solar sails, where a large, highly reflective gossamer film is deployed and the action of Sunlight on the film – the sail – produces a modest amount of thrust. A variant of this idea was advanced some years ago by Robert Winglee. He proposed a system where a magnetic field traps a plasma, and Sunlight acts on the trapped plasma to produce thrust. The thrust on the plasma back-reacts through the trapping magnetic field on the spacecraft's field generator.

Robert Forward, before his death, advanced the "tether" scheme, where a long cable is allowed to dangle in Earth's magnetic field as it orbits Earth. The orbital motion through the magnetic field generates a current, and currents can be applied to the tether to generate forces that can be used for orbital boost and suchlike. Another way to produce propulsion is the "lightcraft" proposal of Liek Myrabo, where an intense laser beam is reflected off of the bottom of a craft designed to ablate the atmosphere in proximity to the craft, producing thrust. Yet another scheme is space elevators, where a carbon nanotube filled strip is extended into space, and vehicles crawl up and down the strip.

---

[2] The reaction chamber here is not a chemical explosion reaction chamber. It is a purely mechanical reaction chamber where the heated propellant is allowed to expand exerting a force on the end of the chamber opposite the nozzle where propellant exits (and exerts no force that would balance the force on the other end of the chamber).

[3] An instructive analog here is a diesel-electric locomotive. The diesel engine does not directly drive the wheels of the locomotive. Rather, it drives an electric generator whose output is used to power electric motors that drive the wheels.

The non-chemical propulsion schemes being explored in the 1950s and 1960s were all shut down by the early 1970s. An economic recession, oil crisis, and the Vietnam War combined with programmatic decisions for NASA – the shuttle and *Skylab* – led to alternative propulsion schemes being shelved for a couple of decades. Only after the mid-1990s were any resources worth mentioning allocated to non-chemical propulsion. And then only the sorts of propulsion schemes being explored 30–40 years earlier got any significant support. Why?

Two reasons. First, aside from the Orion project, none of the earlier schemes involved systems that could be used for access to low Earth orbit (LEO). As such, they did not represent competition to heavy lift chemical rockets. Second, none of those schemes – or, for that matter, any mentioned so far – involved "speculative" physics. That is, as far as the physics involved in these schemes is concerned, there is no question whatsoever whether they will work. They will. Period. Getting them to work is simply a matter of engineering. For these reasons, no real, serious risk is involved should you decide to promote them. Most managers and administrators are deeply risk averse.

However, you may be thinking, we're talking about the mid-1990s and beyond. And Kip Thorne and those energized by his work were doing serious investigations. Miguel Alcubierre, in 1994, had published the "warp drive" metric solution to Einstein's field equations. And the sub-culture of "anti-gravitiers" – those who had been searching all along for a revolutionary propulsion breakthrough involving gravity – had not disappeared.

Actually, revolutionary propulsion got some traction in the mid-1990s, in no small part because Daniel Goldin had been appointed NASA Administrator by the Bush Administration in 1992 and kept on by the Clinton Administration. Goldin, a former Vice President and General Manager for Aerospace and Technology at TRW, knew the aerospace industry well, but did not come from the NASA bureaucracy. Soon after taking over at NASA, he articulated the catch phrase for which he is still known: faster, better, cheaper. Goldin, attuned to the developments in physics, also decided that NASA needed to invest some serious effort in the area of revolutionary propulsion. As he put it at the time, he wanted his grandchildren to be able to study planetary cloud patterns from space – on planets orbiting stars other than the Sun.

The various NASA centers are semi-autonomous. So when Goldin decided NASA should be exploring revolutionary propulsion, he went around to each of the centers and individually tasked them to start such investigations. The culture of NASA wasn't equipped then to deal with this sort of request. Much of the NASA staff thought revolutionary propulsion irrelevant nonsense. Indeed, one of the centers' response to Goldin's tasking was to construct a website where they explained why revolutionary propulsion involving wormhole physics was physically impossible. When Goldin returned to the center some 6 months later and asked what progress had been made on revolutionary propulsion, there was an awkward silence, for there had been none; no one had done anything more than help with the website. Blunt remarks accompanied by fist pounding on the conference table supposedly ensued.

Goldin's revolutionary propulsion program eventually found a home in the Advanced Propulsion Laboratory at the Marshall Spaceflight Center managed by John Cole. Cole, notwithstanding being a chemical rocketeer at heart, helped set up the Breakthrough

Propulsion Physics (BPP) program headed by Marc Millis at NASA's Glen Research Center (in Cleveland, Ohio) and ran interference for it when it was attacked by critics. Millis recruited a number of people with respectable scientific credentials, including a few world-class physicists, and set about identifying and examining the various schemes then on offer for a propulsion breakthrough. A similar, less controversial virtual (web-based) program, the National Institute for Advanced Concepts, was set up to provide modest support for ideas such as Winglee's plasma sail and the space elevator.

The European Space Agency started a small program in the same vein as the BPP project. And the private sector became involved. The president of the Aerospace Division of Lockheed-Martin, convinced that the Skunk Works was living on its reputation of glory days long gone, set up "Millennium Projects," headed by Jim Peoples, who reported directly to him. In addition to being tasked with fairly conventional work – by revolutionary propulsion standards anyway – Millennium Projects was tasked with looking at "absolutely everything available in the field of propulsion." It appears that almost without exception, anyone touting a revolutionary propulsion scheme was invited to Lockheed's Lightspeed facility in Fort Worth, Texas, to present their scheme.

Boeing was a few years behind Lockheed-Martin. But eventually, an internal proposal was generated outlining how Boeing might "partner" with Millis's BPP program and explore revolutionary propulsion schemes using its Phantom Works program in Seattle. The internal proposal, however, was leaked and eventually found its way into the hands of Nick Cook. Cook, a reporter for Janes Defence Weekly, several years earlier, had been recruited by a shadowy figure, presumably associated with the British intelligence community, to find out what the Americans were up to in the area of revolutionary propulsion.

The cause for the British interest seems to have been that in 1994 Bernard Haisch, Alfonso Rueda, and Harold Puthoff had published a paper in *Physical Review A* claiming that inertia could be understood as the action of the "zero point fluctuation" (ZPF) electromagnetic field on electric charges subjected to accelerations. Toward the end of that paper, they had hinted that were their conjecture true, it might be possible to engineer radical, new propulsion systems. The hints occurred in other places, too. But there was no follow-up explaining exactly how this might actually be possible. Instead, there was a stream of papers and conference reports addressing technical details of the inertia origin proposal without application to the issue of rapid spacetime transport.

Now, there are two ways one might account for this silence regarding propulsion. One is that the electromagnetic ZPF conjecture on the origin of inertia is not tenable and so doesn't lead to the hinted revolutionary breakthroughs; or, even if the conjecture is tenable, no revolutionary breakthrough follows from it. The other is that revolutionary breakthroughs do follow from the conjecture, but someone decided that such breakthroughs did not belong in the public domain. It would appear that Cook's mentor thought the latter of these two at least possible, and were it true, the Americans had not kept the British informed of whatever progress had been made. The American treatment of the British during the Manhattan Project would not have been reassuring to the British. So, Cook's mentor aimed him in the direction of the Americans, hoping to get clues to the actual state of affairs. Cook, however, was diverted by some profoundly silly stories about how the Germans in the Nazi era had developed "antigravity" devices.[4] The American end of things had to await this diversion. But eventually Cook did go to the United States and interview those involved with the ZPF conjecture.

Cook recounted all this in his book, *The Hunt for Zero Point*. When Cook got his hands on the Boeing internal proposal, he couldn't resist publishing a report in Janes Defence Weekly about it. Denials ensued.

If Boeing did get involved in a revolutionary propulsion program, they've been much more circumspect about what they have been up to. If Millennium Projects continued more than a few years beyond the turn of the millennium, Lockheed-Martin, too, has been circumspect.

As these developments were taking place, in the background, Robert Bigelow, motel magnate extraordinaire, set up the National Institute for Discovery Science (NIDS) in Las Vegas. And not very long thereafter, he set up Bigelow Aerospace to develop commercial hotel facilities in space using inflatable satellites licensed from NASA. The chief activities of the scientists hired by Bigelow for NIDS seem to have been the investigation of weird happenings on a remote ranch purchased by Bigelow in Utah. This had no impact on advanced propulsion. But it did keep a few in the field employed doing interesting things.

About 5 years after setting up NIDS, Bigelow lost interest in the weird stuff and let the staff of NIDS go. One of the scientists, Eric Davis, got contract work from Franklin Mead (Senior Scientist in the Advanced Concepts and Enigmatic Sciences group at Edwards Air Force Base) to do a study of "teleportation." Star Trek-type teleportation is fiction, of course. But the same end can be achieved with absurdly benign wormholes,[5] so Davis's study became one of wormhole physics. When it was released (as it was judged by the authorities to be unclassified), much opprobrium was heaped on Mead, for having granted the contract, and Davis, for wasting tax dollars on such an irrelevant topic. Davis, not long after this affair, found a permanent position at the Institute for Advanced Studies at Austin,

---

[4] Traced to its origin, this tale started with one Victor Shauberger. Schauberger, forester (not physicist), observed the behavior of trout trying to reach their spawning grounds, leaping tall waterfalls in a single bound. Convinced that it was impossible for trout to do this unassisted, Schauberger concluded that they must be getting an anti-gravity boost in the waters below the falls. What does one find in the waters below falls? Vorticular eddies of course! Obviously, water vortexes cause anti-gravity! So, all one needs to do to create anti-gravity is build "trout turbines" that create water vortexes. Allegedly, Nazi "Aryan" physicists actually built such nonsensical stuff. I have it from a reliable source that he actually once saw a piece of one of these trout turbines. Max Planck, according to Cook, was invited to an audience with Hitler where Schauberger presented his "research" on trout turbines. Planck, according to Cook, unceremoniously walked out as soon as he got the drift of Schauberger's claims. Belaboring the obvious, let us note that Planck was no fool.

[5] Should you want to get from your ship in orbit around a planet to the surface, you merely open an absurdly benign wormhole from the ship to the surface and step through. No need for "Heisenberg compensators," "pattern buffers," and such other improbable rubbish.

the operation set up by Hal Puthoff some years earlier to investigate speculative physics claims.[6]

As far as the BPP and NIAC programs were concerned, neither survived long after the departure of Goldin when the Bush Administration took over in 2000. Goldin was succeeded by Sean O'Keefe. When Michael Griffin took over from O'Keefe as Administrator some years ago, he systematically shut down essentially all of NASA's research efforts, redirecting the 2.5 billion dollars a year thus captured to the Constellation program, a traditional chemical rocket project aimed at a return to the Moon. Griffin is said to have publicly referred to this reallocation of resources as, "draining the swamp." This characterization seems a bit odd, as the Breakthrough Propulsion Physics Program and NIAC were already long gone.

The NASA Administrator is a political appointment. When the Bush Administration was superseded by the Obama Administration, Griffin was replaced by Charles Bolden. Early in the presidential campaign, candidate Obama had remarked that perhaps NASA should be shut down and the resources freed up be redirected to education and other programs. Fifteen gigabucks or so a year, a small part of the federal budget, after all, is not small change. And save for the Hubble Telescope and Mars rovers, much of NASA's activities had the odor of jobs programs – as a friend in grad school days once remarked, NASA and the aerospace community were "welfare for the overeducated."

Faced with the reality of governing, the Obama Administration's problem was how to redirect NASA to more productive goals than doing, in Griffin's turn of phrase, "Apollo on steroids." The answer – the Augustine Commission, a panel of people with expertise in the aerospace sector, chaired by Norman Augustine (former CEO of Lockheed-Martin), charged with assessing the role of humans in space exploration.

Pending the report of the Augustine Commission, the Obama Administration proposed to restore some of the money Griffin had stripped out of NASA's research activities, and the cancellation of the Constellation program could pay for this. Powerful political lobbies were implacably opposed, including some former astronauts. But the Obama Administration tried to stick to its guns. The "compromise" eventually reached upheld the cancellation of Constellation, but provided for continuation of work on the Orion crew capsule (now already having cost 9 gigabucks and counting) and 35 gigabucks over several years to recreate a heavy lift vehicle of the Saturn V type based on the space shuttle main engines. That is, Constellation in almost all but name for all practical purposes on a different time

---

[6] This should not be considered a comprehensive review of advanced propulsion in the period since 1988. I have only included material that relates directly to wormhole physics, and even then, not the literature that can be easily found in traditional searches of scholarly sources on the subject (for example, Matt Visser's *Lorentzian Wormholes*). And, for example, the whole subject of superconductivity in advanced propulsion, starting with the alleged work of Podkletnov, followed by that of Torr and Li, and then Tajmar, et al. is left untouched. John Brandenburg's ideas on "vortex" fields are likewise not addressed, as are the conjectures based on "extended Heim theory." Robert Baker's schemes involving "high frequency gravitational waves" are also not included. From the institutional perspective, I have ignored ISSO, an operation set up by Joe Firmage to further advanced propulsion. The ISSO people were chiefly supporters of zero point schemes, but funded other work like that of James Corum and people at Science Applications Research Associates (SARA). Corum's work eventually found funding through earmarks channeled through the BPP project. Doubtless, there are other conjectures that have been advanced with which I am not familiar.

schedule. A bit more than half of the research money Griffin stripped out for Constellation, however, has been restored to research activities. And NIAC in a new guise has been resuscitated. This time, though, it's being administered out of NASA headquarters.

How did revolutionary propulsion figure into all of this? It didn't. All of the happy talk about advanced propulsion and wormholes on those TV shows (the best of which is, arguably, "Through the Wormhole," hosted by Morgan Freeman) had essentially no impact on NASA at all. You may be wondering, then, why the need for this book?

Well, advanced propulsion didn't go away just because it found no favor in the upper reaches of the government bureaucracy. And those dedicated to advancing advanced propulsion soldiered on. For example, Marc Millis and Eric Davis convinced the editor of an AIAA series on propulsion to publish *Frontiers of Propulsion Science*, a book they edited together. And Tony Robertson and Paul Murad convinced Mohamed el Genk, organizer of a major annual conference called Space Technology Applications International Forum (STAIF), held in early spring in Albuquerque, to sponsor a sub-conference on "New Frontiers." When el Genk retired in 2008, STAIF ceased to exist. So Robertson, largely singlehandedly, created a new conference, called Space, Propulsion, and Energy Sciences International Forum (SPESIF). Recently, Paul Murad and Frank Mead have tried to revive STAIF, held in Albuquerque – without, however, the anti-nuclear pickets that adorned the earlier conferences.

The publication of books and organization of conferences, however, were not the chief motivation for the renewed interest in revolutionary propulsion. After all, routine professional activity rarely produces popular buzz. Buzz is created by charismatic individuals with a message. Who might that be? More than anyone else, Greg Meholic.

Most of the people who try to do revolutionary propulsion in a serious way are so concerned about looking like flakes in public that they studiously eschew any behavior that might suggest that they really think that this stuff can actually be done. They affect a demeanor of hopeful skepticism and are non-committal when asked if anything looks like it might work. Greg, who worked his way up in the aerospace business doing conventional propulsion, has long believed that revolutionary propulsion is the only long-term solution to the problems of space exploration. So, several years ago, he put together a presentation on revolutionary propulsion and started giving it to anyone who was interested.

His inaugural presentation was to the Los Angeles section of the American Institute of Aeronautics and Astronautics. When it was announced, the response was strong enough to require doubling of the usual dinner/meeting facilities. The doubled facilities filled up. A famous astronaut who normally does not attend these sorts of meetings, scheduled to present some awards at the meeting, stuck around for Greg's talk. At an hour and 20 min, the talk was hardly the usual after dinner fare. No one got up and walked out. The response seemed to have been skeptically hopeful. Greg was invited to give the talk in other venues. Seems there is an audience of tech types interested in hearing that it's about time we got serious about advanced propulsion.

It is easy to speak out in support of a human space exploration program. As long as this is understood to be a program limited to the exploration of the Solar System, while such a program might be very expensive, it is a plausible proposal, for we know how to build craft capable of reaching, say, Mars. When the destination is the stars, this is no longer so. We do not know how to build craft capable of reaching even the nearest stars in less than a human lifetime, much less in a few hours, days, weeks, or years. We do not yet know how to build the absurdly benign wormhole generators or warp drives that would enable such travel. Some seem to be convinced that the entire energy output of the Sun for several years would be required to power a wormhole generator. Others are convinced that it is impossible to build stable wormholes in any circumstances. But we do know that that is what must be done.

Before Thorne and his graduate students specified the solution to the problem that must obtain, not even that was known. Given the wormhole specifications generated by Thorne, et al. in the 1980s, in the remainder of this chapter we look at some of the suggestions that have been made regarding how one might go about making absurdly benign wormholes. Or wormholes of any sort.

ABSURDLY BENIGN WORMHOLES

When Morris and Thorne wrote their paper on wormholes and interstellar travel, they presented it as a heuristic for teaching GRT. It was more than that, of course. Nonetheless, they relegated their discussion of absurdly benign wormholes to an appendix; and their remarks on this special class of wormholes were quite brief. So brief, in fact, that they are easily reproduced here[7]:

---

[7] The symbols used in the following excerpt can be identified with the help of Fig. 9.1.

If we allow ourselves to use matter with negative energy density as measured by static observers, $\rho c^2 < 0$, we can confine the exotic matter to an arbitrarily small throat region and thereby obtain an absurdly benign wormhole. An example is

$$b(r) = b_0[1 - (r - b_0)/a_0]^2, \quad \Phi(r) = 0$$
$$\text{for } b_0 \leqslant r \leqslant b_0 + a_0, \tag{A28a}$$
$$b = \Phi = 0 \text{ for } r \geqslant b_0 + a_0. \tag{A28b}$$

We may use the Einstein equations (17)–(19) to tell us what kind of material would be necessary to produce this wormhole: At $b_0 < r < b_0 + a_0$ the material must have

$$\rho(r) = [(-b_0/a_0)/(4\pi Gc^{-2}r^2)][1 - (r - b_0)/a_0] < 0, \tag{A28c}$$

$$\tau(r) = b_0[1 - (r - b_0)/a_0]^2/(8\pi Gc^{-4}r^3), \tag{A28d}$$

$$p(r) = \tfrac{1}{2}[\tau(r) - \rho(r)c^2], \tag{A28e}$$

while at $r \geqslant b_0 + a_0$ spacetime is flat [Eq. (A28b)] and empty, $\rho = \tau = p = 0$. Because $\Phi = 0$ everywhere, if a traveler moves through the wormhole at constant speed v, accelerative forces are nonexistent and tidal forces are bearable so long as the motional constraint of Eq. (50) is satisfied:

$$\left| \frac{\gamma^2}{2r^2} \left(\frac{v}{c}\right)^2 \left(\frac{b'r - b}{r}\right) \right| \lesssim \frac{1}{(10^8 \text{ m})^2}. \tag{A29}$$

This reduces, by virtue of Eqs. (A28), to

$$(v/c)^2 \lesssim a_0 b_0/(10^8 \text{ m})^2 \text{ at } b_0 \leqslant r \leqslant b_0 + a_0. \tag{A30}$$

The total traversal time, so long as $v/c \ll 1$, is

$$\Delta\tau_T \simeq \Delta t \simeq \pi a/v \gtrsim 1 \text{ sec } \sqrt{a_0/b_0}. \tag{A31}$$

Whatever may be the wormhole's circumference $2\pi b_0$, by choosing a_0 arbitrarily small we confine the exotic matter to a region of arbitrarily small thickness $\Delta l = \pi a_0$ and volume $4\pi^2 b_0^2 a_0$, and we ensure that it can be traversed with comfort arbitrarily quickly.

Reprinted from Michael Morris and Kip S. Thorne, *American Journal of Physics*, vol. 56, pp. 395–412, (1988). © 1988, American Association of Physics Teachers.

As we've noted from the outset, there are two problems with this class of solutions to Einstein's field equations. One is the negativity of the rest mass of the matter used to support the throat against collapse. The other is the amount of exotic material required, that is, a Jupiter mass in a structure of small dimensions. Those are the chief problems. But they are not the only problems.

Even if we assume that we can lay our hands on the requisite amount of exotic matter and confine it in a structure of suitable dimensions, there is the problem of how the induced wormhole forms. In general terms, there are two possibilities. One is that wormhole induction by the exotic matter causes a tear in spacetime before the forming wormhole reconnects (God knows where!) with some distant location in spacetime. In this scenario, the tearing of spacetime occurs because the topology of the spacetime is changed by the wormhole. The other is that wormhole induction produces a smooth deformation of spacetime, so no topology change accompanies wormhole formation.[8]

Both of these scenarios are problematical. In the first, when the tear occurs, a singularity is formed at the edge of the tear. The laws of physics break down at singularities, so we have no way to investigate, theoretically, how this process works in any detail. We might assert, as did Thorne many years ago, that the classical singularity is really governed by the laws of quantum mechanics. So this process, presumably, is governed by the laws of quantum gravity.

The problem is that no widely accepted theory of quantum gravity exists. Super stringers, nonetheless, are convinced that their theory encompasses GRT as a "low energy, effective" theory of gravity. But superstring theory is not background independent. It models gravity as a process mediated by the transfer of gravitons in a flat background spacetime. Should this be true, it seems unlikely that any process can produce the change in the spacetime structure required to create a wormhole. GRT, however, is background independent – so making wormholes is in principle possible. But if that's right, evidently, superstring theory cannot be correct, at least as regards gravity.

The other candidate theory for a quantum theory of gravity is "loop quantum gravity." In this theory spacetime itself is quantized. So, when topology change is induced by producing a tear in spacetime, while quanta of spacetime are exposed at the edge of the tear, no singularity appears as the spacetime quanta present at the edge are finite. How this affects the process of reconnection in topology change, however, is not presently known. In the case of smooth deformation of spacetime in wormhole formation, we are faced with the problem of Hawking's chronology protection mechanism. As Thorne pointed out long ago, though, the flaring of wormhole throats produces a defocusing of any path through the wormhole, so even if a Closed Timelike Curve forms as a smooth deformation wormhole with time shifted mouths is created, Hawking's mechanism may not cause the destruction of the wormhole.

---

[8] A commonplace illustration of the topology involved in these scenarios is found in making a cup or mug from a lump of clay. We can smoothly deform the lump to create the cup without punching any holes in the material, no matter what the size or shape of the cup. No topology change has taken place. When we put a handle on the cup, as long as we make no holes in the clay, the topology is unaffected. If we want a traditional handle on the cup, however, we must make a hole in the lump. Now the material that makes up the cup is no longer connected in the way it was before the handle was added. By punching the hole in the clay to make the handle, we have changed the topology of the material.

These considerations are at best "academic" in the absence of any prospect of actually building a wormhole generator. Should that prospect be realized, however, likely experiment will guide theory and vice versa as devices are developed. Until experiment and theory reach that point, though, there are some observations that have been, and can be made about wormhole physics.

THE VACUUM

The vacuum – the quantum vacuum in particular – has been associated with traversable wormholes from the publication of Morris and Thorne's paper. Before their paper was published, it was widely assumed that the "positive energy theorem" was true. This "theorem" was the bald assertion that negative energy was physically impossible.

Several ways of looking at this were in common currency. One was that the idea that negative energy was analogous to asserting that temperatures less than absolute zero were possible. That is, that states of less than no motion were possible – clearly an impossibility. A more subtle argument was that were negative energies possible, since all systems left to their own devices seek the state of minimum energy, they would be unable to reach a state of stable equilibrium, as there would always be a lower energy state available.

This problem had already been encountered in Dirac's electron theory, which has negative energy solutions. Dirac dealt with this problem by proposing the vacuum to be a "negative energy electron sea," a population of negative energy electrons that supposedly filled all of the negative energy states in the vacuum. Positrons, the anti-particles of electrons, in this view are "holes" in the negative energy electron sea created when one of the negative energy electrons is boosted into a positive energy state, creating (transiently) an electron-positron pair. The positrons predicted by Dirac's theory were detected early in the 1930s, but quickly came to be regarded as positive energy particles in their own right, rather than holes in a negative energy electron sea.

Dirac's theory is a particular example of the problems associated with negative energy called the issue of being "bounded from below." As long as negative energy states are excluded, there is no problem of being bounded from below, because there is always a well-defined minimum energy – zero. Or, in the case of quantum mechanics, the "zero point state" that is always a state, small one hopes for the vacuum, of positive energy. If negative energies are allowed, there is no lower bound, and if you assume that systems seek the lowest possible energy, serious problems ensue, as there is nothing to stop a system from seeking an infinitely negative energy state.

This problem has been in the background of gravity for a very long time. The reason why is that gravitational potential energy is negative, owing to the force of Newtonian gravity being attractive. You have to do work to separate two gravitating objects in close proximity, which means that the gravitational interaction has conferred negative energy on the objects when they are in a bound system. When James Clerk Maxwell, inventor of electrodynamics, tried to extend his formalism to encompass gravity, this was the problem that stymied him. The way the negativity of gravitational energy is dealt with is to use what later came to be called "gauge invariance."

In electrodynamics it is well-known that the scalar electric potential can be globally rescaled by an additive constant without changing any of the electric fields or their interactions (hence "invariance" because the rescaling changes nothing of physical signif-icance). Why? Because electric fields, the real manifestations of electromagnetism, are not the potentials themselves. They are gradients of the potentials. And a constant added to the potential, if globally applied, disappears in the gradient operation since the derivative of a constant is zero. Maxwell, of course, knew this. But he found it deeply implausible that a trick such as this had to be applied to make the negative gravitational potentials always manifest themselves in terms of positive energies – especially since the additive constant would have to be infinity since negative energies are not bounded from below. Indeed, this led him to abandon his efforts to write a field theory of gravity like his theory of electrodynamics.

When Thorne found that in order to stabilize a wormhole throat negative energy would be required, knowing the standard story about the positive energy theorem and the alleged impossibility of negative energy, he asked himself if all that were really true? If negative energy really is impossible, then there will never be any wormholes or warp drives. Classical systems, gravity notwithstanding, did not seem promising candidates for real negative energy, so Thorne looked to quantum mechanical systems. In quantum mechan-ics there is a well-defined state of minimum energy that can be taken as the zero energy state – the state of so-called zero point fluctuations. It follows that if any means can be found to reduce the energy in some region of spacetime below this state of zero point fluctuations, the energy density in that region would have to be genuinely negative. Is this possible? Yes, it is. Thorne appealed to an even then long known effect in quantum systems called the Casimir effect (after Hendrick Casimir, one of its discoverers in the late 1940s).

What Casimir noted was that should you take two plane parallel perfectly conducting plates separated by a small distance, you will find that they experience a force of "attraction." Why? Because perfectly conducting metal plates do not permit the presence of electric fields with components in the direction of their surfaces at their surfaces. Were such fields present, they would induce currents in the surface of the metal that would cause the redistribution of the electric charges present so as to cancel those fields.

This "boundary condition" on electric fields between the plates limits electric fields present between the plates to those that have zero components in the direction of the surfaces at the surfaces. Now, in the unconstrained vacuum presumably all "modes" of electromagnetic fields are present. That is, there are photons of all possible frequencies (and thus wavelengths) present, as there is nothing to preclude them from (transiently) flitting into existence from nothing spontaneously.[9] But between the plates this is not so because of the limitation imposed by the boundary condition at the plates' surfaces. The only photons that can exist between the plates are those with half wavelengths and their integral multiples equal to the distance between the plates, as shown in Fig. 6.1. All others are excluded.

Since most low energy, long wavelength photons are excluded, and they are the most common photons in the zero point fluctuation spectrum (as they, having low energy, last the longest), the result is that the total energy density between the plates is less than the

---

[9] This is explained in a little more detail in the context of the cosmological constant problem below.

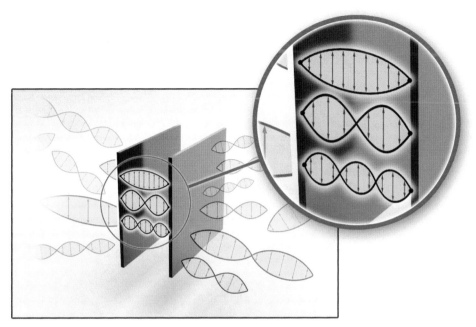

Fig. 6.1 An illustration of the Casimir effect. Outside the plates, photons of all wavelengths, frequencies and thus energies are possible. Between the plates, this is not so. Since some types of photons present outside the plates cannot exist between, the energy between the plates is lower than the energy outside. Since the outside energy density is, by definition, zero, between the plates it must be negative

energy density in the vacuum outside of the plates. The Casimir force is a consequence of the (virtual) photon pressure on the exterior of the plates not being balanced by an equal outward pressure due to the photons between the plates. More importantly for our purposes, as Thorne pointed out, the energy density between the plates being less than the unconstrained zero point energy means that the energy density between the plates is genuinely negative.

Nowadays, pretty much everyone allows that Thorne was right. When Thorne made this argument, he was not looking for a mechanism to make wormholes. Rather, he was chiefly interested in dispelling the notion that negative energy was physically impossible. The association between wormholes and the Casimir effect, however, has been perpetuated because it is very difficult to think of other ways to create negative energy that might be used to stabilize wormhole throats.

Indeed, even a cursory search of the web for wormholes leads one to several sources that claim that "zero point energy" or "zero point fields" are the path to making real wormholes. Some have even suggested that the problem of advanced propulsion is figuring out how to expand the distance between a pair of Casimir plates to macroscopic dimensions while maintaining the negative energy density of microscopic dimensions. Since the dependence of the Casimir effect on plate separation is well-known to go as the inverse fourth power of the distance, such conjectures should not be taken seriously. But because of the problem of bounding from below that seems to bedevil classical systems, it seems that it may be that negative energy is only to be found in quantum systems where the bounding problem can be sidestepped.

Now, you may be thinking, "Ah ha! It may not make any sense to try to increase the size of Casimir cavities to get more negative energy. But what if we make a structure that is filled with zillions of Casimir cavities? Each cavity may not contain much negative vacuum energy, but when all of the zillions of cavities are summed...." The problem with this scheme is that the maximum negative energy you can create in each of the cavities is limited by the amount of energy in the zero point field at the suppressed frequencies. That energy is not infinite. Indeed, the distribution of the energy of virtual photons in the quantum vacuum is well-known.[10] The "spectral energy density" goes as the cube of the virtual photon frequency. It has to be this distribution. Why? Because that is the only distribution that is invariant under transformations from one inertial frame of reference to another moving with non-zero velocity with respect to the first frame of reference. If this Lorentz invariance is violated, then SRT is wrong, as it would always be possible to single out some preferred frame of reference. We know as a matter of fact that that is impossible.

In addition to the issue of the spectral energy density of vacuum fluctuations, there is the problem of their total energy density. It's called the "cosmological constant" problem. The problem is that if you do a straightforward calculation of how much energy should reside in the vacuum due to all of those fluctuating zero point fields, you get an idiotically high number for the energy density. *Decades* of orders of magnitude too high. Indeed, unless you introduce a "cutoff" that suppresses all photons with energies higher than some specified limit, the energy density of the vacuum turns out to be infinite.[11] The fact of experience is that the energy density of the vacuum is almost exactly zero. In general terms, this is an outstanding anomaly, and one presumes that it will eventually find some reasonable resolution.

For those advocating the electromagnetic zero point fluctuation origin of inertia proposed by Haisch, Rueda, and Puthoff in the mid-1990s, it was a more critical issue. The energy density of the vacuum had to be very much higher than nearly zero in order that there be sufficient electromagnetic energy in the vacuum to produce the inertial reaction forces putatively being explained. But at the same time, the energy in the vacuum

---

[10] The energies of these transient photons are governed by the Heisenberg uncertainty relationship for energy and time, which says that $\Delta E \times \Delta t = \hbar$, where $\hbar$ is Planck's constant divided by 2π, a very small number. From this it follows that low energy vacuum fluctuation photons last longer. This suggests that low energy photons are more prevalent than higher energy photons. Nonetheless, since the energy per photon scales with the frequency, more energy resides in the higher frequency photons. And the distribution of energy by frequency in the zero point field is dictated by the constraint of Lorentz invariance, which demands scaling with the cube of the frequency.

[11] How one chooses this cutoff is in some measure a matter of taste. The one used by Haisch, Rueda, and Puthoff initially was the "Planck frequency." Planck units are constructed with the constants of nature: the speed of light in vacuum, Planck's constant, the constant of gravitation, and so on. See below for more on Planck units. The Planck time turns out to be 10^{-43} s. The Planck frequency is, roughly, the inverse of the Planck time – a very large number, but not infinity.

could not act as predicted by standard theory, for the vacuum energy would then curl the universe up into a little ball. Our reality simply would not exist.

Initially, the advocates of zero point inertia suggested, invoking the gauge invariance argument mentioned above, that only gradients of the vacuum fields mattered; that the energy of the field could be stipulated as arbitrarily high without the usual physical consequences. When that failed to attract adherents, Haisch and Rueda proposed that the vacuum could act on other things, but not on itself. This proposal didn't attract adherents either. Energy, after all, is energy, and energy is a source of gravity. It has real consequences.

The zero point fluctuation explanation of inertia has other serious problems. For example, it predicts that the inertial masses of the proton and neutron should be very different. In fact they are almost exactly the same. These problems notwithstanding, it's worth noting that if the energy density of the vacuum is as high as the vacuum fluctuation aficionados want, then the Casimir cavity array approach to creating enormous amounts of exotic "matter" may have a future. Lots of vacuum energy is there to be suppressed by cavities of suitable geometry. The really neat thing about all this is that it can be tested by experiment. For example, you could make up the parts for an array of Casimir cavities – two sheets of metal and a sheet of material with an array of holes in it so that when sandwiched between the metal plates forms the array of cavities. You would then simply weigh the parts before and after assembly. If there's a lot of energy in the vacuum, the assembled parts should weigh more than they did before assembly.[12] But if we stick to the view that energy in the vacuum has the consequences dictated by standard theory, then the total energy density observed and the spectral distribution of that energy preclude making wormholes with arrays of Casimir cavities. The energy available to be suppressed would simply be too meager to have much effect.

WHAT'S OUT THERE IN THE VACUUM?

An even more fundamental problem attends the quantum vacuum than those we've considered so far. As Peter Milonni and others showed about 40 years ago, quantum electrodynamics does not – on the basis of the Casimir effect at least – demand that the quantum vacuum be filled with anything at all. That is, quantum electrodynamics can be consistently interpreted as without any zero point vacuum fluctuations of the electromagnetic field at all. In this view, the Casimir force is a result of direct interactions between the particles in the plates, so one can still claim that there is an effective negative energy density in the space separating the plates. However, there are no vacuum fluctuations *with independent degrees of freedom* between the plates. If this sounds suspiciously familiar, it should. It is just Wheeler-Feyman "action-at-a-distance" theory that we encountered in

---

[12] Since the energy in the cavities is negative, you might think the assembled part should weigh less than when they are unassembled. But the Equivalence Principle requires that the inertial mass of the negative energy be negative too. So when the Earth's gravity acts on it, it accelerates downward, just like positive mass and energy. By the way, this also suggests methods of energy generation. But that would draw us too far afield from our topic of interest.

Chap. 2 in our discussion of Mach's principle applied to quantum systems. Apart from the fact that this sort of interpretation of quantum systems is consistently possible, is there any reason why you might want to affect this position? Well, yes. It's that little issue of the "cosmological constant problem."

Vacuum fluctuations are spontaneous processes. That is, they are not "caused" by any external stimulus. The photons of the zero point fluctuation electromagnetic field simply pop up into existence, last a brief interval, and disappear back into nothingness. Photons of all possible energies participate in this process. Since the energies of photons are related to their frequencies and wavelengths, directly and inversely, respectively, this means that photons of all frequencies and wavelengths are present in the zero point electromagnetic field. But not in equal number and with infinitely high energy, for were that the case, the vacuum would have infinite energy density, and Einstein's field equation of GRT tells us that were that the case, the universe would be shriveled up into a singularity. And it would never have expanded into the world that we know and love.

SPACETIME FOAM

In the early days of traversable wormhole physics, some speculated that it might be possible to motor around in one's spacecraft and harvest such exotic matter as one might find lying around here and there. This approach presumes that there is exotic matter out there waiting to be harvested. This is an extremely dubious proposition. Were there naturally occurring exotic matter, and were it fairly common, one might reasonably expect that there should be some here on Earth. None has ever been reported.

You might think that exotic matter with negative mass, since it is repelled by the positive mass matter that makes up Earth and its environs, if ever present locally, would have been driven away, explaining why we don't see the stuff around us all the time. This is a common mistake. Negative mass matter, notwithstanding that it is repelled by the positive mass matter that makes up Earth, nonetheless moves *toward* Earth because its inertial mass, like its gravitational mass, is negative, too (as demanded by the Equivalence Principle). So its mechanical response to a force in some direction is to move in the direction opposite to that of the force.

Speaking in terms of Newtonian physics, when negative mass matter is pushed away from Earth by the repulsive force of its gravity, it responds by moving toward Earth. Richard Price wrote a very nice paper in the *American Journal of Physics* (Volume 61. pp. 216–217) on the behavior of negative mass matter back in 1993. Moreover, this approach is bedeviled by another problem. Even if you could harvest a Jupiter mass of exotic matter, you would have to find a way to compact it all into a structure of very modest dimensions. To characterize this problem as "challenging" is to slightly understate the case.

In light of the problems of the harvesting scenario, it seems as though the only realistic methods of making stargates are going to depend on finding a way to transform some modest amount of pre-existing stuff into a Jupiter mass of exotic matter in situ so that the compaction problem is averted as the exotic matter is "created" already compacted. This means that the exotic matter we seek to make our stargate must, in some sense, already be present in latent form in the world as we find it. And we must find a way to expose the already present exotic mass. We set aside the proposition of creation of exotic matter *ex nihilo,* as that is an egregious violation of the principle of local momenergy (momentum plus energy) conservation. We've already seen that the quantum vacuum of standard relativistic quantum field theory holds out no hope of the creation of the gargantuan amount of exotic matter we require. So, the question would seem to be: Is there another view of matter and the vacuum that holds out any hopeful prospect?

As it turns out, there is another view of reality that may be relevant here. It's casually referred to as "spacetime foam." It's a conflation of two ideas that have been around for many years, both of which we've encountered repeatedly so far. The first idea is that there are fundamental limits to the stuff that makes up physical reality. Back around the turn of the twentieth century, several people noted that the constants of nature can be combined in various ways to give back numbers with the dimensions of mass, length, and time in particular, and other dimensions of interest, too.[13] The person who most assiduously advocated this idea was Max Planck, founder of quantum theory in his work on blackbody radiation. Planck's fundamental relations are:

$$l_P = \left(\frac{G\hbar}{c^3}\right)^{1/2} \approx 10^{-33} \, cm \tag{6.1}$$

$$t_P = \left(\frac{G\hbar}{c^5}\right)^{1/2} \approx 5 \times 10^{-44} \, s \tag{6.2}$$

$$m_P = \left(\frac{c\,\hbar}{G}\right)^{1/2} \approx 10^{-5} \, gm \tag{6.3}$$

Taking the scale of quantum theory to be that of the microscopic interactions of matter and radiation, it is obvious that the Planck scale is of a very different order. The masses of elementary particles are decades of orders of magnitude smaller than the Planck mass. And the distance and time scales of quantum theory are decades of orders of magnitude larger than the Planck distance and time. Evidently, the Planck scale has little or nothing to do with everyday quantum mechanics. What does it have to do with? Well, maybe the putative theory of quantum gravity, as mentioned above in our discussion of topology change.

The second idea involved in spacetime foam is the wormhole notion. The first exact solution of Einstein's field equations – that for a spherically symmetric massive object – was found almost immediately by Karl Schwarzschild. When the object is assumed to contract under the action of its gravity, as the collapse proceeds a wormhole is transiently formed. The wormhole doesn't last long enough for anything interesting to happen.

---

[13] This idea seems to have first occurred to the Irish physicist G. Johnstone Stoney some 30 plus years before Planck. Since he didn't have Planck's constant to work with, he used the atom of electricity – the electric charge of the ionized hydrogen atom and, with a minus sign, the electron when it was later discovered. As a result, his "natural" units differ from Planck's by a factor of the square root of the "fine structure constant" $[\alpha = e^2/\hbar c]$. The fine structure constant is a dimensionless number roughly equal to 1/137.

Fig. 6.2 A hyperspace diagram of an Einstein-Rosen bridge

Einstein was convinced that singularities were nonsense, so in the mid-1930s, with his collaborator Nathan Rosen, he proposed to deal with two problems at once. The first problem was singularities, and the second was the infinite self-energy of electrons when they are assumed to go to zero radius. They proposed that electrons should be viewed as the patching together of two wormholes, creating a so-called "Einstein-Rosen bridge." A hyperspace representation is displayed in Fig. 6.2. The neat thing about these structures is that if you assume that the electric field threading the wormhole is responsible for keeping the wormhole open, you can get rid of the notion of electric charge. As John Wheeler would later say, "charge without charge."

By the mid-1930s it was clear that reality was more complicated than could be accounted for with the simple Einstein-Rosen bridge model of the electron, and physicists generally ignored the model. Even then, Einstein was largely marginalized because of his unwillingness to hew to the orthodox view of quantum mechanics advocated by Niels Bohr and his followers. But in the 1950s, John Wheeler resuscitated the wormhole model of electrons as part of his theory of "geometrodynamics." Wheeler put together Planck's ideas on fundamental properties of reality captured in Planck units with GRT and the wormholes of Einstein and Rosen to assert that at the Planck scale spacetime consists of a foam of transient wormholes flitting into and out of existence as quantum fluctuations at the scale of quantum gravity. Given Wheeler's stature and penchant for catchy phrases and

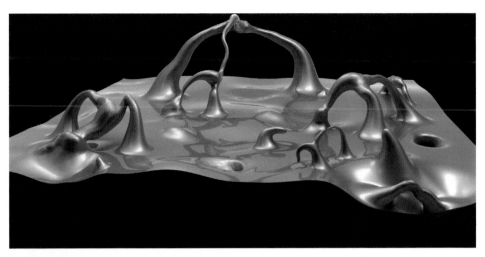

Fig. 6.3 A time slice hyperspatial depiction of the hypothetical "foam" structure of spacetime at the scale of the Planck length – 10^{-33} cm. All of this structure fluctuates wildly as time goes on

ideas, spacetime foam created a fair amount of buzz in the early 1960s. A hyperspatial snapshot of what the foam might look like is shown in Fig. 6.3. But it didn't lead to the theory of quantum gravity even then long sought.

Back in the early days of traversable wormhole physics, faced with the problem of dealing with quantum gravity for topology changing wormhole formation, Thorne chose to invoke the spacetime foam of geometrodynamics. Thorne's way to avoid topology change was to assert that the way to make a traversable wormhole was to extract a pre-existing wormhole from the hypothetical Planck-scale quantum spacetime "foam" where spacetime consists of a frothing sea of microscopic wormholes that flit into and out of existence in conformity with Heisenberg's energy-time Uncertainty relationship. The transient energy conservation violation enabled by the Uncertainty relationship at the Planck scale turns out to be surprisingly large – about 10^{-5} g – the Planck mass mentioned above. By unspecified means, a microscopic wormhole is "amplified" to macroscopic dimensions. Should you want to make a time machine, one of the 2 mouths of the wormhole can then be taken on a "twins paradox" trip at high speed relative to the other mouth. Alternatively, one of the mouths can be placed near the event horizon of a black hole, where time slows markedly, for some reasonable duration. This produces a time shift between the mouths as the traveling or horizon-proximate mouth ages more slowly than the non-traveling or normal space mouth.

At the most profound level, Thorne's proposal suffers from a fundamental problem: there is no generally accepted background independent quantum theory of gravity and there is no evidence whatsoever that spacetime at the Planck scale consists of a foam of transient microscopic wormholes. (Experiments, however, intended to detect predicted effects of spacetime foam are presently being contemplated, and one is being executed at Fermilab using matched Michelson interferometers. See: Michael Moyer, "Is Space Digital?", *Scientific American,* **306**, no. 2, pp. 31–37 [2012].) Appealing to spacetime foam is a way of bringing in quantum gravity without burdening oneself with such a theory whose formalism is so conspicuously absent. Without that quantum theory of gravity and evidence for spacetime foam, there is no reason to believe that this proposal has any chance of working.

Nonetheless, the belief that a quantum theory of gravity will eventually be created, and that it will justify belief in the existence of spacetime foam, is widespread. So we ask, if spacetime foam does exist and consists of microscopic transient wormholes, is there any prospect of amplifying one into a stargate? And if so, how might this be done?

The customary way of making things larger is to blow them up. This is usually done by adding energy, sometimes explosively, to the object to be enlarged. It is worth noting that this process works for blowing things up *in* spacetime. Whether it will work for blowing up spacetime itself is another matter. The size of the wormholes of the putative quantum spacetime foam presumably are about the Planck length large – that is, about 10^{-33} cm across. This is about 20 orders of magnitude smaller than the classical electron radius and 18 orders of magnitude smaller than the diameter of nuclei. How do you blow up something so fantastically small? By smoking everything in its environs along with it. The two tools at our disposal to attempt this are the Relativistic Heavy Ion Collider (RHIC) and the Large Hadron Collider (LHC). Taking these devices (optimistically) to put about 10 TeV per interaction event at our disposal, we find an equivalent mass that gives us about 10^{-20} g. This is roughly 15 orders of magnitude less than the mass of our transient wormholes to be ablated – hardly enough to perturb a wormhole, much less blow it up. Moreover, the transient wormholes exist for only 10^{-43} s, making the temporal interaction cross-section hopelessly small. So, given present and foreseeable technology, this scheme is arguably impossible. It seems most doubtful that anyone who has success-fully made traversable wormholes would have pursued this avenue to stargates.

If amplifying quantum spacetime foam is an impossible scheme, at least in the absence of a quantum theory of gravity, are all possible microscopic wormhole schemes irrefragably flawed? Not necessarily. Recall, Wheeler's motivation for introducing the concept of microscopic wormholes was not to create the concept of spacetime foam. It was to make the structure of electrons and other electrically charged elementary particles into wormholes threaded by self-repulsive electric fields to stabilize them so that electric charge could be eliminated as a fundamental physical entity. Where foam wormholes only exist fleetingly, electrons and quarks exist forever.

So it would seem that instead of wasting our time on some scheme to amplify the vacuum that we assume to have some structure that has never been detected, we should focus our attention on electrons instead. They, at least, have been detected. From the energetic point of view, this course of action seems much more promising, as the mass of the electron is 10^{-27} g, some seven orders of magnitude *smaller* than the interaction energy available at the RHIC and the LHC. Of course, amplification by ablation of electrons or other elementary particles into mesoscopic wormholes has never been observed at either the RHIC or the LHC, or at any other accelerator for that matter, so evidently if electrons are to be used to make wormholes, some more subtle process than simply blowing them up by slamming other stuff into them will be required. Absent an explicit scheme for ablating electrons into wormholes and given the non-observation of wormholes in the collision debris at accelerators, however, we set this speculation aside.

SOFTENING SPACETIME

If the solution to the problem of making traversable, absurdly benign wormholes is not to be found in the quantum vacuum or the spacetime foam of putative quantum gravity, is a solution to be found at all? The nature of the problem is easy to identify in terms of Einstein's GRT field equations. In simple terms, those field equations say:

$$\textbf{Geometry} = \textbf{coupling constant} \times \textbf{``matter sources''} \qquad (6.4)$$

The source of the problem is the "coupling constant." The coupling constant has a well-known value: $8\pi G/c^4$. G in cgs units is 6.67×10^{-8}, and c is 3×10^{10} cm/s; so the value of the coupling constant turns out to be 2×10^{-48}, an exceedingly small number. Seemingly, given the value of the coupling constant, the only way to produce the geometrical distortions needed for a traversable wormhole is to assemble gargantuan sources – the Jupiter mass of exotic matter that we've referred to again and again in this discussion. Another approach, however, has recently been suggested by Jack Sarfatti (a capable, colorful physicist from the San Francisco area). His suggestion is that instead of focusing on the sources needed to compensate for a very small coupling constant, we should concentrate on the coupling constant itself.

Coupling constants are an essential part of all physical theories. They are the things that transform both the magnitude and dimensions of one type of thing into another to which it is causally related. Coupling constants are normally just that: constants. That is, they have the same numerical value for all observers. In technospeak, they are Lorentz scalars, or scalar invariants. But coupling *coefficients* need not necessarily be constants. For example, the energy density E of the electromagnetic field is given by:

$$E = \frac{1}{8\pi}(\textbf{D} \bullet \textbf{E} + \textbf{B} \bullet \textbf{H}), \qquad (6.5)$$

where **E** and **H** are the electric and magnetic field strengths respectively, and **D** and **B** are the electric displacement and magnetic flux respectively. These quantities obey the so-called "constitutive" relations:

$$\textbf{D} = \varepsilon\textbf{E}, \qquad (6.6)$$

$$\textbf{B} = \mu\textbf{H}, \qquad (6.7)$$

where ε and μ are the dielectric permittivity and magnetic permeability respectively.[14] The permittivity and permeability, in addition to characterizing the media, are, of course, coupling coefficients.[15] But they are not constants, as they depend on the properties of any material media that may be present where one is computing the energy density of the

---

[14] In all but the simplest media, the permittivity and permeability are tensors and everything is much more complicated. But simple situations suffice to make the point of this argument.

[15] When no material medium is present, these coefficients take on their values for free space, and those are constants.

electromagnetic field. So the equation for the energy density in the electromagnetic field becomes:

$$E = \frac{1}{8\pi}\left(\varepsilon\mathbf{E}^2 + \mu\mathbf{H}^2\right),\tag{6.8}$$

and we have coupling coefficients that are not constants.

Now, Sarfatti notes that the speed of light is different in material media than in a vacuum. Usually, this is put in terms of the index of refraction, n, where:

$$c = c_0/n,\tag{6.9}$$

Where c_0 is the vacuum speed of light and the index of refraction is related to the permittivity and permeability:

$$n = \sqrt{\varepsilon\mu}.\tag{6.10}$$

If we assume that the speed of light that appears in the coupling constant in Einstein's equations is the speed in media, rather than the vacuum value, the equations become:

$$\mathbf{G} = \frac{8\pi G n^4}{c_0^4}\,\mathbf{T},\tag{6.11}$$

And now we have a way to soften spacetime because if we choose a medium with a large index of refraction, we can make the coupling coefficient much larger than it would be were the speed of light in the coefficient that for a vacuum.

How big can n be? Well, normally n lies between 1, the value for the vacuum, and at most a few thousand, for materials with high dielectric constants. But with the advent of superconductors, and especially Bose-Einstein condensates, very much higher indexes of refraction have been achieved. The speed of light in these materials can be reduced to a few cm/s, and even stopped entirely. A speed of 3 cm/s corresponds to an index of refraction of 10^{10}. Substituted into Eq. 6.11, the coupling coefficient becomes 40 orders of magnitude larger than the customary coupling constant of GRT. If Sarfatti is right, Bose-Einstein condensate superconductors should dramatically soften the stiffness of spacetime. Is he right? Almost certainly not. But we can make Bose-Einstein condensate superconductors, so we can find out. Perhaps we will be fantastically lucky, and his conjecture will pan out.

You may be thinking: Well, even if Sarfatti is right about superconductors and coupling coefficients, it won't matter because the energy we need for wormholes and warp drives is negative, and superconductors don't change positive energy into negative energy. Is there anything that can pull off this trick? Depends on who you talk to. Back in the 1960s it occurred to some folks that it might be possible to have materials with negative indexes of refraction. They worked through the theory that such materials would have to have. The materials came to be called "metamaterials" (because they go "beyond" normal materials). Metamaterials have either ε or μ, or both, negative. They display the sort of behavior shown in Fig. 6.4. Their study has become very trendy in the last decade or so. If you look at Eq. 6.8 above, you'll see that if the negativity of either ε or μ, or both, is true,

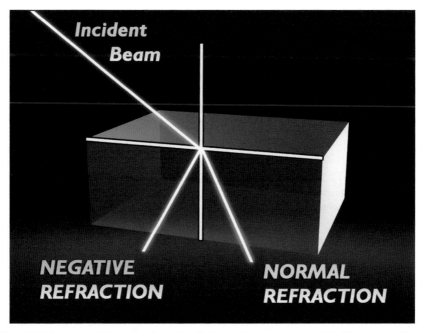

Fig. 6.4 Diagram of the refraction of a light ray by a metamaterial. Where in ordinary refraction the direction of the light ray in the material is bent toward the normal to the surface, in a metamaterial it is bent away from the normal, as shown here

it would appear that the energy density of the electromagnetic field can become negative. Since this possibility presented itself in the age of adherence to the positive energy theorem, theorists looked around for a way to explain why negative energy would not occur for electromagnetic waves in metamaterials.

The solution to the problem of negative energy in electromagnetic waves in metamaterials involves details of the propagation of those waves. Consider a well-known feature of real waves. Namely, real waves come in "packets" of finite length. A consequence of the finite length of real waves is that they do not have an exactly defined frequency (or wavelength). Indeed, the shorter the length of the wave train (the packet), the less well-defined its frequency (and wavelength). Actually, we can put this a bit more precisely by invoking the branch of mathematics invented by a French engineer of the early nineteenth century, Joseph Fourier.

Fourier, you might recall, showed that any periodic function – even if there is only one period to consider – can be represented as the sum of some suitable combination of simple (sine and cosine) periodic functions, the frequency of each of those functions being some multiple (that is, harmonic) of the fundamental frequency corresponding to the lowest period of the function. When we now consider a packet of electromagnetic waves, we can view the packet as a "group" of waves that can be decomposed into a Fourier series of pure waves with relative phases such that the pure waves just add up to the "group" wave. The pure waves of the decomposition are called "phase" waves. The interesting thing about this is that the phase waves that sum to the group wave need not travel at the same speed as the group wave. Indeed, the phase waves need not travel in the same direction as the group wave. And since the phase waves do not individually carry real energy or momentum, and are thus unable to transfer information from one place to another, they can travel at any velocity – including faster than the speed of light in a vacuum.

Normally, the phase waves that together make up the group wave of a packet travel in the same direction as the group wave. What the early investigators of negative indexes of refraction and metamaterials found was that they could account for the negativity of the permittivity and permeability of metamaterials not as indicating the presence of a negative energy density. Rather the negativity was taken to indicate that the phase wave propagation direction in a metamaterial reverses, so the phase waves move in the opposite direction to that of the group waves.

If this is universally true, Sarfatti's conjecture has no hope of working, for exoticity is required to stabilize wormhole throats. Sarfatti's way to deal with this problem is to assert that the metamaterial to be used in a device implementing his scheme must display negative properties at or near zero frequency. Why? Because if it does so, you can't use the phase wave explanation for the negativity, and it must represent a real negative energy density. Note that Eq. 6.8 applies for all frequencies, including zero. Note, too, that if this scheme is actually to work, you'd want a DC effect in any event, for you'd want to be able to just switch on a system and not have a lot of propagating radiation floating around in it. For these reasons, Sarfatti specifies that the superconducting material that slows the speed of light to a snail's pace also be a metamaterial at low or zero frequencies so that the negativity of the energy in any electromagnetic field in the material will necessarily be negative. Will this scheme actually work? Probably not. But it has the merit of being one that can be explored experimentally, so we can find out whether it will work.

WHERE DOES ALL THIS LEAVE US?

We seem to be faced with a difficult, perhaps insuperable, problem. The customary view of the quantum vacuum has it seething with energy. But that seething energy is inconsistent with what we know of gravity, for if it were real, spacetime as we know it wouldn't exist. And even were a way around this problem found, we'd be no better off, for the structures that it alleges to present are too hopelessly small and short-lived to be engineered into real, macroscopic devices. The conjecture that elementary particles might be miniature wormholes that can be ablated to macroscopic dimensions is likewise not promising, for in decades of running of high energy accelerators, ablation of an elementary particle into a mesoscopic wormhole has never been observed.

Coming at the problem by attacking the coupling constant in Einstein's equations does not look promising either. Nonetheless, the quantum vacuum and superconducting metamaterial proposals can be tested by experiments, distinguishing them from many of the conjectures and proposals in the advanced propulsion business. And if no other way can be found, they merit serious investigation.

Faced with this bleak outlook, we may ask: Is there any other scheme on offer – based on plausible physics – that provides a way to make stargates? Yes! We'll explore this scheme in the following chapters.

7

Where Do We Find Exotic Matter?

In the last chapter we've seen that Kip Thorne's work on wormholes was widely recognized as having a direct bearing on the problem of rapid spacetime transport. But the requirements he and his collaborators suggested were and remain so daunting that almost no one has paid much attention to them when trying to devise realistic advanced propulsion systems. The single scheme recently proposed that addresses the issues squarely, Jack Sarfatti's "softening" scheme, is based on physical assumptions that almost certainly are not correct. But his proposal is based on serious physics that can be tested in the laboratory. Should we be fabulously lucky, his proposal may turn out to have merit.

If the source material needed to create wormhole throats cannot be achieved by fudging with the coupling constant in Einstein's gravity field equations and making zero frequency metamaterials, then, by a simple process of elimination, the Jupiter mass of exotic matter we need must be found in the sources of the field *per se*. Since no exotic matter is lying around to be harvested (and compacted), if absurdly benign wormholes are to be made, then the normal everyday matter we find around us must contain exotic matter. Somehow it must be possible to transform normal matter into the gargantuan amount of negative mass matter required to make wormholes. And if this is to be technically realistic, it must be possible to effect this transformation using only "low" energy electromagnetic fields.

The idea that some modest amount of normal matter might be transformed into some idiotically large amount of exotic matter may seem hopelessly quixotic. Certainly, if the theory of the nature and structure of matter of mainstream physics – the Standard Model – is true, then this is arguably impossible. But, as Peter Milonni pointed out in the fall of 1992, it is worth noting that the presence of ridiculous amounts of exotic matter in normal matter is a feature, albeit rarely mentioned, of the Standard Model. The Standard Model is a generalization of the theory of Quantum Electrodynamics (QED), and negative mass is already built into that theory.

Recall, in general terms, the motivation for that theory and how it works. The problem that bedeviled quantum theory and classical electrodynamics, from at least the 1920s onward, was that of "divergences" – computed quantities in the theory that are infinite. (In GRT, the analogous things are "singularities.") Infinity may be an interesting concept to contemplate. But when you get infinity for, say, the mass of an elementary particle, you know something is wrong with your theory.

J.F. Woodward, *Making Starships and Stargates: The Science of Interstellar Transport and Absurdly Benign Wormholes*, Springer Praxis Books, DOI 10.1007/978-1-4614-5623-0_7,
© James F. Woodward 2013

THE MASS OF THE ELECTRON

The fundamental divergences in the classical and quantum mechanical theories of matter arise because of infinite self-energy. This occurs in even the simplest case, that of the electron. Until the mid-1920s, this was not a matter of critical concern. After the discovery of the electron (by Thomson in 1897) and relativity theory, Einstein's second law especially (in 1905), Lorentz quickly developed a theory of the electron. He envisaged it as a sphere (in a reference frame where it is at rest) of electrically charged dust.

Like electric charges, of course, repel each other. So the constituent particles of the dust that make up electrons should repel each other. Unless some other interaction takes place to balance the electrical repulsion of the dust for itself, electrons cannot exist. Lorentz simply assumed the existence of some mechanism that would hold the particles of dust together once assembled. After all, electrons are observed facts of reality. So something must hold them together. Henri Poincaré, who had also contributed to the development of relativity theory, lent his name to these presumed forces – so-called "Poincaré stresses."

Now, work must be done to assemble the particles of electrically charged dust into our spherical electron with some small radius. And Einstein's second law tells us that the energy expended in the assembly process, which can be regarded as being stored in the electric field of the assembled electron, will contribute to the mass of the electron. This mass is due to the "self-energy" of the charge distribution – the amount of work that must be done to assemble the charge distribution.

The electromagnetic contribution to the mass of the electron is easily calculated. That calculation shows that the self-energy depends on the radius of the charge distribution, indeed, as:

$$E = \frac{e^2}{r_0} \tag{7.1}$$

where E is the energy, e the electronic charge, and r_0 the radius of the charge distribution. Gaussian units, the traditional units of field theory of yesteryear, are used here.[1] You don't need to do the formal calculation to see that Eq. 7.1 is reasonable. The force between two charged particles is proportional to the product of the charges, and if the charges have the same magnitude, this will just be the square of either charge. So e^2 is to be expected. And the force between pairs of charged dust particles depends on the inverse square of the separation distance. So when these forces are integrated [summed] as the separation distance decreases, the result will be inverse first power – the distance dependence of Eq. 7.1.

---

[1] Nowadays it is fashionable to use either SI units or "natural" units. With natural units one or more of the constants of nature, like Planck's constant and/or the speed of light in a vacuum, are set equal to one. In the case of SI units, one carries around a lot of factors involving π and small integers, and the values of things scale to human size objects rather than the small constituents of real material structures. With natural units, the magnitude of things as we usually measure them get transformed into unintuitive values owing to the very large or small values of the constants of nature. Gaussian units avoid both of these issues.

Now, the radius of the electron, until it is actually measured, can be pretty much anything you want. Lorentz, motivated by the principle of simplicity, proposed that the radius of the electron should be exactly that which would make the mass of the electron entirely due to its electrical self-energy. That is, we set the rest energy mc^2 equal to the electrical self-energy:

$$mc^2 = \frac{e^2}{r_0}. \qquad (7.2)$$

Using Eq. (7.2) to calculate the radius r_0 turns out to be 2.82×10^{-13} cm and is called the "classical electron radius." Until the mid-1920s most everyone just assumed that electrons might really look like this. There was a potential problem, though. If you assumed that electrons were point particles, their self-energies would be infinite because the electron radius is zero, and zero divided into anything is infinity.[2] Infinite self-energy means infinite mass, and that's ridiculous. At the time there was no reason to believe that the radius was zero, so the physicists of the early 1920s were not concerned. But hints of trouble were on the horizon.

Niels Bohr laid out his quantum theory of the hydrogen atom in the early twentieth century, shortly after Ernest Rutherford discovered the nuclear structure of atoms. Bohr's insight was to see that electrons in stable orbits around atomic nuclei had to satisfy the condition that their orbital angular momenta were integral multiples of Planck's constant (which has dimensions of angular momentum). This condition leads to electron energies for the stable orbits with differences that correspond exactly to the energies of the photons of the spectrum of the hydrogen atom.[3]

A few years later, the "fine structure" of the hydrogen spectrum – the splitting of the "lines" of the spectrum into several closely spaced lines when a magnetic field is applied to the radiating atom – had been successfully explained by Arnold Sommerfeld by assuming that electrons in orbit experience relativistic mass increase due to their orbital motion and applying the Wilson-Sommerfeld quantization rules to elliptical as well as circular orbits. A common, dimensionless factor occurs in Sommerfeld's equations – the fine structure constant we have already encountered. In QED the fine structure constant turns out to be the coupling constant for the electromagnetic field and its sources. With a value of 1/137, the fine structure constant is decades of orders of magnitude larger than the coupling constant in Einstein's gravity field equations.

Problems with Bohr's theory arose in the early '1920s as the shell theory of atoms more complicated than hydrogen was developed. First it was found that half-integer quantum numbers had to be used to get theory and observation to coincide. And, in a turn of phrase by Wolfgang Pauli, a "two-valuedness" appeared. This was especially obvious in the

---

[2] Purists will point out that division by zero is "undefined." But for situations like these, when one lets the denominator go to zero, the quotient obviously goes to infinity as the radius goes to zero.

[3] Recall that the relationship between the wave properties of the electromagnetic field and photon energy is $E = h\nu$, where h is Planck's constant and ν is the frequency of the wave.

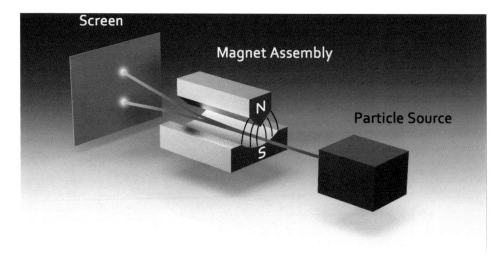

Fig. 7.1 A schematic diagram of the apparatus used by Otto Stern and Walther Gerlach to study silver atoms in a strong magnetic field with a non-vanishing gradient. The gradient was expected to smear out the beam of atoms at the detection screen. In fact, two distinct spots formed on the screen

results of the Stern-Gerlach experiment where a beam of electrically neutral silver atoms passed through an inhomogeneous magnetic field, as shown in Fig. 7.1.

Now, an electron circulating around a nucleus in orbit produces a small circular current, and circular currents create magnetic moments. The magnetic moment of a circular current will try to align itself with an externally applied magnetic field, as shown in Fig. 7.2. If the field is uniform, however, the force produced on the moment is a torque and will not make the moment align with the field. Rather, it will make the moment precess around the direction of the field. If the field has a gradient, though, the force on one of the equivalent poles of the dipole moment will be greater than the other, and the dipole moment will experience a net force in the direction of the gradient of the field.

The problem with the Stern-Gerlach experiment results was that the atoms had been chosen so that the expected total magnetic moment of the orbiting electrons was zero. So, if the only magnetic moments present were due to the orbital motion of the electrons in the atoms, the atoms in the beam should not have been deflected by the inhomogeneous magnetic field. But they were – into the two spots on the screen. Evidently, something in the atoms had a magnetic moment, and its interaction with the field was quantized with two allowed values.

The two-valuedness of the Stern-Gerlach results was explained in the mid-1920s by Samuel Goudsmit and George Uhlenbeck by assuming that electrons spin. Actually, electron spin had been proposed a bit earlier by Ralph Kronig. But when Kronig told Wolfgang Pauli of his idea, Pauli ridiculed it. Pauli pointed out that an electron with the classical electron radius must spin with a surface velocity roughly 100 times that of light in order to produce the measured magnetic moment. Goudsmit and Uhlenbeck learned of this

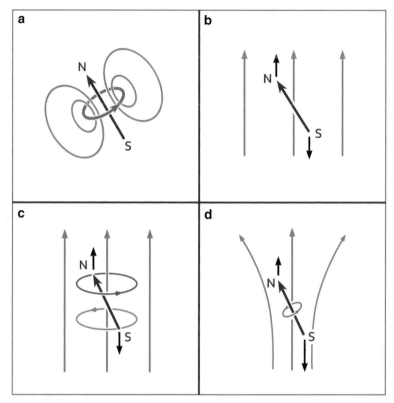

Fig. 7.2 A spinning electron, like a small electric current loop, produces a dipole magnetic field (*Panel A*). A uniform external magnetic field (*Panel B*), applies a torque to the dipole equivalent to the current loop. A simple dipole would align itself with the magnetic field. But since the current loop/spinning electron has angular momentum, its response to the torque is to precess in the direction of the magnetic field (*Panel C*). If the magnetic field has a gradient, the magnitude of the force on one of the poles of the dipole is greater than the force on the other pole, and the dipole moves toward the net force (*Panel D*)

defect of their idea from Lorentz, whom they asked to look at their work, but not until after Paul Ehrenfest, Uhlenbeck's graduate supervisor, submitted their paper on electron spin to be published. So electron spin made it into the professional literature notwithstanding that it was known to conflict with SRT.

Another problem of quantum mechanics in the mid-1920s was that the replacement theory for Bohr's "old" quantum theory, only then created by Schrodinger (wave mechanics) and Heisenberg (matrix mechanics), was not formulated in relativistically invariant form. Paul Dirac solved the problem of relativistic invariance and spin. Dirac's theory of electrons and their interaction with the electromagnetic field, however, treats the electron as a point particle. This is problematic for two reasons. First, if you are a literalist, it is difficult to see how a point particle can have spin that produces a magnetic moment. Magnetic moments are produced by circular currents, and point particles, with zero radius,

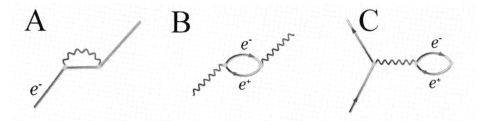

Fig. 7.3 The Feynman diagrams for the lowest order corrections for mass and charge renormalization (see Sakurai, *Advanced Quantum Mechanics*). *Wavy lines* represent photons, *non-wavy lines* electrons. The first diagram is for virtual photon emission and reabsorption by an electron, the chief contributor to the electron's self-energy. The second shows the spontaneous production of a virtual electron-positron pair by a photon. The third diagram is for the production of the virtual charges that dress the electron by polarization in charge renormalization

cannot be real circular currents. This was dealt with by identifying electron spin as an "internal" or "intrinsic" property.

A far more important and fundamental problem with the Dirac electron is that with zero radius, it has infinite self-energy, and thus infinite mass. Real electrons, however, obviously do not have infinite mass. Quantum theory made this "divergence" less severe than it was in the classical electron theory. But the divergence did not go away. Oh, and Dirac's equation for electrons has negative energy solutions.

Working through a solution to infinite electron self-energy took more than 20 years.[4] The procedure for dealing with divergences eventually constructed is "renormalization." It is predicated on the assumption that the observed mass and charge of the electron (and other fundamental particles) is not the same as their "bare" masses and charges. The bare mass and charge of an electron are "dressed" by adding the energy (and charge) of surrounding quantum fields to the bare value to get their observed values. Put in pictorial terms using Feynman "diagrams", this looks like Fig. 7.3. Electrons are constantly emitting and absorbing "virtual" photons that make up the electric field of the charge. These photons are "virtual" because they do not convey any energy or momentum to other charges. But they have energy – the energy stored in the electric field viewed classically.

The energies of virtual photons do not respect the usual energy conservation law. Rather, they are limited by the Heisenberg uncertainty relation involving energy and time. If the lifetime is very short, the energy of a virtual photon can be very high – higher indeed than the rest energies of the source electron. Virtual photons with energy greater than twice the rest energy of the electron can spontaneously decay into electron-positron pairs (that recombine before the virtual photon is reabsorbed by the electron that emitted it). These transient virtual electron-positron pairs, during their brief existence, are

---

[4] The story of these developments has been told many times in many ways. Of the accounts I have read, I like that of Robert Crease and Charles Mann, *The Second Creation* the best.

polarized by the electric field of the electron, and their polarization charge must be added to the bare charge of the electron to get the observed charge. Accounting for the effects of virtual photons and electron-positron pairs on the masses and charges of electrons is the renormalization program.

The first person to figure out this procedure seems to have been Ernst Stueckelberg in 1943. But the paper on it he submitted to the *Physical Review* was refereed by Gregor Wentzel, who turned it down. Stueckelberg did not pursue the matter. After World War II papers laying out the procedures of renormalization were published by Julian Schwinger, Sin-itiro Tomonaga, and Richard Feynman. Schwinger and Feynman's methods were shown to be mathematically equivalent by Freeman Dyson. Wentzel approached Stueckelberg with a proposal to revise and publish his paper to make an assertion of priority. Stueckelberg turned him down. Schwinger, Tomonaga, and Feynman eventually won the Nobel prize for this work.

Dirac never accepted renormalization as a truly legitimate mathematical procedure; and Feynman occasionally referred to it as a "dippy" process. But it worked. And when Gerard t'Hooft extended the method to fields other than electrodynamics in the early 1970s, the Standard Model was born. But in the late 1960s there was open talk of replacing QED with some other type of field theory that was not beset by the divergences dealt with by renormalization. String theory was invented at this time, one of the explicit motivations being that strings were not points and thus free of the divergences associated with point charges.

The mass contributed to the total mass of the electron by the cloud of virtual photons, by Einstein's second law, is positive. So the bare mass of the electron must be less than the observed mass. How much less? Well, if the virtual photons have energies greater than twice the electron mass so they can make the electron-positron pairs needed for charge renormalization, then the bare mass of the source electron *must* be negative. In principle, one can always keep things finite by introducing a "cut-off": the assertion that for whatever reason, energies larger than some specified value do not occur.

Popular cut-offs have been the energy equivalent of the mass of this or that particle, the energy dictated by the Heisenberg uncertainty relations to this or that length (for example, the Planck length), and so on. For the renormalization procedure to be convincing, though, it must return a finite mass for the electron (and other particles) even if the allowed virtual photon energy becomes infinite. And the procedure does this. So, formally, the bare mass of the electron is negative and infinite, as Milonni commented in a conversation with me back in 1992. And if the Standard Model has merit, we live in a sea of stuff with infinite negative mass disguised by clouds of virtual photons, electrons, and positrons.

You're probably thinking to yourself, "So what? Even if we could find a way to expose the bare masses of a bunch of electrons (and other particles), since it is infinite, we'd destroy the universe. And besides, it's preposterous to suggest that a way might be found to turn off the quantum fluctuation processes that generate the virtual photons of the electric field." Indeed. There is no quantum mechanical process that would make it possible to turn off the production of virtual photons of an electric charge that are its electric field. Were there such a process, it would violate electric charge conservation. And were there such a process, we could be sure that there was something wrong with the theory, as it would expose an infinite exotic bare mass. What might that defect be? Well, the obvious candidate defect is that the Standard Model does not encompass gravity, and, as we have seen in our discussion of Mach's principle, gravity is intimately involved in the origin of mass and inertia. So whether we are interested in electrons as wormholes or large repositories of negative mass, we arguably need an electron model that explicitly includes gravity.

THE STRUCTURE OF ELEMENTARY PARTICLES

The concerns about the inclusion of gravity in models of the electron, for the purposes of advanced propulsion, emerged in the wake of the uncovering of the always negative "wormhole" term in the Mach effect equation in 1991. The prospect of transiently inducing mass fluctuations by accelerating things while their internal energies change raised two issues. The first was: Are there any limitations on the accelerations that can be applied to real objects? The second was: Can the wormhole term effect be used to produce large amounts of exotic matter?

In the matter of a limitation on the acceleration an object can be subjected to, it turns out that there is a limit that depends on how large the object is. It arises because the speed at which parts of the object can communicate with each other is limited to light speed. The resulting relationship between the proper acceleration a_0 and length l in the direction of the acceleration is:

$$a_0 = \frac{c^2}{l} \tag{7.3}$$

If we take the Compton wavelength as the characteristic length for elementary particles,[5] we get:

$$a_0 \leq \frac{m_0 c^3}{h} \tag{7.4}$$

where m_0 is the particle rest mass and h is Planck's constant. Substitution of the values for the electron shows that even by this standard, the maximum accelerations ultimately tolerable by real objects are so large that this is not a matter of serious concern.

In the matter of the application of Mach effects to make large amounts of exotic matter, the situation is not so simple. In the circumstances of interest, we need an electron model that includes gravity. It turns out that electron modelling has been an off-beat cottage industry practiced by a small band of quite capable physicists beginning with Lorentz and Einstein, and continuing to the present day. We've already encountered one of the early

---

[5] This length is actually much larger than the sizes of elementary particles by at least several orders of magnitude. It comes from the Compton effect, where X-ray photons are scattered from electrons, the photon wavelengths being increased in the scattering process because the recoiling electrons carry away some of the incident photon energy. The Compton wavelength, in terms of the rest mass m of the electron and Planck's constant h, is h/mc. Since m is the rest mass, the Compton wavelength is a universal constant.

attempts to model electrons as something other than the point particles of QED – the Einstein-Rosen bridge wormhole model proposed in 1935. No one paid much attention to the model.[6]

A few years later, Dirac wrote a paper on classical models of the electron. His reason for writing about classical electron models rather than the electron of QED was that he figured that if the problem of infinite self-energy could be solved for classical electrons, a path to eliminating the divergences of QED would be made evident. Of course, he didn't solve the problem. But he identified some of the really weird features of the process of radiation reaction. And he also wrote down the "Lorentz-Dirac equation" that governs these processes. Subsequent modeling efforts, sometimes ingenious and fascinating, haven't enjoyed much more success than these early efforts.

Were you looking into electron modeling in the early 1990s, you would have been helped by a review article written by Paul Wesson that included the subject.[7] And had you been an experimentalist chiefly interested in figuring out how to do advanced propulsion, you probably would have wanted to find out what was going on in quantum gravity, for even then it was common knowledge that superstring theory purported to encompass gravity and GRT. String theory and the precursor to loop quantum gravity were at the cutting edge of theoretical physics in the early 1990s. In the eyes of many, they still are. And the Standard Model is a construct created by literally hundreds, if not thousands of very smart theoretical physicists.

An experimentalist chiefly interested in advanced propulsion isn't single-handedly going to create a major extension to, or replacement for any of these theories. But, if extraordinarily lucky, he or she might be able to find a way to examine the issues of interest involving the role of gravity in elementary particles. In any event, such an experimentalist has no choice if failure is to be avoided. Either the person tries, risking failure, or doesn't try, accepting certain failure. You might guess that if such a simple route to address the role of gravity in elementary particles existed, it would be found in the electron modeling literature. You'd be mistaken. Neither does it lie in the string theory literature. In string theory gravity is a background dependent field, so this isn't really very surprising. As we've remarked several times already, a background dependent gravity theory, if the background spacetime has real physical meaning, doesn't admit real wormholes.

What we seek is found in a review article by Abhay Ashtekar in the proceedings of the Texas Symposium on Relativistic Astrophysics that took place in 1989.[8] Ashtekar had proposed a program of "new Hamiltonian variables" for GRT not long before the '89 Texas Symposium. It promised to be a "non-perturbative" approach to quantum gravity and held out the hope of avoiding the divergences that plagued all attempts at a theory of quantum gravity up to that time.

---

[6] In my reading of the electron modeling literature years ago now, I once ran across a quote attributed to Einstein that went, roughly, "Every Tom, Dick, and Harry thinks he knows what electrons are. They are mistaken." This seems to be a widely shared view in the electron modeling community.

[7] Wesson, P., "Constants and Cosmology: The Nature and Origin of Fundamental Constants in Astrophysics and Particle Physics," *Space Sci. Rev.* 1992 **59**:365–406.

[8] Ashtekar, A , "Recent Developments in Quantum Gravity," *Annals. N.Y. Acad. Sci.* 1989 **571**:16–26.

Early in his talk/paper Ashtekar employed a heuristic example to make the point that non-perturbative techniques are mandatory if one hopes to get realistic results. His example is the self-energy of a sphere of electrical charge. He used "natural" units where c is set equal to 1 and notation, which we replicate for ease of comparison with his own words excerpted in the addendum at the end of this chapter. If one ignores gravity, one gets for the mass of the charged sphere:

$$m(\varepsilon) = m_o + \frac{e^2}{\varepsilon}, \tag{7.5}$$

where m_o is the bare mass of the charge, e its electrical charge (of the electron), and ε the radius of the distribution. The electrical self-energy, the second term on the right hand side of this equation, diverges as the radius of the distribution goes to zero.

Ashtekar then noted that gravity is ignored in Eq. 7.5. To include gravity, we must add a term for the gravitational self-energy. Adding this term makes sense, since the bare mass will have an interaction analogous to the electrical interaction that produces the electrical self-energy. Including this term changes Eq. 7.5 to:

$$m(\varepsilon) = m_o + \frac{e^2}{\varepsilon} - \frac{Gm_o^2}{\varepsilon}. \tag{7.6}$$

Gravitational self-energy is negative, whereas electrical self-energy is positive, so the signs of the two self-energy terms are different. Separately, the self-energy terms go to infinity as the radius of the charge distribution goes to zero. But if these terms are "fine tuned," since they are of opposite sign, they can be made to cancel. Now we have a mass for our sphere of charge – our electron – that is finite for zero radius, so it might seem that all that is required to solve the divergences of electrodynamics is to include gravity with the right amount of bare mass. But this is not a relativistic calculation.

The situation is different when GRT is taken into consideration, Ashtekar noted. In general relativity, since everything couples to gravity, including gravity itself, the mass in the gravitational self-energy term becomes the total mass, rather than the bare mass. Accordingly, Eq. 7.6 gets replaced by:

$$m(\varepsilon) = m_o + \frac{e^2}{\varepsilon} - \frac{Gm^2}{\varepsilon}. \tag{7.7}$$

This equation is quadratic in $m(\varepsilon)$. Solving this equation for m involves nothing more than high school algebra, in particular, application of the "quadratic formula." That solution has two roots, the positive root being:

$$m(\varepsilon) = -\frac{\varepsilon}{2G} + \frac{1}{G}\sqrt{\left(\frac{\varepsilon^2}{4} + Gm_o\varepsilon + Ge^2\right)}. \tag{7.8}$$

Since the bare mass in Eq. 7.8 is multiplied by the radius of the charge distribution, in the limit as ε goes to zero we have

$$m(\varepsilon = 0) = \frac{e}{\sqrt{G}}, \tag{7.9}$$

a finite result obtained without any fine tuning at all. Aside from one little problem, this calculation is an electron modeler's dream come true — a simple calculation that gives the electron mass in terms of fundamental constants with no fudge factors present at all. The little problem is that the value of the mass is decades of orders of magnitude larger than that of real electrons. Sigh.

For Ashtekar's heuristic purposes, getting the wrong mass was irrelevant. The calculation served very neatly to make the points he thought important. Chief among the points he wanted to make was that had one done a perturbation expansion in powers of Newton's constant to compute an approximation to the mass – perturbation expansions being a feature of the technique of renormalization – each of the terms in the resulting series would have diverged, notwithstanding that the non-perturbative solution is finite. Moreover, this turns out to be the case for all gravity theories that are background independent, like GRT. The inference Ashtekar wanted to emphasize was that only non-perturbative methods have any hope of yielding divergence-free solutions in the case of background-free quantum theories of gravity.

Ashtekar pointed out that the above calculation was shown to be an exact solution of Einstein's gravity field equations by Arnowitt, Deser, and Misner in the 1960s. The solution presumably does not return a realistic mass for electrons because it is too simple. For example, it ignores quantum mechanical effects like spin. But the background independence of GRT means that we *must* use a model like that of Arnowitt, Deser, and Misner (ADM) to explore the behavior of matter in the presence of gravity if we want to get finite, reasonable results. So our question is: Can something like the ADM model be constructed to get a realistic mass for electrons? And if this can be done, will we find an *accessible* exotic bare mass that can be used to make stargates?

EXOTIC BARE MASS ELECTRONS

Actually, in the early 1990s, the first question was a bit more practical: Where does one find the ADM papers on electrical self-energy? We're talking about the pre-Google age — indeed, the pre-Internet age. And Ashtekar, evidently assuming his audience to be familiar with the ADM work, didn't bother to give explicit references to the papers in question. By a quirk of fate, I had copies of some papers by Lloyd Motz that did have the needed references.[9] And it did not take long to ascertain that Ashtekar's characterization of their work was correct.

---

[9] Lloyd, a friend of the family [my mother was an astronomer who had met Lloyd at Harvard when she was a grad student there in the 1930s], had given them to me when I was a grad student. Mom had sent me to visit with Lloyd in his attic office in Pupin Hall at Columbia University hoping, I suspect, that Lloyd would counsel me with good sense. At that time Lloyd was a staid senior professor of astronomy and the author of texts on astronomy and cosmology. Little did Mom know that Lloyd was also an electron modeler and quantum gravitier who was delighted to regale me with his work on the problem. In addition to the copies of his papers, published and unpublished, he showed me a hand-written postcard from Abdus Salam complimenting him on his work on gravity and fundamental particles. An absolutely delightful afternoon.

You may be thinking, "So what? The ADM calculation gives the wrong mass for the electron." Yes. But ADM, and for that matter everyone else who had worked on the problem, had not taken into account the possibility that the bare mass of the electron might be *negative*. Remember the positive energy theorem?

There was another feature of the ADM model that got my attention: In Gaussian units, the ADM electron mass is almost exactly a factor of c^2, larger than the actual electron mass. This may strike you as just an odd coincidence. And perhaps it is. But if you know the stuff about Mach's principle in Chap. 2, this coincidence is suggestive. The square of the speed of light, in relation to Mach's laws of inertia, is just the locally measured invariant value of the total gravitational potential due to all the gravitating stuff in the universe. And Mach's second law of inertia says that the masses of things are caused by the gravitational action of all of that stuff on local objects. Perhaps you're thinking, "Ah! But the dimensions are wrong. If you divide the ADM electron mass by the square of the speed of light, you don't get a mass." Yes, that's right. We'll get to that.

The ADM model was first published in 1960.[10] It is the *exact* general relativistic solution – in isotropic coordinates – for a sphere of electrically charged dust with total charge e and bare mass m_o. Rather than try to create a new model of the ADM type, we ask if the ADM model itself can be adapted to our purpose by the simple substitution of a negative bare mass for the electron. We start by writing the ADM solution in slightly different notation and units (i.e., we do not set $c = 1$) than those used by Ashtekar (and here above):

$$m = m_o + \frac{e^2}{Rc^2} - \frac{Gm^2}{Rc^2}, \tag{7.10}$$

where R is the radius of the cloud of dust with charge e and bare (dispersed and non-interacting) mass m_o. The solution for zero radius is:

$$m = \pm\sqrt{\frac{e^2}{G}}. \tag{7.11}$$

In order to turn the solution of Eq. 7.10 into a realistic model of the electron, the first thing we must do, suggested by the Standard Model, is assume that the bare mass of the electron is *negative*, not positive. Moreover, we must also assume that the radius of the charge distribution is *not* zero, for zero radius, irrespective of the bare mass, returns Eq. 7.11, and that's not realistic. The question then is: What do we use for the bare mass of the electron? Well, the charged leptons are simple. They only interact via the electromagnetic interaction.[11] So, it would seem a reasonable assumption that they consist

---

[10] Arnowitt, R., Deser, S., and Misner, C.W., Gravitational-Electromagnetic Coupling and the Classical Self-Energy Problem, *Phys. Rev.* 1960 **120**:313–320 and Interior Schwartzschild Solutions and Interpretation of Source Terms, *Phys. Rev.* 1960 **120**:321 – 324.

[11] Note, however, that the essentially exclusive decay mode of the muon is an electron accompanied by a pair of neutrinos mediated by a W boson.

only of dispersed dust particles of electric charge. Since only electrical and gravitational energies contribute to the ADM solution of Eq. 7.11, it appears that the constituent dust should obey:

$$dm_o = \pm de/G^{1/2}. \tag{7.12}$$

This amounts to the assumption that the gravitational self-energy contribution to the bare mass for the infinitesimal electrical charges that are the dust is not non-linear. That is, since the electrical charge of the dust is infinitesimal, Eq. 7.10 is applicable. Summing over the dispersed dust to get the bare mass gives the ADM solution for the total mass of the dispersed dust:

$$m_o = \pm\sqrt{\frac{e^2}{G}} \tag{7.13}$$

Arguably, this is reasonable, as there is no gravitational self-energy of assembly in the dispersed dust. The presumed negativity of the bare mass dictates that the negative root be chosen in Eq. 7.13 when substitutions for m_o are made in other equations.

Substitution of the negative root of the radical in Eq. 7.13 for m_o isn't the only modification of the ADM model we must make when we assume that the bare mass of electrons is negative. When we change the signs of charges such as an electric charge, that is all we need to worry about when we then look at the dynamical consequences. But gravity is different. When we change the sign of gravitational charge – that is, mass – we also change the sign of the inertial mass that figures into dynamical equations independently of gravity *per se*. That is, negative mass matter, exotic matter, figures into Newton's second law with a minus sign in front of it. So, if a force produces an acceleration in one direction, but acts on exotic matter, the matter moves in the opposite direction.

This means, for example, that the repulsive electrical force produced by like charges results in what appears to be an attractive force if the masses of the charges are negative, as the repulsion results in the charges moving toward each other. This is required by the Equivalence Principle. Consequently, the negativity of the inertial mass of the dust that accompanies it, assuming that it is gravitationally negative, reverses the normal roles of gravity and electricity in the dust. The repulsive electrical force effectively becomes attractive, and the attractive gravitational force effectively becomes repulsive. That means that the signs of the self-energy terms in Eq. 7.10 must be reversed. That is, Eq. 7.10 must be replaced by:

$$m = m_0 - \frac{e^2}{Rc^2} + \frac{Gm^2}{Rc^2}. \tag{7.14}$$

Solution of this equation is affected from the rearrangement:

$$m^2 - \frac{Rc^2}{G}m + \frac{Rc^2}{G}m_0 - \frac{e^2}{G} = 0. \tag{7.15}$$

Application of the quadratic formula gives:

$$m = \frac{Rc^2}{2G} \pm \sqrt{\left(\frac{Rc^2}{2G}\right)^2 - \left(\frac{Rc^2}{G}m_0 - \frac{e^2}{G}\right)}. \tag{7.16}$$

Just as before the sign reversal to account for the Equivalence Principle, when R goes to zero, Eq. 7.11 is recovered because all of the terms on the right hand side are zero at zero radius save for e^2/G. And the sign under the radical of this term is positive. So, using Eq. 7.13 for the dust bare mass is unaffected. This substitution produces:

$$m = \frac{Rc^2}{2G} \pm \sqrt{\left(\frac{Rc^2}{2G}\right)^2 \pm 2\frac{Rc^2}{2G}\sqrt{\frac{e^2}{G}} + \frac{e^2}{G}}. \tag{7.17}$$

The main radical can be factored and its square root taken to give:

$$m = \frac{Rc^2}{2G} \pm \left(\frac{Rc^2}{2G} \pm \sqrt{\frac{e^2}{G}}\right). \tag{7.18}$$

We can still take $R = 0$ and get back Eq. 7.13. But here we are not interested in getting the mass for a zero radius particle with the ADM mass, be that mass positive or negative. We want two terms that can be differenced to yield a very small positive mass if possible. Happily, this can be done. Reading from the left, we take the first root positive and get two terms on the RHS of Eq. 7.18. Since the charged dust bare mass is negative, we take the second root negative, and we thus get for m:

$$m = \frac{Rc^2}{G} - \sqrt{\frac{e^2}{G}}. \tag{7.19}$$

If the first term on the RHS of Eq. 7.19 is very slightly larger than the second term – which will be the case when the dust lies about at its gravitational radius (defined by $2Gm_0/Rc^2 \approx 1$ in this case) – a very small positive mass for the electron results.[12]

You may be thinking about Poincaré stresses at this point and wondering about a critical feature that any remotely realistic electron model must possess: stability. If your electron model depends on a charge distribution with finite radius, but the distribution at that radius is unstable, you're not talking about a realistic electron model. So, the question is: Is the negative bare mass ADM electron model stable?

Astonishingly, the answer to this question is yes. The reason why is because of the sign reversal of the self-energy terms brought about by making the bare mass negative. And it depends on the electrical interaction being linear and the gravitational interaction being

---

[12] As a technical matter, the absolute value of m_0 is used in the gravitational radius, as that is a positive quantity for both positive and negative mass objects.

non-linear. Non-linearity of the gravitational interaction means that when the radius of the dust decreases a bit, the gravitational interaction becomes stronger than were it linear. Since that interaction is effectively repulsive, that means that the repulsion will be stronger than the effective attraction of the electromagnetic interaction, and the equilibrium radius will tend to be restored. Similarly, when the dust expands beyond the equilibrium radius, the gravitational interaction will become a little weaker than the electromagnetic interaction, and the dust will tend to contract back to that radius. So no interactions beyond the gravitational and electromagnetic are required to construct stable electrons. No Poincaré stresses. ...

This is all very neat. But what does it have to do with making stargates? Note that latent in the negative bare mass ADM electron is the second term on the RHS of Eq. 7.19 – an ADM mass of exotic matter. From the point of view of making stargates, the obvious question is: Is there any way to expose the latent exotic masses of electrons by suppressing the first term on the right hand side of Eq. 7.19? To answer this question, we need Mach's principle.

ADM ELECTRONS AND THE UNIVERSE

Back in Chap. 2 we saw that the Machian nature of GRT and its foundational Einstein Equivalence Principle (EEP) require that at every point in spacetime there is a gravitational potential present that is a locally measured invariant equal to the square of the vacuum speed of light. The ADM electron, and, for that matter, all other local solutions of the GRT field equations that assume asymptotically flat spacetime are constructed in the spacetime where c and φ are locally measured invariants and $\varphi \approx c^2$. Gravitational potential energy, as required by the EEP, may not be localizable in GRT, but that does *not* mean that gravitational potential energy can be ignored. Indeed, we find that the total gravitational potential energy, E_{grav}, of a body with mass m is just:

$$E_{grav} = m\varphi = mc^2. \tag{7.20}$$

The obvious generalization of this observation is that the masses of material objects in the universe arise because of their gravitational potential energy whose source is the rest of the matter in the universe. Thus we see that Mach's principle not only identifies the source of inertial reaction forces as the gravitational action of chiefly distant matter, it also identifies the origin of mass-energy itself as gravitational potential energy.

We now turn to the issue of how the universe couples to the ADM negative bare mass electron. The only quantity in Eq. 7.19 that isn't either a constant or locally measured invariant is R. It seems that we must express R in terms of gravitational potential(s) in order to achieve our objective. We proceed by first writing down energy E_e of the electron as determined by distant observers:

$$E_e = mc^2 = m\varphi_u, \tag{7.21}$$

where we have put a subscript u on φ to identify the potential as arising from chiefly cosmic matter.

Next we consider the gravitational energy of the charge distribution from the perspective of an observer located near to the distribution, but not in the distribution (as the gravitational potential seen by an observer in the distribution, owing to the locally measured invariance of the potential, is just c^2). Our nearby observer will see the bare mass m_o and the potential in the dust φ_i, which has two contributions, the potential due to distant matter and the potential due to the bare dust itself. E_e for the nearby observer will be $m_o\varphi_i$, and that, by energy conservation, is equal to $m\,c^2$, so:

$$m_o = m\frac{c^2}{\varphi_i}. \tag{7.22}$$

Now the gravitational radius of m_o by definition is $R = 2Gm_o/c^2 = 2Gm/\varphi_i$, so substituting for R in Eq. 7.19 and with a little algebraic manipulation we get:

$$m = -\frac{\sqrt{\frac{e^2}{G}}}{\left[1 - \frac{2c^2}{\varphi_i}\right]} = -\frac{\sqrt{\frac{e^2}{G}}}{\left[1 - \frac{2c^2}{\varphi_u + \varphi_b}\right]}. \tag{7.23}$$

Note that we have written the potential in the dust as the sum of its constituent parts, the potential due to the universe and the potential due to the bare mass. That is:

$$\varphi_i = \varphi_u + \varphi_b. \tag{7.24}$$

When φ_u and φ_b are almost exactly equal and opposite with a difference of order unity, as they are in this case since φ_b is negative owing to the negative bare mass of the charged dust, the denominator of the RHS of Eq. 7.23, up to a factor of order unity, becomes $\sim -c^2$ and the actual mass of the electron is recovered. The presence of φ_i in Eq. 7.23 insures that the dimensions of the electron mass are correct.

Electron modelers who want an exact calculation of the electron mass in terms of fundamental constants without adjustable parameters will likely be disappointed by the negative bare mass ADM model. In a sense, the subtraction procedure of renormalization has been replaced by an analogous subtraction procedure involving the universal and bare mass gravitational potentials, and the bare mass potential can be viewed as an adjustable parameter. But speaking physically, this makes sense as the internal gravitational potential must be present in any model that includes gravity.

We now remark that could we find a way to screen our electron from the gravitational potential due to the rest of the universe, the denominator would become of order unity and the *exotic* bare mass of the electron – 21 orders of magnitude larger than its normal mass and negative – would be exposed. Do this to a modest amount of normal stuff, and you would have your Jupiter mass of exotic matter to make a traversable stargate – *if the negative bare mass ADM model of elementary particles is a plausible representation of reality.*

There is an outstanding problem for the ADM model as presented to this point. If we take the minimum energy to be zero and assume that the dust distribution will seek minimum energy, then it will settle to a radius that makes the electron mass zero. If we assume that the energy is not bounded from below at zero, the dust will collapse to a point, and we will have electrons with ADM masses, positive or negative makes no difference. Reality is made up of small positive mass electrons. Something is missing. Quantum mechanics. We address this in the next chapter.

ADDENDUM

Excerpt from Ashtekar.

Let us now bring in general relativity. The key idea here is that everything couples to gravity including gravity itself. Therefore, in the expression of the gravitational self-energy, we have to replace m_0 by m. The resulting equation,

$$m(\epsilon) = m_0 + \frac{e^2}{\epsilon} - \frac{Gm^2}{\epsilon},$$

is quadratic in $m(\epsilon)$ and thus has two roots. Let me just appeal to physical requirements and choose the positive root:

$$m(\epsilon) = -\frac{\epsilon}{2G} + \frac{1}{G}\sqrt{\frac{\epsilon^2}{4} + Gm_0\epsilon + Ge^2},$$

which, in the limit as ϵ tends to zero, yields a finite result,

$$m(\epsilon = 0) = \frac{e}{\sqrt{G}}.$$

Note that we did not have to fine-tune any of the parameters. If we had done a perturbation expansion in powers of Newton's constant, as is clear from the formula for $m(\epsilon)$ above, each term in the series would have diverged even though the result is perfectly finite. Can this argument be made rigorous? This was achieved by Arnowitt, Deser, and Misner already in the 1960s using the exact framework of general relativity. Of course, the model itself is too simple (in particular, it ignores all quantum effects) to provide realistic values of mass of observed particles. However, it does suggest that general relativity has certain "built-in" regulating mechanisms that, unfortunately, are lost if we insist on using perturbation expansions in powers of Newton's constant. A detailed examination shows that this is a feature shared by all theories in which there is no background space-time metric. More precisely, the Hamiltonian structure of theories changes dramatically once the metric itself becomes dynamical, and it is this change that is at the root of the "regulating mechanism".

Reprinted from Abhay Ashtekar, "Recent Developments in Quantum Gravity," *Annals of the New York Academy of Sciences*, vol. 571, pp. 16–26 (1989) with permission of John Wiley and Sons.

8

Making the ADM Electron Plausible

When Rutherford discovered the nuclear structure of atoms by bombarding gold foil with alpha particles starting in 1909, he delayed publication of his results, for it was obvious to him that there was a serious problem with the nuclear atoms implied by his results. The electrons presumed to orbit the nuclei would be accelerated in their motion, and classical electrodynamics dictates that accelerating electric charges radiate electromagnetic waves. The energy carried away by the waves would quickly dissipate the orbital energy of the electrons, which would consequently spiral into the nuclei. And matter as we know it would cease to exist.

After a disastrous sojourn at the Cavendish with Thomson in the fall of 1911, Niels Bohr came to work with Rutherford in Manchester in the spring of 1912. They hit it off famously, and Bohr learned of Rutherford's alpha particle scattering results. Quantum "theory" already existed, having been created by Planck to address the "ultraviolet catastrophe" of blackbody radiation, used by Einstein to explain the photo-electric effect, and applied to the low temperature behavior of solids by Einstein and others. Line spectra produced by heated gases had also extensively been studied since their discovery in the nineteenth century.

People had tried to invent schemes where spectra could be explained by the quantum hypothesis, since the frequencies of the spectral lines were discrete, suggesting that in light of Einstein's conjecture on the relationship between frequency and "photon" energy,[1] the energy of electrons in atoms might be quantized. But the quantum hypothesis, despite its successes, had not gained widespread traction in the physics community. Acceptance of truly new ideas is always very difficult and chancy at best.

Bohr's profound insight on learning of the nuclear atom from Rutherford was the realization that the thing quantized in atoms is not energy. Rather, electron orbital angular momentum is the thing constrained by quantum mechanics. The ADM electron, as a sphere of electrically charged dust, isn't exactly like the atom, of course. But it is close enough to invite the use of quantum mechanics to deal with the energetic issues involved.

---

[1] Photons weren't called photons until 1926 when Gilbert N. Lewis invented the name.

J.F. Woodward, *Making Starships and Stargates: The Science of Interstellar Transport and Absurdly Benign Wormholes*, Springer Praxis Books, DOI 10.1007/978-1-4614-5623-0_8,
© James F. Woodward 2013

And in any event, we must deal with electron spin if the ADM model is to be made plausible.[2]

HOW REALISTIC ARE ADM ELECTRONS?

The negative bare mass ADM electron model we have considered in Chap. 7, attractive though it may be for making stargates, is not realistic. It is a Planck scale object that lacks the properties that arise from spin – angular momentum and a magnetic moment. The fundamental problem for all electron models that have faced the issue of spin has been known from the time of Goudsmit and Uhlenbeck. If the electron is viewed as sphere of charge with a "classical" radius, then its equatorial surface velocity must be about 100 times the speed of light c to account for its magnetic moment. If the surface velocity is to be kept at c or less, the radius must be of the order of the Compton wavelength of the electron, that is, about 10^{-11} cm. Scattering experiments, however, show no structure to the electron down to a scale of 10^{-16} cm or less.

Classical models that make the radius of the charge distribution smaller than the Compton wavelength face another criticism. They at least appear to violate the constraints of the Uncertainty Principle. Using the position-momentum Uncertainty Principle, and noting that $E \approx pc$, it is easy to show that the confinement region of an electron Δx dictated by the Uncertainty Principle is:

$$\Delta x \approx \hbar / mc, \tag{8.1}$$

where $\hbar$ is Planck's constant divided by 2π. When the electronic mass is substituted into Eq. 8.1, the minimum confinement distance turns out to be about 10^{-11} cm. – the Compton wavelength of the electron. This is hardly surprising, inasmuch as h/mc is the definition of the Compton wavelength. So, both relativity and quantum mechanics seem to require that electrons not be smaller than the Compton wavelength. But the fact of observation is that electrons are exceedingly small objects orders of magnitude smaller than the Compton wavelength.

Undaunted by reality, electron modelers have pressed on. Recent examples of this sort of work are those of Burinskii and Puthoff.[3] Burinskii has long advocated the Kerr-Newman solution of Einstein's field equations (for an electrically charged, rotating source) as a plausible representation of an extended electron. Part of the appeal of the Kerr-Newman solution is that it automatically returns a gyromagnetic ratio of 2, rather

---

[2] All this and more was known before the publication of Making the Universe Safe for Historians: Time Travel and the Laws of Physics [MUSH; *Found. Phys. Lett.* 1995 **8**:1–39] early in 1995. But an attack on the problems in 1995 was thwarted by an algebraic error that misdirected the investigation into blind alleys. For years, whenever I'd think of the problem, I quickly dismissed it and went on to other matters.

[3] Burinskii, A., Kerr's Gravity as a Quantum Gravity on the Compton Level, Arxiv:gr-qc/0612187v2, 2007; Kerr Geometry as Space-Time Structure of the Dirac Electron, Arxiv:0712.0577v1, 2007; Regularized Kerr-Newman Solution as a Gravitating Soliton, Arxiv:1003.2928v2, 2010. Puthoff, H. E., Casimir Vacuum Energy and the Semiclassical Electron, *Int. J. Theor. Phys.* 2007 **46**:3005–3008.

than 1, an achievement attributed to the Dirac electron.[4] The problem with the model is the size of the ring singularity: the Compton wavelength of the electron.

Were we intent on suggesting a replacement for the Standard Model of elementary particles and their interactions, we might be interested in exploring Kerr-Newman analogs of the negative bare mass ADM solution of Einstein's field equations. Such an undertaking, however, is not required, since we are only interested in seeing if the ADM solution can be adjusted to include spin in a plausible way. After all, our interest is in making stargates, not creating a new theory of matter. Should this be possible, it is arguably reasonable to suppose that electrons might be exceedingly small spinning structures with negative bare masses that can be exposed if a way can be found to gravitationally decouple them from the action of the bulk of the matter in the universe.

The only obvious generalization of the ADM solution of Einstein's field equations (with the self-energy terms' signs reversed to account for the negative bare mass) that leaves it a simple quadratic equation that can be solved for m in the usual way is:

$$m = m_o - \frac{Ae^2}{Rc^2} + \frac{Gm^2}{Rc^2}, \tag{8.2}$$

where A is a constant to be determined.[5] Following the procedures spelled out in Chap. 7, and taking account of the fact that angular momentum conservation requires that we associate angular momentum with the dispersed bare mass of our electrons, we find:

$$m = \frac{Rc^2}{G} - \sqrt{\frac{Ae^2}{G}}. \tag{8.3}$$

As long as we can find a way to express the angular momentum and its associated energy of rotation and magnetic self-energy in terms of e^2/R, then all we need do is choose A appropriately to incorporate these quantities into our solution.

ANGULAR MOMENTUM AND MAGNETIC MOMENT

If we are willing to make a simple approximation, including quantized angular momentum and magnetic-self energy turns out to be much simpler than one might expect. The approximation is that the electronic charge, rather than being a sphere, is a ring of radius R circulating with velocity v. Viewed at the scale of the electron, this may seem an unreasonable assumption. But viewed at the scale of everyday stuff, electrons look like points.

---

[4] Before Dirac published his theory of the electron, this anomaly in the gyromagnetic ratio was dealt with by Thomas, who showed it to be a relativistic effect arising from choice of frame of reference in the barycenter of the atom. It is known as the "Thomas precession." Since Dirac's theory is relativistic from the outset, this is automatically included in his theory.

[5] Constant coefficients can also be introduced for the other terms in the equation if needed. However, in this case only A will be required.

And the electric field of a point is the same as that of a sphere. So approximating a ring current for electrical purposes as a sphere or point is arguably reasonable as long as we do not inquire about the details of the source and fields at the scale of several times the radius of the ring current or less.

We start by considering the mechanical energy of rotation, E_{rot}, of the charged dust. This is just $\frac{1}{2}I\omega^2$, I being the moment of inertia of the ring and ω its angular velocity. For the simple case of a circular ring spinning around its axis of symmetry, E_{rot} is just $\frac{1}{2}mv^2$, where v is the speed of circulation of the ring current. To put this in terms of e^2/R we first invoke the Bohr quantization condition on angular momentum: $mvR = \frac{1}{2}n\hbar$, where n is an integer. The Bohr condition gives $m = \frac{1}{2}n\hbar/vR$, so:

$$E_{rot} = \hbar\, nv\,/\,4R \tag{8.4}$$

This gives us the R dependence, and it turns out to be the dependence we require. We can use the fine structure constant [$\alpha = e^2/\hbar c$] to show the explicit dependence on e^2. We solve the definition α for $\hbar$ and substitute the result into Eq. 8.4, getting:

$$E_{rot} = \frac{n\,v}{4\alpha c}\,\frac{e^2}{R}. \tag{8.5}$$

If v were $\approx c$, the coefficient of e^2/R would be especially simple. We know, however, that cannot be the case, for if it were true, the predicted angular momentum of the electron would be many orders of magnitude too small. To recover reality, v must be very much larger than c. The principle of relativity, however, requires that v be less than or equal to c.

Despite the seeming defect of violation of relativity because of a highly superluminal v, at least quantization is implicit in Eq. 8.5 via the presence of the fine structure constant that depends on Planck's constant and the principal quantum number n that comes from the Bohr quantization condition. We defer comment on the *apparent* violation of relativity until after discussion of magnetic self-energy.

Turning now to the magnetic self-energy associated with the loop of current created by the spinning electric charge, we note that it is a commonplace of both classical and quantum physics that the angular momentum and magnetic moment of a current are intimately related, their ratio being a constant called the "gyromagnetic," or "magnetogyric" (depending on how prissy the physicist speaking happens to be) ratio. This is a consequence of the fact that the charge to mass ratio of elementary particles is fixed, and the magnetic moment of a current loop is defined as $\mu = evR/2\,c$. This being the case, it stands to reason that if the mechanical angular energy can be expressed as a function of e^2/R, it should be possible to do the same for the magnetic self-energy. An approximate calculation of the magnetic self-energy that shows this is:

$$E_{mag} \approx \frac{1}{8\pi}\int_R^\infty H^2 dV, \tag{8.6}$$

where H is the magnetic field intensity and dV is a volume element that is integrated from $r = R$ (exceedingly small) to infinity. Equation 8.6 is recovered from the general definition of the energy density of the electromagnetic field, adapted to the special circumstances of a simple ring current. Now H for a simple dipole source is $\approx \mu/r^3$ and $dV = 4\pi r^2 dr$. Note that μ here is the dipole moment of the current ring, not the permeability of the vacuum. Carrying out the integration of Eq. 8.6 gives us:

$$E_{mag} \approx \frac{1}{6}\frac{\mu^2}{R^3} = \frac{1}{24}\left(\frac{v}{c}\right)^2\frac{e^2}{R} = A_{mag}\frac{e^2}{R}, \tag{8.7}$$

where we have used the definition of the magnetic moment to get the energy in terms of e. As in the case of mechanical energy of rotation, the magnetic self-energy can also be expressed in terms of e^2/R. But where the mechanical energy depends on v/c, the magnetic self-energy depends on the square of that quantity. And if v is larger than c, then the magnetic self-energy dominates the contributions to A. That is, $A \cong A_{mag}$.

ISOTROPIC COORDINATES AND THE "COORDINATE" SPEED OF LIGHT

We now turn to the issue of the superluminality of v required by the observed values of the angular momentum and magnetic moment of the electron. The exceedingly small size of the ADM electron, even with spin included, requires that $v >> c$ be the case to recover the observed values of the angular momentum and magnetic moment. How can that be?

Well, recall that the ADM solution is done in "isotropic" coordinates. In these coordinates the speed of light is the same in all directions point-by-point (hence "isotropic") and the coordinates give space the appearance of flatness. But in fact, for a compact or point source the radial coordinate may be highly distorted by the source at or near the origin of coordinates.[6] For example, as Arthur Eddington had already pointed out in his *Mathematical Theory of Relativity* published in the early 1920s, the relationship between the usual r of the Schwarzschild solution and r_i of the isotropic coordinates for the same solution is:

$$r = \left(1 + \frac{GM}{2c^2 r_i}\right)^2 r_i, \tag{8.8}$$

where M is the mass of the source. More importantly, the speed of light in isotropic coordinates differs from its locally measured invariant value. For distant observers it becomes:

$$c_{obs} = \frac{\left(1 - \frac{GM}{2r_i c^2}\right)^2}{\left(1 + \frac{GM}{2r_i c^2}\right)^3}\, c. \tag{8.9}$$

---

[6] Eddington, A. S., *The Mathematical Theory of Relativity* (Cambridge, Cambridge University Press, 2nd ed., 1960). See especially p. 93.

c_{obs} is commonly called the "coordinate" speed of light. If M is positive, as, for example, it is for a black hole, then as one observes light propagating near the event horizon (the gravitational radius) from a distance, the speed appears to drop, going to zero at the horizon. And not only does c go to zero at the horizon of a black hole. As we noted back in Chap. 1, time stops there as well. These are well-known features of black holes.

As long as the mass of the source of the gravity field is positive, all is well. Should the mass of the source be negative, however, things are not the same. Instead of the speed of light decreasing in the vicinity of, say, a negative mass star as measured by distant observers, it speeds up. As a result, instead of getting the usual prediction of GRT for the deflection of starlight, the light from distant stars is deflected away from the negative mass star. Accordingly, stars viewed around such a negative mass star during a total eclipse will appear to move toward, rather than away from, the eclipsed star.

Back in the mid-1990s, John Cramer, Robert Forward, Michael Morris, Matt Visser, Gregory Benford, and Geoffrey Landis[7] appreciated this distinction and suggested that negative mass astronomical objects could be sought by looking for an interesting light signature. At the time a search for MACHOS (massive compact halo objects) was under way. What was sought was the transient brightening of background stars as they were occulted by MACHOS employing gravitational lensing in their orbital motion about the galactic center. Cramer et al. realized that were such objects (which were called GNACHOS, for gravitationally negative astronomical compact halo objects) made of exotic matter, the occultation light curves would show a distinctive feature by which they could be identified. Owing to the deflection of light away from such objects, a cone of light from the distant star should form around the GNACHO, leading to a double peak in the light curve with no light from the star between the peaks. Alas, no such light curve has been reported in surveys of this sort. But what about electrons?

If M is negative, as is the bare mass of our ADM electron, Eq. 8.9 returns a very different result: *the speed of light, for distant observers, appears to increase as the gravitational radius is approached, going to infinity at the gravitational radius* (as the denominator of the coefficient of c on the right hand side in Eq. 8.9 goes to zero). If our ring of electronic charge circulates very near to its gravitational radius – as it must for the small observed mass of the electron to be recovered – then while measured locally v is always less than c, distant observers will measure v to be orders of magnitude larger than c.

The negative bare mass of our ADM electrons thus provides a natural explanation for the feature of classical and semi-classical electron models that has confounded these models from the outset. $v \gg c$ as measured by us distant observers is *not* a violation of the principle of relativity when the bare mass of the electron is negative. It is expected. Electron spin thus need not be treated as an "intrinsic" quality of electrons. Classicists may find this result appealing.

---

[7] Cramer J., et al. *Natural Wormholes as Gravitational Lenses*. Physcial Review D. volume D15, pp. 3117–3120 (1995).

SPIN AND THE MASS OF THE ELECTRON

The inclusion of spin in the ADM electron as sketched here changes things so that the solution obtained using Mach's Principle and energy conservation in the previous chapter does not return the electron mass in terms of interest. To calculate the electron mass, as before, we proceed by substituting for R in the first term on the RHS of Eq. 8.3. But instead of using energy conservation and the energies measured by near and distant observers of the dust cloud, we use the Bohr quantization condition to express R in terms of the other quantities in that relationship. That is, we use $R = \frac{1}{2}n\hbar/mv$. Substitution and a little algebra yields:

$$\sqrt{\frac{G}{Ae^2}}\,m^2 + m - \sqrt{\frac{G}{Ae^2}\,\frac{n\hbar c^2}{2Gv}} = 0. \tag{8.10}$$

Equation 8.10 has the solution:

$$m \cong \frac{1}{2}\sqrt{\frac{Ae^2}{G}}\left[-1 \pm \left(1 + \frac{1}{2}\frac{n\hbar c^2}{2Ae^2v}\right)\right], \tag{8.11}$$

and choosing the root that leads to the cancellation of the ADM bare mass by the first two terms on the RHS,

$$m \cong \frac{1}{8}\sqrt{\frac{Ae^2}{G}\,\frac{n\hbar c^2}{Ae^2v}}. \tag{8.12}$$

Since we know that $v >> c$ (as measured by us distant observers), it follows that A will be dominated by the magnetic self-energy term, so from Eq. 8.7 we can take:

$$A \cong \frac{1}{24}\left(\frac{v}{c}\right)^2. \tag{8.13}$$

Substitution for A in Eq. 8.12 and more algebra yields:

$$m \cong 3\,\frac{c^3}{v^2}\,\frac{n\hbar}{e^2}\,\sqrt{\frac{e^2}{24\,G}}. \tag{8.14}$$

Using the definition of the fine structure constant, this expression can be stated simply in terms of the ADM mass as:

$$m \cong n\sqrt{\frac{9}{24}}\left(\frac{c}{v}\right)^2 \alpha^{-1}\sqrt{\frac{e^2}{G}}. \tag{8.15}$$

The fine structure constant is present in Eq. 8.15 because we used the Bohr quantization condition to express R in Eq. 8.3, so Eq. 8.15 is semi-classical.

THE LEPTON MASSES

Armed with Eq. 8.15, we might want to know the (mass)-energy spectrum that follows from the choice of n equal to values other than one. We might expect to get back the masses of the heavier charged leptons. The situation, however, is a bit more complicated than simply putting n values greater than one in Eq. 8.15. The reason is that the muon and tau, like the electron, are all spin one half particles. So their total angular momenta are the same, while their energies differ.

Evidently, we are dealing with more than one principle quantum number. The principle quantum number that records total angular momentum must be the same for all of the charged leptons. So we ask: How does the energy of our ADM electron depend on the value of a principle quantum number other than that of recording total angular momentum? As it turns out, the calculation involved was carried out by Asim O. Barut many years ago.[8] Barut started by remarking on the well-known, but unexplained, relationship between the electron and muon:

$$m_\mu = \left(1 + \frac{3}{2\alpha}\right) m_e, \qquad (8.16)$$

claiming it could be derived on the basis of the magnetic self-interaction of the electron. (Barut's one-page paper is included at the end of this chapter as an addendum.) In a footnote he did a simple calculation to show that the magnetic self-energy should depend on a principal quantum number as n^4.

Considering an electric charge moving in a circular orbit in the field of a magnetic dipole, Barut noted the equation of motion is $F = ma = mv^2/R = e\mu v/R^3$ and the Bohr quantization condition is $mvR = n\hbar$. He solved the quantization condition for R and substituted into the equation of motion, getting $v = n^2\hbar^2/e\mu m$. Squaring this equation gives: $v^2 = n^4\hbar^4/e^2\mu^2m^2$, so the kinetic energy of the charge in orbit is: $mv^2/2 = n^4\hbar^4/2 e^2\mu^2m$ and energy quantization goes as n^4. A charge in circular orbit around a dipole is equivalent to the magnetic self-interaction of a ring current, so the energy quantization dependence of a ring electron on n should be the same as that calculated by Barut. Taking $n = 1$ to be implicit in Eq. 8.16, Barut conjectured that the mass of the τ should be:

$$m_\tau = m_\mu + \frac{3}{2\alpha}n^4 m_e, \quad n = 2. \qquad (8.17)$$

or, more generally expressed,

$$m_{lep} = m_e \left(1 + \frac{3}{2\alpha}\sum_{n=0}^{n_{lep}} n^4\right), \qquad (8.18)$$

[8] Barut, A. O., The Mass of the Muon, *Physics Letters B* 1978 **73**:310–312; B; and Lepton Mass Formula, *Physical Review Letters* 1979 **42**:1251.

where $n = 0$ for the electron, $n = 1$ for the muon, and $n = 2$ for the tau lepton. The charged lepton masses computed with Barut's formula are surprisingly accurate. Barut did not know of the negative bare mass ADM electron model modified to include spin. Presumably, he assumed that something like it must exist, for it was obvious to him that the magnetic self-energy of the electron must dominate the electromagnetic self-energy. That being the case, he thought it natural to view the muon and tau as excited states of the electron. However, quoting Glashow, Barut remarked, "We have no plausible precedent for, nor any theoretical understanding of, this kind of superfluous replication of fundamental entities [the charged leptons]. Nor is any vision in sight wherein the various fermions may be regarded as composites of more elementary stuff. . . ."

The spin modified negative bare mass ADM model provides a physical model that arguably underpins Barut's conjecture. Note, however, that the principal quantum number 1 is assigned to the electron in Eqs. 8.10, 8.11, 8.12, 8.13, 8.14, and 8.15, whereas it is assigned to the muon in Barut's calculation. Evidently, the ns of the two calculations, both principal quantum numbers, are not the same. Since the electron, muon, and tau are all spin one half particles, the n implicit in Eqs. 8.10, 8.11, 8.12, 8.13, 8.14, and 8.15 must be one, and the same value for all the charged leptons as they all have the same angular momenta: $\frac{1}{2}\hbar$. Notwithstanding that Barut invoked the Bohr quantization condition in his argument, his n must assume a range of values to account for the different mass-energies of the charged leptons.

It is tempting to pursue old quantum theory and invoke Wilson-Sommerfeld quantization conditions to explain the charged leptons as splitting of a ground state due to structural differences in the ring current. However, such speculations exceed our purpose, which is to show that a simple, plausible theory of elementary particles including gravity is possible, and that the theory shows that an enormous, but finite, amount of exotic matter resides in the bare masses of those elementary particles – bare masses that would be exposed if a way to decouple some local matter from the gravitational influence of distant matter can be found. Arguably, we have done what we set out to do.

THE UNCERTAINTY PRINCIPLE

What about the argument about energy and confinement size based on Heisenberg's Uncertainty Principle? Well, if you believe, with Einstein and his followers, that the Uncertainty Principle is a statement about our ability to measure reality, rather than an assertion about the inherent nature of reality, you won't have a problem with the negative bare mass ADM electron. After all, how big something is, is not the same thing as how accurately you can measure its position. However, you may think the Uncertainty Principle is an assertion, with Bohr and his legion of followers, about the inherent nature of reality – that an object with a size less than the Compton wavelength of the electron must have energy given by the Uncertainty Principle, far in excess of the electron rest energy – and you may have a problem with all this. You may think it impossible to build stargates. Ever. You may be right. But in this section of this book, we will assume that Einstein and his followers are right.

ADDENDUM

Asim O. Barut's paper on the masses of the charged leptons.

VOLUME 42, NUMBER 19   PHYSICAL REVIEW LETTERS   7 MAY 1979

Lepton Mass Formula

A. O. Barut

Department of Physics, The University of Colorado, Boulder, Colorado 80309
(Received 28 March 1979)

A recently proposed mass formula for the muon is extended to τ and other heavy leptons. It is postulated that a quantized magnetic self-energy of magnitude $\frac{3}{2}\alpha^{-1}n^4 M_e$, where n is a new quantum number, be added to the rest mass of a lepton in the chain e, μ, τ, $\cdots$, with $n=1$ for μ, $n=2$ for τ, etc. I predict $M_\tau = 1786.08$ MeV, and for the next lepton $M_\delta = 10\,293.7$ MeV.

Recently I have suggested that the mass formula for the muon, $M_\mu = (\frac{3}{2}\alpha^{-1} + 1)M_e$, can be derived on the basis of magnetic self-interaction of the electron.[1] The radiative effects give an anomalous magnetic moment to the electron which, when coupled to the self-field of the electron, implies an extra magnetic energy. Assuming that this energy of a charge in the field of a magnetic moment is quantized and is successively added to the rest energy, we arrive at the following set of mass values:

(i) M_e,

(ii) $M_\mu = M_e + M_{\text{magnetic}}$ $(n = 1)$,

(iii) $M_\tau = M_\mu + M_{\text{magnetic}}(n = 2)$, etc.

The magnetic energy of a system consisting of a charge and a magnetic moment quantized according to Bohr-Sommerfeld procedure implies quantized energies

$$E_n = \lambda n^4,$$

where n is a principal quantum number.[2] Determining the proportionality constant λ from the muon-mass formula $(n=1)$, we obtain

$$M_\tau = M_\mu + \tfrac{3}{2}\alpha^{-1}n^4 M_e \quad (n=2)$$
$$= (M_e + \tfrac{3}{2}\alpha^{-1}1^4 M_e) + \tfrac{3}{2}\alpha^{-1}2^4 M_e$$
$$= 1786.08 \text{ MeV}.$$

The three very recent experimental determinations[3] of the τ mass agree with this value extremely well.

The next lepton with $n = 3$ would have a mass

$$M_\delta = M_\tau + \tfrac{3}{2}\alpha^{-1}3^4 M_e = 10\,293.7 \text{ MeV}.$$

It is possible that although the Bohr-Sommerfeld quantization is approximative, the final result might be exact as was the case in Bohr-Sommerfeld derivation of the Balmer formula. There-

fore, the hypothesis which I advanced should be considered of a heuristic nature towards the development of a more complete theory. On the other hand, the occurrence of the fermion chain e, μ, τ, ..., is a novel phenomenon for which we have so far no theory in order to derive a mass formula from first principles: "We have no plausible precedent for, nor any theoretical understanding of this kind of superfluous replication of fundamental entities. Nor is any vision in sight wherein the various fermions may be regarded as composites of more elementary stuff. No problem is more basic than the problem of flavor, and it has been with us since the discovery of muons. Sadly, we are today no closer to a solution."[4]

---

[1] A. O. Barut, Phys. Lett. 73B, 310 (1978).
[2] Consider a charge moving in circular orbits in the field of a fixed magnetic dipole $\vec{\mu}$. The equations of Bohr-Sommerfeld quantization are

$$mv^2/r = e\mu v/r^3 \text{ and } mvr = n\hbar.$$

From the second of these equations we have $r = n\hbar/mv$, which we insert into the first equation to obtain $v = n^2\hbar^2/e\mu m$, or $v^2 = n^4\hbar^2/e^2\mu^2 m$. Hence quantized energies are proportional to n^4.
[3] R. Brandelik *et al.* [Phys. Lett. 73B, 109 (1978)] gave a value of $m_\tau = 1807 \pm 20$ MeV. W. Bacino *et al.* [Phys. Rev. Lett. 41, 13 (1978)] gave a value of $m_\tau = 1782^{+2}_{-7}$ MeV, now revised to $m_\tau = 1782^{+3}_{-4}$ MeV [see, e.g., G. J. Feldman, in *Proceedings of the Nineteenth International Conference on High Energy Physics, Tokyo, Japan, August 1978*, edited by S. Homma, M. Kawaguchi, and H. Miyazawa (Physical Society of Japan, Tokyo, 1979), p. 786]. W. Bartel *et al.* [Phys. Lett. 77B, 331 (1978)] gave a value of $m_\tau = 1787^{+10}_{-18}$ MeV, now revised to $m_\tau = 1790^{+7}_{-10}$ MeV [see the rapporteur's talk by Feldman cited above].
[4] S. L. Glashow, Comments Nucl. Part. Phys. 8, 105 (1978).

9

Making Stargates

If you have paid attention to the semi-tech, popular media that deals with things such as astronomy and space, you'll know that there has been a change in the tone of the commentary by capable physicists about serious space travel in the past 5 years or so. Before that time, though Thorne's and Alcubierre's work on wormholes and warp drives (respectively) was quite well-known, no one seriously suggested that building such things might be possible within any foreseeable future. The reason for this widely shared attitude was simply that within the canon of mainstream physics as it is understood even today, no way could be imagined that might plausibly lead to the amassing of the stupendous amounts of exotic matter needed to implement the technologies. Within the canon of mainstream physics, that is still true. The simple fact of the matter is that if the technology is to be implemented, our understanding of physics must change. The kicker, though, is that the change must be in plausible ways that actually have a basis in reality. Fantasy physics will not lead to workable technologies no matter how much we might want that to be so.

The material you have read through in the chapters leading up to this one purports to be such a structure of arguably plausible physics. It does not rest on one profound insight that radically alters the way in which we understand reality. That is, there is no single key "breakthrough" that enables one to see how simply to build stargates. No one is going to make a few trips to their local home improvement and electronics stores, and go home to construct a wormhole generator in their garage as a weekend project. There are several insights, but none of them are of the revolutionary sort that have radically altered our perception of reality, like the theories of relativity, or the foundational insights that led to the creation of quantum mechanics. The only driving notion behind all of this is the realization that inertia is the physical quantity that must be understood if we want to get around spacetime quickly; and that is hardly a profound insight appreciated by only a few. It has been obvious to all interested parties for decades, if not longer. And the key to understanding inertia – Mach's principle – has been known since Einstein introduced it shortly after publishing his first papers on GRT. True, it almost immediately became a topic of confusion and contention and has remained so to this day. But no one can say that the tools needed to address the origin of inertia have either not existed, or been so shrouded in obscurity that they have been inaccessible to anyone interested in the problem.

J.F. Woodward, *Making Starships and Stargates: The Science of Interstellar Transport and Absurdly Benign Wormholes*, Springer Praxis Books, DOI 10.1007/978-1-4614-5623-0_9,
© James F. Woodward 2013

Quite the contrary. Mach's principle and the question of the origin of inertia has at times been one of the central issues of debate in the gravitational physics community over the years. Though at other times this hasn't been the case. For example, the editors of Richard Feynman's *Lectures on Gravitation* in their preface felt it necessary to explain to readers why Feynman, lecturing on the topic around 1960, made references to Mach's principle. From the publication of Dennis Sciama's paper "On the Origin of Inertia" in 1953, to the work of Derek Raine, Julian Barbour, and Bruno Bertotti in the mid- to late 1970s was the heyday of contentious debates about Mach's principle. Andre K. T. Assis sparked a bit of a revival in the 1990s, but it was short-lived. After the publication of the proceedings of the 1993 Tübingen conference organized by Julian Barbour and Herbert Pfister, some follow-up papers, notably those of Wolfgang Rindler, Herman Bondi and Joseph Samuel, and Ciufolini and Wheeler's book *Gravitation and Inertia* in 1995, interest in the origin of inertia again lapsed.

Given how fundamental inertia and its measure, mass, is – it is a feature of *all* physical processes – this lack of interest seems odd at best. Abraham Pais's comment on the "obscure" nature of mass and inertia in the theory of particles and fields, quoted at the outset of Chap. 2, recently has been echoed by Frank Wilczek:

> As promised, we've accounted for 95 % of the mass of normal matter from the energy of [rest] massless building blocks, using and delivering on the promise of Einstein's second law, $m = E/c^2$. Now it's time to own up to what we've failed to explain.
>
> The mass of the electron, although it contributes much less than 1 % of the total mass of normal matter, is indispensable. And yet, we have no good idea (yet) about why electrons weigh what they do. We need some new ideas. At present, the best we can do is to accommodate the electron's mass as a parameter in our equations – a parameter we can't express in terms of anything more basic.

Since starships and stargates will not be built with theories that are unable to characterize the nature of mass and inertia, perhaps the excursions into physics that lie outside of the Standard Model and its popular extensions is warranted. Let's recapitulate where those excursions have taken us by way of review.

REVIEW OF CHAPTERS

In Chap. 1 we saw that Galileo, in the early seventeenth century, introduced the principles of inertia, relativity, and equivalence – the conceptual foundations of all modern mechanics. Owing to his belief in the Copernican version of cosmology, however, he did not assert these principles as "universal," that is, applying everywhere and every time. Newton corrected this limitation of Galileo's version of the principles, building them into his theories of mechanics and gravitation. Newton, however, defined space and time as independent and absolute, making the creation of the *theory* of relativity impossible within his system.

Working under the influence of Ernst Mach and in the heyday of classical electrodynamics, Einstein corrected this shortcoming of Newton's system, realizing that the principle of relativity *required* that the speed of light be a "constant" – that all observers, no matter what their inertial motion might be, measure the same numerical value for the speed of light (in a vacuum), insuring that the speed of light could not be used to identify some preferred inertial frame of reference as more fundamental than all others. This "invariance" of the speed of light, Einstein realized, mixed up space and time in such a way that he could show that more than rest mass alone had inertia. In particular, energy as well as normal matter has inertia; and the measure of inertia, mass, was therefore equal to the non-gravitational energy present divided by the square of the speed of light, or $m = E/c^2$, Einstein's second law in Frank Wilczek's turn of phrase.

In the absence of gravity, Einstein's second law was as far as we could go in our attempt to understand inertia and its measure mass-energy. Einstein used three "principles" in his construction of general relativity theory (GRT): the Einstein Equivalence Principle (EEP); the principle of general covariance (that the laws of physics should take the same form in all frames of reference); and Mach's principle (the "relativity of inertia," or that cosmic matter should determine the inertial features of local physics).

The principle that speaks to our issue of interest, Mach's principle, was the only one of the three that Einstein appeared to fail to incorporate in GRT. GRT, nonetheless, introduced some important changes from SRT with relevance for the issue of inertia, notwithstanding that the foundation of GRT is the local applicability of SRT and accordingly that might not seem to be the case. The Equivalence Principle reveals gravity to be universal because when you write down the equation of motion for the Newtonian gravity force (that is, set it equal to $m\mathbf{a}$), the passive gravitational mass and inertial mass cancel out, showing that all massive objects, regardless of their mass or composition, respond to gravity with the same acceleration. Since this is so, gravity as a force can be replaced with a suitable geometry where inertial frames of reference accelerate so that bodies we normally think of as in "free fall" under the action of gravity are in fact moving inertially. The implementation of this idea requires curved spacetime to accommodate local concentrations of mass-energy, and the gravitational "field" becomes the curvature of spacetime, rather than being something that exists in a flat background spacetime.

Although the speed of light measured locally remains an invariant in general relativity theory, the curvature of spacetime introduced to deal with local concentrations of mass-energy makes the speed of light measured by observers at locations other than where they are no longer invariant. Generally, the speed of light depends on whether a gravity field, a local distortion of spacetime by a local concentration of mass-energy, is present and how strong it is. Since gravity does this, and by the EP gravity and accelerations are equivalent, this raises the question of whether accelerations change the measure of the speed of light. For non-local determinations, the same sorts of changes in c produced by gravity must follow. But the Champney experimental results using the Mössbauer effect and a high speed centrifuge, and the local applicability of SRT, show that the *locally measured* value of c remains invariant, even in gravity fields and accelerating frames of reference. That is, in the turn of phrase of Kip Thorne, space and time are relative. The speed of light is absolute.

The universality of gravity, that it satisfies the EP, is what makes possible the substitution of non-Euclidean geometry for a customary field representation. This leads to the notion that gravity is a "fictitious" force. But gravity is not the only fictitious force. In addition to gravity, inertial and Coriolis forces have the same property.

Coriolis forces really are fictitious. They arise from viewing inertial motion in an accelerating frame of reference. So the acceleration of objects viewed in that frame are artifacts of the observer's non-inertial motion, and no forces actually act on the viewed objects. But, while inertial forces are fictitious, as they satisfy the EP, unlike Coriolis forces, they are real forces. They are "real" because they are the reaction forces to real, non-gravitational forces exerted on objects. Non-gravitational forces cannot be eliminated by some suitable choice of geometry, so while in principle inertial forces can be eliminated by geometry, in fact they usually only appear in local circumstances when a non-gravity force acts. For this reason, they do not lend themselves to the "rubber sheet" analogy that can be used to illustrate the geometrical effect of mass-energy on spacetime for gravity as it is normally understood.

The "fictitious" nature of inertial forces suggests that they might be understood as the gravitational action of the whole universe on local things when they are acted upon by non-gravity type forces. Historically, Einstein framed the issue a little differently. He asserted the "relativity of inertia," the notion that the inertia of local things should be determined by the amount and distribution of matter – everything that gravitates – contained in the universe. Since very little was known about the amount and distribution of matter in the universe in the teens and twenties of the twentieth century, that this might work, and actually be the case, was conjectural at best in his time. After repeated attempts to incorporate Mach's principle (as he called it), including the introduction of the "cosmo-logical constant" term in his field equations, after learning of Mach's disavowal of relativity theory before his death in 1916, in 1922, Einstein simply asserted that the universe must be closed, insuring that only the matter contained therein could be respon-sible for inertia – if anything was responsible for inertia.

Dennis Sciama revived interest in Mach's principle in the early 1950s using a vector field theory for gravity, a theory that he initially thought different from general relativity theory. Eventually Sciama's vector theory was shown to be just an approximation to general relativity. The simplicity of the vector formalism combined with idealized assumptions about the distribution of matter in the universe (homogeneity and isotropy, and ignoring Hubble flow) made it easy to show that inertial reaction forces could be accounted for with gravity, especially for a universe that is spatially flat at cosmic scale with its concomitant "critical" cosmic matter density. Those conditions lead to the condition that the total scalar gravitational potential of the universe – GM/R, where G is Newton's constant of gravitation, M the mass of the universe, and R its radius – is just equal to the square of the speed of light. This makes the coefficient of the acceleration of a test body in Sciama's gravelectric field equation one, and thus the force it exerts on the test body is exactly the inertial reaction force.

The mass and radius of the universe are presumably functions of time, so it was not obvious in the 1950s and 1960s that Sciama's calculation really meant that inertial reaction forces were caused by gravity. Two considerations, however, did suggest this might be the case. First, if critical cosmic matter density really is the case, general relativity told everyone that that condition would persist as the universe evolved, so $GM/R = c^2$ would remain true even if it were just an accident at the present epoch.

A more powerful argument was provided by Carl Brans in the early 1960s. In the early 1920s, Einstein had addressed Mach's principle in some lectures he gave at Princeton University. In a calculation he did for those lectures he claimed to show that if "spectator" masses were piled up in the vicinity of some local object, the gravitational potential energy they would confer on the local object would change its mass. Brans showed that Einstein was mistaken about this, for it would be a violation of the Equivalence Principle.

Indeed, the only way the Equivalence Principle can be satisfied is if gravitational potential energy is "non-localizable," that is, spectator masses in the vicinity of objects do not change the masses of local objects. The non-localization condition is what distinguishes the Einstein Equivalence Principle from the other versions of this principle. This does not mean, however, that the total scalar gravitational potential can be taken to be anything you choose. It does mean that the cosmic value is the value measured everywhere in local measurements. And since the WMAP survey results show the universe to be spatially flat at the cosmic scale, this means that inertial reaction forces really are caused by gravity.

Mach's principle – the gravitational origin of inertial reaction forces – comes at a price. The most obvious issue is that inertial reaction forces are instantaneous. But Sciama's calculation shows that they are a radiative phenomenon, for they have the characteristic distance dependence of radiative interactions, and signals in general relativity propagate at light speed or less. This problem is compounded by the fact that the interaction does not have an exact counterpart in electrodynamics. Inertial reaction forces depend on the gravimagnetic vector potential, which in turn depends on mass-energy charge currents. Since the mass of the constituents of the universe is positive, accelerating objects see non-zero mass-energy charge currents all the way to the edge of the universe. In electrodynamics, where the mean electric charge density at cosmic scales is zero (because there is as much positive as negative charge out there), there are no cosmic scale net electric charge currents. So there are no electromagnetic inertia-like effects observed.

How does one deal with a radiative interaction with cosmic matter that occurs instantaneously? Two ways seem the only ones possible. One is to assert that inertial effects are actually not radiative effects at all, notwithstanding Sciama's calculation. If they are attributed to initial data constraint equations, then instantaneity is not a problem, as such equations are elliptic and thus act instantaneously throughout spacetime. This approach, adopted in the 1990s by John Wheeler and others, has the curious feature that the entire future history of the universe is already written into the initial data. The other approach, advocated by Fred Hoyle and Jayant Narlikar and others, implements Wheeler-Feynman action-at-a-distance (or "absorber") theory, where both retarded and advanced wave exist, making instantaneous communication with the distant future possible while preserving the appearance of purely retarded interactions. This, of course, means that the future is in some real sense already out there, for things that happen in the present (and happened in the past) depend on its existence. It's complicated.

If you choose the action-at-a-distance representation with its radiative interaction, there's another problem. In standard field theory, the lowest order radiative solution of the field equations when all of the sources have the same sign (positive or negative) is the quadrupole component in a standard multipole expansion of the field. The problem is that gravitational radiation from local sources is incredibly weak. The idea that inertial reaction forces might arise from the reaction to the launching of quadrupole gravity waves as customarily understood is just ridiculous. Sciama's calculation (and others like it) sidestep this problem. But if you insist on assuming that quadrupole radiation must be what causes the radiation reaction to produce inertia, then the quadrupole must involve some mass of the scale of the mass of the universe in addition to the mass of the accelerating local object.

Other issues beset Mach's principle and the origin of inertia. Best known perhaps is the relationalist/physical debate. In no small part, most of these issues trace their origin to the fact that Mach's principle has been conceived chiefly as an issue of cosmology and held hostage thereto. If we are to exploit Mach's principle in the quest for starships and stargates, however, we need a local operational definition, for although we can hope to control our local circumstances, there is no chance that we will ever be able to manipulate the whole cosmos.

With this in mind, we can identify two laws of inertia that we attribute to Mach, Einstein, and Sciama as they put the issue on the table in the first place:

- First law: $\varphi = c^2$ locally always; or, inertial reaction forces are due to the gravitational action of causally connected "matter," where matter is understood as everything that gravitates.
- Second law: $m = E/\varphi$, or the mass of an entity (isolated and at rest) is equal to its non-gravitational energy divided by the locally measured total gravitational potential.

The second law follows from the first law, but note that it is not simply a restatement of the first law. The first law is that required to insure in a Sciama-type calculation that inertial reaction forces are due to gravity. As long as $\varphi = c^2$, it doesn't matter what the exact numerical values of φ or c are. So, in a gravity field, or accelerating reference frame, c might have some value other than 3×10^{10} cm/s, but inertial reaction forces would still be due to gravity.

The second law, suggested by the first, constrains matters by making the masses of objects depend on their gravitational potential energies. In effect, it has the gravitational field of cosmic matter play a role analogous to that attributed to the Higgs field in relativistic quantum field theory. It also fixes the locally measured value of c to be the same in *all* frames of reference, for the mass-energies of objects do not depend on whether they are moving inertially, or accelerating. The result of the Champney experiment tells us that the clock rate measured by a co-accelerating observer is independent of any acceleration combined with the fact that accelerations do not induce length contractions and assures us that the speed of light locally measured by an accelerating observer is the same as that for inertial observers. That is, quoting Thorne again, "space and time are relative. The speed of light is absolute."

Armed with the Mach-Einstein-Sciama laws of inertia, in a form applicable to local phenomena, we have a sufficient grasp of the nature of inertia to be able to look for ways to make starships and stargates. But the laws of inertia, by themselves, are not enough. If you are lucky enough to blunder onto the formalism for relativistic Newtonian gravity, interesting possibilities present themselves. Putting Newtonian gravity into relativistically invariant form leads to the introduction of time dependent terms in the familiar Newtonian equations. In particular, you get:

$$\nabla \bullet \mathbf{F} + \frac{1}{c}\frac{\partial q}{\partial t} = -4\pi\rho. \tag{9.1}$$

where ρ is the matter density source of the field $\mathbf{F}$, and q is the rate at which gravitational forces do work on a unit volume. The term in q in this equation appears because changes in gravity now propagate at the speed of light. It comes from the relativistic generalization of force, namely, that force is the rate of change in proper time of the four-momentum, as discussed in Chap. 1.

Can we treat the term in q as a transient source? After all, we can subtract that term from both sides of the equation, making it a source. Physically speaking, whether something gets treated as a field, or a source of the field, is not a simple matter of formal convenience. q is not $\mathbf{F}$, so transferring the term in q to the source side wouldn't obviously involve treating a field as a source. But q may contain a quantity that should be treated as a field, not a source. In the matter of rapid spacetime transport, this question has some significance because if the time-dependent term can be treated as a source of the gravitational field, then there is a real prospect of being able to manipulate inertia, if only transiently.

The way to resolve the issues involved here is to go back to first principles and see how the field equation is evolved from the definition of relativistic momentum and force. When this is done, taking cognizance of Mach's principle, it turns out that it is possible to recover not only the above field equation but also a classical wave equation for the scalar gravitational potential – an equation that, in addition to the relativistically invariant d'Alembertian of the potential on the (left) field side, has transient source terms of the sort that the above equation suggests might be possible. *But without Mach's principle in the form of the formal statement of the laws of inertia, this is impossible.*

The procedure is straightforward. You assume that inertial reaction forces are produced by the gravitational action of the matter in the universe, which acts through a field. The field strength that acts on an accelerating body – written as a four-vector – is just the inertial reaction four-force divided by the mass of the body. That is, the derivative with respect to proper time of the four-momentum divided by the mass of the body. To put this into densities, the numerator and denominator of the "source" terms get divided by the volume of the object. In order to get a field equation of standard form from the four-force per unit mass density, you apply Gauss' "divergence theorem." You take the four-divergence of the field strength. Invoking Mach's principle judiciously, the field and source "variables" can be separated, and a standard field equation is obtained.

$$\nabla^2\varphi - \frac{1}{c^2}\frac{\partial^2\varphi}{\partial t^2} = 4\pi G\rho_0 + \frac{\varphi}{\rho_0 c^2}\frac{\partial^2\rho_0}{\partial t^2} - \left(\frac{\varphi}{\rho_0 c^2}\right)^2\left(\frac{\partial\rho_0}{\partial t}\right)^2 - \frac{1}{c^4}\left(\frac{\partial\varphi}{\partial t}\right)^2, \qquad (9.2)$$

or, equivalently (since $\rho_0 = E_0/c^2$ according to Einstein's second law, expressed in densities),

$$\nabla^2\varphi - \frac{1}{c^2}\frac{\partial^2\varphi}{\partial t^2} = 4\pi G\rho_0 + \frac{\varphi}{\rho_0 c^4}\frac{\partial^2 E_0}{\partial t^2} - \left(\frac{\varphi}{\rho_0 c^4}\right)^2\left(\frac{\partial E_0}{\partial t}\right)^2 - \frac{1}{c^4}\left(\frac{\partial\varphi}{\partial t}\right)^2. \qquad (9.3)$$

where ρ_0 is the proper matter density (with "matter" being understood as everything that gravitates) and E_0 is the proper energy density. The left hand sides of these equations are just the d'Alembertian "operator" acting on the scalar gravitational potential φ. "Mach effects" are the transient source terms involving the proper matter or energy density on the right hand sides.

The terms that are of interest to us are the transient source terms on the right hand sides in these equations. We can separate them out from the other terms in the field equation, getting for the time-dependent proper source density:

$$\delta\rho_0(t) \approx \frac{1}{4\pi G} \left[\frac{\varphi}{\rho_0 c^4} \frac{\partial^2 E_o}{\partial t^2} - \left(\frac{\varphi}{\rho_0 c^4} \right)^2 \left(\frac{\partial E_0}{\partial t} \right)^2 \right] \qquad (9.4)$$

where the last term on the right hand side in the field equation has been dropped, as it is always minuscule. The factor of $1/4\pi G$ appears here because the parenthetical terms started out on the field (left hand) side of the derived field equation. If we integrate the contributions of this transient proper matter density over, say, a capacitor being charged or discharged *as it is being accelerated*, we will get for the transient total proper mass fluctuation, written δm_0:

$$\delta m_0 = \frac{1}{4\pi G} \left[\frac{1}{\rho_0 c^2} \frac{\partial P}{\partial t} - \left(\frac{1}{\rho_0 c^2} \right)^2 \frac{P^2}{V} \right] \qquad (9.5)$$

where P is the instantaneous power delivered to the capacitor and V the volume of the dielectric. If the applied power is sinusoidal at some frequency, then $\partial P/\partial t$ scales linearly with the frequency. So, operating at elevated frequency is desirable. Keep in mind here that the capacitor must be accelerating for these terms to be non-vanishing. You can't just charge and discharge capacitors and expect these transient effects to be produced in them. None of the equations that we have written down for Mach effects, however, show the needed acceleration explicitly. Writing out the explicit acceleration dependence of the Mach effects is not difficult. We need to write the first and second time-derivatives of the proper energy in terms of the acceleration of the object. The mathematics is straight-forward, though some important technical details need to be taken into account.

The mass fluctuations predicted by Eqs. 9.4 and 9.5 are the Mach effects. Examination of Eq. 9.5 makes plain that the two terms are normally of very different magnitudes owing to the way the inverse square of the speed of light enters into their coefficients. But their coefficients also contain the proper matter density in their denominators, so if that can be driven to zero using the first, larger term on the right hand side of Eq. 9.5, both coefficients can be driven toward infinity. Since the coefficient of the second, normally much smaller term is the square of the coefficient of the first term, it will blow up much more quickly than the first, making the second term dominate. It is always negative. So a fluctuating mass with a negative time-average should result. Since the first term is normally very much larger than the second, and since it is required to create the circumstances where the second term becomes important, experiments to check on the reality of these predicted effects have, and should, focus on the first term.

Detection of first-term Mach effects is done by producing stationary forces – thrusts – in systems where the mass fluctuation can be acted on by a second force, periodic at the frequency of the mass fluctuation. When the relative phase of the mass fluctuation and second force is adjusted so that the force acts in one direction as the mass fluctuation reaches its maximum, and the opposite direction when the fluctuation is at minimum, the result is a stationary force acting on the system. The momentum imparted by this force is compensated for by a corresponding momentum flux in the gravinertial field that couples the system to chiefly distant matter in the universe. Although looking for such thrusts only confirms, if found, the existence of the first Mach effect, since one cannot be present without the other, thrust effects are indirect confirmation of the second effect that is needed to construct stargates.

Experimental work (Chaps. 4 and 5) designed to look for Mach effects accompanied theoretical work from the outset. For many years, the full import of the condition of acceleration in the prediction of effects was not fully appreciated. Accordingly, when it was, sometimes inadvertently, designed into test systems, interesting results were found. When proper account of the condition was not taken, results were equivocal at best. Other issues relating to the availability of technology and the level of support for the work were complicating factors in developments. Some of these were "external" in the sense that they involved people not directly associated with CSU Fullerton. This was especially the case in the matter of attempted replications of the work at CSU Fullerton. Eventually, Nembo Buldrini called attention of those interested to the "bulk acceleration" issue, and the haphazard treatment thereof in the design of experiments stopped.

The devices eventually settled on as those most promising for the development of Mach effect thrusters were stacks of disks of lead-zirconium-titanate (PZT) 1 or 2 mm thick, 19 mm in diameter, and 19 mm in length when assembled. The stacks were clamped between a brass reaction disk 9 mm thick and aluminum cap 4.5 mm thick, both being 28 mm in diameter, secured with 6, 4–40 stainless steel socket head machine screws. These assemblies were mounted on a thin aluminum bracket bolted into a small, mu metal lined, aluminum box that acted as a Faraday cage. This was mounted on the end of a very sensitive thrust balance suspended with C-Flex flexural bearings and equipped with coaxially mounted galinstan liquid metal contacts for power transfer to the device being tested. These contacts were designed to minimize torques that might act on the thrust balance beam when power was flowing to the test device.

The mount on the end of the thrust balance beam was designed so that the direction in which the test device pointed could be rotated in a plane perpendicular to the beam by loosening a single nut. Perhaps the most important protocol used to test the thrust being measured was to reverse the direction of the test device to see if the thrust in question reversed direction, too. A number of other tests were also done to make sure that observed thrust signals did not arise from mundane or spurious causes. With some optimization, it proved possible to get these devices to produce thrusts on the order of 10 microNewtons or more – signals easily seen in real time in single cycles. The measured thrusts, that match predictions based on idealized operation to better than order of magnitude, clearly carry the signature of the first term Mach effect in operation.

Thorne's work on wormholes and the publication of the warp drive metric by Alcubierre in the late 1980s and early 1990s (Chap. 6) changed the field of advanced propulsion in a fundamental way. Before the mid-1990s, almost no attention was paid to seriously advanced propulsion. The only non-chemical propulsion scheme that got any attention at all was "electric" propulsion, the iconic realization thereof being "Deep Space One." Even the schemes that had been formulated in the 1950s and 1960s that might prove advances on the chemical had been ignored in the heyday of the space shuttle, a program that had limited space exploration to low Earth orbit and become in the eyes of some little more than a jobs program for an agency that had lost its way.

Wormholes and warp drives, notwithstanding that there seemed no plausible chance that they might be practical in the foreseeable future, had reignited interest in the quest for human exploration of deep space. NASA administrator Daniel Goldin tried to get the bureaucracy working toward that goal, with only very modest success. Physics that might enable revolutionary advances in propulsion were explored. One group, convinced that "zero point" fluctuations in the quantum vacuum were real sought to show that were that true, propulsive advantages might be enabled. Others dreamed of accessing the hypothetical quantum spacetime foam that putatively might exist at the Planck scale – many orders of magnitude smaller than even the tiniest entities accessible with present and foreseeable techniques. Other schemes were proposed; for example, Jack Sarfatti's proposal to "soften" spacetime using superconductors and warp it by having the superconducting materials also be metamaterials in which he hoped negative energy might be induced. Aside from Sarfatti's scheme, none of the proposals even hinted at how their goals might be achieved.

Faced with the Jupiter mass of exotic matter requirement for both practical wormholes and warp drives, it is evident that if they are to be realized, a way must be found to transform the constituents of reality as we find it into the sought-after stuff (Chap. 7). Although quantum electrodynamics suggests that vast reservoirs of exotic matter lie hidden in the "bare" masses of normal matter, the theory gives no indication whatsoever that the bare masses of elementary particles can be exposed and put to use. The extension of quantum electrodynamics – the Standard Model of relativistic quantum field theory – modeled on quantum electrodynamics, since it too is based on the renormalization program, give no indication that exotic bare masses of elementary particles can be exposed. If a program to expose gargantuan exotic bare masses of elementary particles is to succeed, it must in some way be fundamentally different from the Standard Model.

Should you go looking for some alternative to the Standard Model, and you are informed about general relativity and Mach's principle, the obvious thing to note is that the Standard Model does not include gravity. This is hardly secret knowledge or fringe fantasizing. The universally acknowledged outstanding problem of modern field theory is reconciling general relativity and quantum mechanics – in the putative theory of quantum gravity. Candidates for a theory of quantum gravity already exist, of course: superstring theory and loop quantum gravity. Neither of these theories, however, hold out any prospect of enabling the exposure of enormous exotic bare masses.

Curiously, a widely unappreciated electron model created in 1960 by Arnowitt, Deser, and Misner (ADM) – an exact solution of Einstein's equations for a spherical ball of electrically charged dust with an arbitrary bare mass included – considered in conjunction with Mach's principle holds out the promise of a realistic, semi-classical electron model that includes the features sought for making wormholes and warp drives.

The ADM model was created in the heyday of the "positive energy theorem," which prohibits the existence of negative mass. If you believe this theorem, then the ADM electron model returns a fascinating but completely unrealistic solution for the mass of the electron. As Abhay Ashtekar noted in a Texas symposium talk in 1989, the ADM general relativistic model gives back a finite mass for the electron without invoking perturbation expansions and renormalization at all. Indeed, the bare mass of the electron, whatever it may be, doesn't even appear in the solution. So you can make it anything you want. The problem with the model is that the ADM mass for the electron, equal to the square root of the electric charge squared divided by the gravitation constant, in CGS units is 21 orders of magnitude larger than the observed mass. Interestingly, this is just the magnitude of the square of the speed of light.

Thorne, in his exploration of wormholes, changed the positive energy dogma by showing that negative mass is a fact of our reality. This opens the way to explore whether the ADM model can be used to discover whether normal matter might be operated upon to expose its presumed gargantuan exotic bare mass. Care must be taken in such an investigation to account for the fact that the Equivalence Principle requires that if any of the active, or passive, gravitational masses, or the inertial mass of an object is negative, then all of the others must be negative, too. In addition to this precaution, account must also be taken of the actions of the local matter *and* cosmic matter in constructing the model. This done, the negative bare mass ADM model is found to have a solution that in principle can return a realistic mass for the electron. And it also shows that the bare mass of the charged dust assumed to make up electrons is the ADM mass – but a negative ADM mass.

Moreover, implicit in the solution is the fact that if could we find a way to isolate some normal matter from the gravitational influence of the universe, the bare masses of the screened elementary particles would be exposed. This solves the assembly problem – putting together the constituents of the Jupiter mass in a structure of modest dimensions – a few tens of meters at most.

The fly in the negative bare mass ADM ointment is that the minimum energy solution of the equation for the electron mass is either zero or negative and large, depending on how you define the energy minimum.[1] It's not the observed, small positive mass of experience. What's missing? Spin and quantum mechanics, not quantum gravity, plain old quantum theory. But there are problems. To get a small mass for the electron, the electrically charged dust must lie very near to its gravitational radius, and that turns out to be near the Planck length – about 10^{-33} cm. Even at the "classical" electron radius, some 20 orders of magnitude larger, in order to get back the magnetic moment of the electron, it must be spinning with a surface velocity more than 100 times the speed of light (which is why such models were abandoned in the mid-1920s). How could an object with Planck length dimensions possibly be realistic?

Negative bare mass. The ADM model is constructed in isotropic coordinates. For positive mass objects, the speed of light near a local matter concentration decreases for distant observers in these coordinates. (Light stops at a black hole horizon as viewed by distant observers.) When the local mass concentration is negative, however, the speed of light measured by distant observers increases near the gravitational radius, going to infinity.

---

[1] If one takes the absolute value of the energy, then zero is the minimum energy. The physical reason for doing this is that negative energy, like positive energy, has real physical consequences. The energy minimum can be defined as the state where energy has the smallest physical consequences, and that is the closest state to zero energy, not some ridiculously large negative value.

Evidently, if the electrically charged dust circulates near to its gravitational radius, it can do so at less than the speed of light observed locally but highly superluminally as observed from a distance, and the measured magnetic moment can be generated by an exceedingly small structure without violating the relativistic light speed limit.

The question remains: Can spin, angular momentum, and magnetic moment be incorporated into the negative bare mass ADM model? Yes. Both the angular momentum and magnetic moment can be put into a form with the same radial dependence as the electrical self-energy. So the only modification needed is the addition of a constant coefficient for this term. Quantum theory enters analogously to the way Bohr dealt with orbital motion of the electron in the hydrogen atom. The quantum condition is that the angular momentum is quantized in integral multiples of Planck's constant. This solves the minimum energy problem, since the lowest energy state is a half quantum of angular momentum, not zero. The energy of the electron in this model turns out to be dominated by the energy in the magnetic field due to the electron's spin – as realized must be the case many years ago by Asim Barut (and Lorentz before him). And Barut, though he didn't have the negative bare mass ADM model to underpin his conjectures, provided a bonus – the energy quantization formula for the charged leptons when the energy is dominated by the magnetic moment. All of this without quantum gravity, or a unified field theory. And we know what must be done if we want to make traversable absurdly benign wormholes.

STARSHIPS VERSUS WORMHOLES

So far we have talked about starships and wormholes as if they put essentially the same demand on resources – a Jupiter mass of exotic matter. If we assume that the Alcubierre metric must be achieved to make a starship, then this equating of starships and wormholes is justified. But we should consider the possibility that other methods of making starships that put less stringent demands on our resources and technology might be possible. The method that comes to mind in this connection is the "negative mass drive." It consists of a craft with some mass M and separate drive mass $-M$ arranged to be held at some small distance from our inhabited craft.

If you analyze the forces acting in such a situation, you will discover that the two masses move off in the line from the negative to the positive mass, and they accelerate spontaneously. They keep accelerating indefinitely. You might think energy and momentum conservation are violated by such behavior, but they aren't. Why? Because the mass of the accelerating system is zero. Relativistic considerations eventually come into play. But such a system should be able at least to achieve some non-negligible fraction of the speed of light. And aside from the cost, whatever it might be, of maintaining the exoticity of the driver mass, no energy at all is required to achieve these high velocities. It's not warp drive. But you can build credible starships this way.[2]

---

[2] John Brandenburg has argued for this approach, pointing out that effectively one is flattening spacetime in the vicinity of the craft, making it gravinertially invisible to the rest of the universe. A tall order, but peanuts compared with trying to make a wormhole.

So, the minimum requirement for a starship would seem to be the production of some large amount of exotic matter. Enough to cancel the mass of some vehicle with a mass on the order of thousands of kilograms. How much exotic matter do we require to drive the formation of a wormhole? A lot more – the Jupiter mass – which we now know to reside cloaked from view in normal matter. That was the lesson of Chap. 7. Recall that what we found for the negative bare mass ADM electron was that:

$$ m = -\frac{\sqrt{\frac{e^2}{G}}}{\left[1 - \frac{2c^2}{\varphi_i}\right]} = -\frac{\sqrt{\frac{e^2}{G}}}{\left[1 - \frac{2c^2}{\varphi_u + \varphi_b}\right]}. \tag{9.6}$$

where m is the electron mass, e its charge, G the gravitation constant, c the speed of light, and φ the gravitational potential. Note that we have written the potential in the dust (subscript i) as the sum of its constituent parts, the potential due to the universe (subscript u) and the potential due to the bare mass (subscript b). That is:

$$ \varphi_i = \varphi_u + \varphi_b. \tag{9.7}$$

When φ_u and φ_b are almost exactly equal and opposite, with a difference of order unity, as they are in this case since φ_b is negative owing to the negative bare mass of the charged dust, the denominator of the RHS of Eq. 9.6, up to a factor of order unity, becomes $\sim -c^2$ and the actual mass of the electron is recovered. The presence of φ_i in Eq. 9.6 insures that the dimensions of the electron mass are correct. Now, if φ_u can be reduced to a small number or zero, then the exotic bare masses of the particles that make up some object are exposed, and this turns out to be our sought after Jupiter mass for objects of suitable dimensions.

The question then is: Is there any way, with a modest expenditure of energy and time, to expose the exotic rest masses of an object by suppressing φ_u? One might think that simply producing enough exotic matter to compensate for some local positive matter might do the trick. But that is a mistake. It gives back the negative mass drive just mentioned. It does not expose the bare masses of local elementary particles. To do that, we need enough exotic matter to create a local gravitational potential equal to $-\varphi_u \approx -c^2$. But the non-linearity of Eq. 9.4 and the presence of the rest mass density in the denominators of the coefficients suggests that only a local compensating negative mass that drives ρ_0 to zero might be sufficient to trigger a process that gets us our Jupiter mass of exotic matter without the expenditure of gargantuan amounts of energy.

How does this relate to the stuff needed for our traversable wormhole? Well, from Morris and Thorne's 1988 paper we know that the line element (or metric if you prefer) for a general spherical wormhole is:

$$ ds^2 = -e^{2\Phi}c^2 dt^2 + \frac{dr^2}{1 - b/r} + r^2\left(d\vartheta^2 + \sin^2\vartheta d\varphi^2\right). \tag{9.8}$$

Φ is the "redshift" function and b the "shape" function of the wormhole (see Fig. 9.1). The minimum traversability conditions require that Φ^2 be everywhere finite – that is, no horizons – and that b not involve excessive curvature. For an absurdly benign wormhole

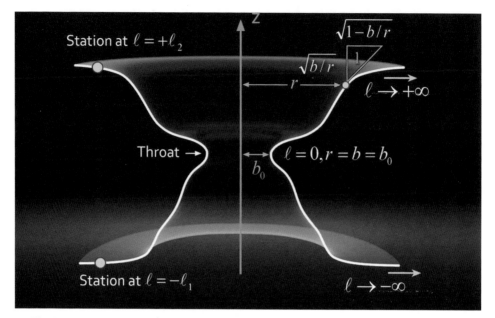

Fig. 9.1 After the figure in Morris's and Thorne's paper on wormholes where the various quantities in their equations are illustrated

the exotic matter is confined to a thin layer of thickness a_o around the throat with radius b_o. The volume of the exotic matter is $4\pi b_o^2 a_o$ and its density is:

$$\rho = -\frac{b_o c^2}{4\pi G r^2 a_o} \left[1 - (r - b_o)/a_o\right]. \tag{9.9}$$

From the density and volume we find that the mass of the matter supporting the throat is:

$$M = -\frac{\pi b_o c^2}{2G}, \tag{9.10}$$

and,

$$-\frac{2GM}{\pi b_o c^2} = 1. \tag{9.11}$$

Aside from the minus sign and factor of π in the denominator, this is just the horizon condition for a Schwarzschild black hole. It follows that we need to at least transiently make enough exotic matter to form an absurdly benign wormhole in order to expose the bare masses of the normal matter initially in the throat to transform it into exotic matter to stabilize the throat. So we know where the exotic matter we need lies, but we have to transiently create the same amount to expose it to stabilize the wormhole we want to make. Doesn't sound very promising, does it?

MAKING WORMHOLES

To make a wormhole, the proper matter density of the matter in the region that is to become the throat must first go to zero, and then become enormously negative. The only hope of doing this lies in the transient effects in the Mach effect equation, Eq. 9.4.

Let's say that we want to drive a state of exoticity by applying an AC signal to a capacitative element in a tuned LC resonant circuit in the laboratory – *taking care to make sure that the requisite acceleration of the capacitor also is present*. To achieve net exoticity, we will need to drive a mass density fluctuation that is as large as the quiescent density of the material absent the AC signal (and acceleration). In all but two of the experimental projects reported in Chaps. 4 and 5, since only a thrust effect was sought, no attempt was made to drive the devices tested to the point where exoticity might be manifested. In those circumstances, the variation of ρ_o in the denominators of the coefficients of the transient terms could be ignored. When going after exoticity is the goal, this is no longer possible. For as ρ_o goes to zero, the coefficients blow up, becoming singular when $\rho_o = 0$.

Equation 9.4 is non-linear. It does not have an analytic solution. If you want to examine its behavior, numerical techniques must be employed. They need not be elaborate, for after all, we are not engineering a real starship or stargate here. (That is left as a homework problem for the interested reader.) We first write the proper matter density in Eq. 9.4 as the proper energy density divided by the square of the speed of light and note that the proper energy density has two components: the quiescent contribution present in the absence of mass fluctuations, and that due to the energy density that accompanies the mass fluctuations. We assume that the energy density fluctuation driven by our applied AC signal is sinusoidal. When we compute the first and second time-derivatives of the proper energy density, the quiescent term drops out, as it is constant, and we get:

$$\frac{\partial^2 \rho_o}{\partial t^2} = -\frac{E_d \omega^2}{c^2} \sin \omega \, t \tag{9.12}$$

$$\left(\frac{\partial \rho_o}{\partial t}\right)^2 = \left(\frac{E_d \omega}{c^2}\right)^2 \cos^2 \omega \, t \tag{9.13}$$

where E_d is the amplitude of the energy density driven by the AC signal and ω is the angular frequency of that signal. We now define:

$$k_1 = \frac{E_d \omega^2 \varphi}{4\pi G c^4} \tag{9.14}$$

$$k_2 = \frac{(E_d \omega \varphi)^2}{4\pi G c^8} \tag{9.15}$$

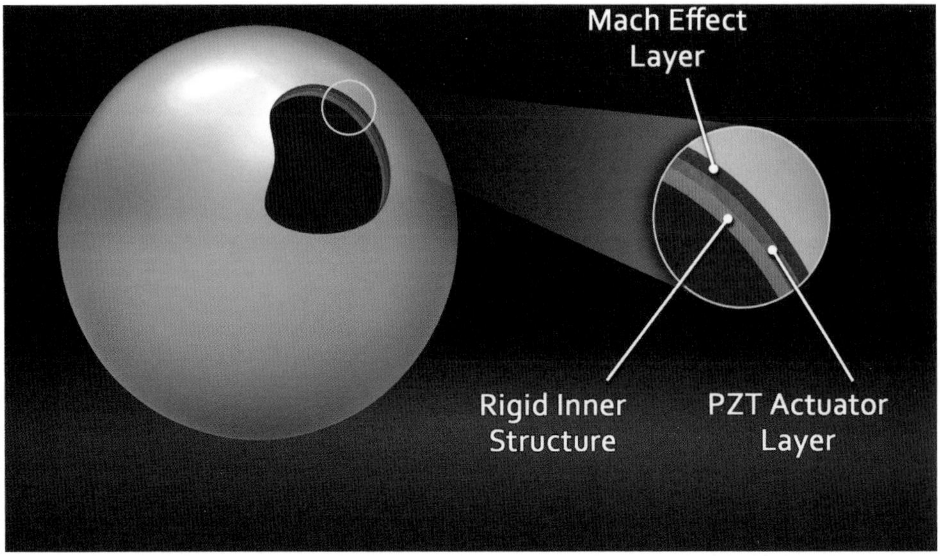

Fig. 9.2 A schematic of a simple spherical Mach effect wormhole generator. A rigid interior structure supports a ferroelectric actuator that provides the acceleration of an outer layer wherein internal energy changes are driven, promoting the Mach effect

We can now write down the equation for the total density:

$$\rho = \rho_o - \frac{k_1}{\rho_o}\sin \omega t - \frac{k_2}{\rho_o^2}\cos^2 \omega t \qquad (9.16)$$

To solve this equation we specify some initial values of ρ_o, k_1, k_2, ω, and t, and compute the value of ρ. We then choose a time step size, $\delta t\,(t+\delta t)$, and using the value of ρ computed in the previous step for the new value of ρ_o, compute a new value of ρ for this time step. This procedure is iterated *ad nauseam*. As long as ρ_o doesn't become exactly zero, the results remain finite and reasonable. Only at $\rho_o = 0$ and its very immediate vicinity are the results questionable. But we can improve the results of the integration in this vicinity by using a very small step size.

Now that we have the Mach effect equation in a useful form for constructing wormholes, namely, Eq. 9.16, to ground our discussion we imagine the qualitative features of a very simple wormhole generator. It is a multilayer sphere some 5–10 m in diameter, as shown in Fig. 9.2. If you prefer to think in terms of starships, you can imagine the multilayer structure to be the ring surrounding the vehicle in Fig. 9.3.[3]

---

[3] Alas, imagining the starship *Enterprise* from the *Star Trek* TV shows and movies will not do. When the models for the series were designed, no one had even a clue as to how warp drive might work, so they winged it. The saucer body (ripped off from flying saucers of pop culture) and power nacelles (envisioned as keeping devices producing lethal levels of radiation away from the crew, no doubt) do not have the geometry needed to deal with the exotic matter we now know required to make a starship.

Fig. 9.3 A ring-style wormhole generator ranged around a starship in the act of forming a wormhole to enable quick transit back to Earth. Nembo Buldrini calls this figure, "Going Home"

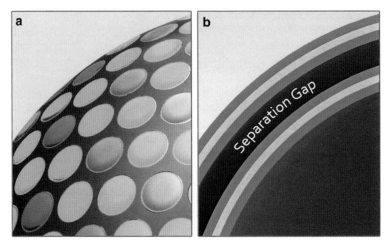

Patches arranged and separated for optimum phase delay.

Concentric layers spaced for light travel time at operating frequency.

Fig. 9.4 Two possible dispositions of Mach effect elements arranged to take advantage of the phase delay between the elements arising from the light speed propagation of the effects of one on the other

Ring or sphere, the layering of the structure is straightforward. The innermost layer will be quiescent from the electro-mechanical point of view. Its purpose is to be the matter whose exotic bare mass is exposed when the outer layers of the structure are driven so as to gravitationally decouple it from the rest of the universe. As such, in a sense, it doesn't really matter too much what this layer is made of save for one consideration − it should be rigid so that it acts in bulk as the reaction mass for the outer layers. That is, it is the analogue for this device of the reaction masses used in the test devices described in Chaps. 4 and 5 (Fig. 9.4).

The next innermost layer in our structure will be the electromechanical actuator layer, presumably made of PZT or some similar substance optimized to produce the acceleration of the outermost layer where we seek to drive the Mach effects. There are a number of technical details that go into the design of this part of the structure. One is whether to use materials that rely on the piezoelectric (first order in the frequency of the applied voltage) or electrostrictive (second order) effects.

Piezoelectric materials must be polarized and are susceptible to depolarization, whereas electrostrictive materials do not display that behavior. Another is the thermal dissipation of the material at the design operating frequency. Inspection of the coefficients of the trigonometric terms in Eq. 9.16 shows that you will want to operate at the highest possible frequency. Practically speaking, since "ionic" response produces the largest effects in the materials of interest, that means working in the 2–3 GHz range. As for the outermost layer, where the Mach effects are driven, you will want materials in which the largest internal energy fluctuations can be driven. That means materials with the highest possible dielectric constants.

Turning away from practical concerns, we now ask: What we have to do to turn our sphere or ring into a wormhole generator? You might think that we could just turn things on, get the system cycling, and wait for wormhole formation to develop. But there is no reason to believe that this will be the sort of cumulative process that would support such a scenario. This is a process that, given the proposed configuration of the system, must be accomplished in a single cycle of the applied AC voltage. So, after turning everything on at low power, we must jack the voltage up quickly to some supposed threshold where wormhole formation proceeds. When is that? When the coefficients of the trigonometric terms in Eq. 9.16 become singular.

For the moment, let us assume that we can do this – that is, that the power available to us is sufficient to drive ρ_o to zero in Eq. 9.16. An important detail calls for our attention at this point. Not only does the sine of ωt go to zero as ωt passes through zero and 180°, its sign reverses, too. So the first transient term can be both positive and negative in a single cycle. We will want to engineer things so that $\sin \omega t$, initially positive along with ρ_o, reverses sign by passing through $\omega t = 0$ as ρ_o passes through 0 so that the sign of the first transient term remains negative through this part of the transition cycle.

Whether $\rho_o = 0$ at exactly the same instant as $\omega t = 0$ is not a matter of critical concern, for $\cos^2 \omega t = 1$ when $\omega t = 0$, and since this term is negative and dominates the equation when $\rho_o = 0$, exoticity will prevail. And if we get everything just so, the magnitude of that exotic state is formally infinite. But we do not want that condition quenched by the first transient term changing sign and becoming large and positive immediately thereafter.

How realistic is all this? Let's run some numbers.

Let's say that $k_1 = 100$. We take the power frequency of the AC driving signal to be 1 MHz. This constrains the power density $[E\omega]$ to be 1.2 kw/cm^3. k_2 is constrained to the value 2.1 $\times 10^{-16}$. For laboratory situations where we are dealing with a few cubic centimeters of material or less, these values are easily attainable. But this value of k_1 requires a megawatt of power per cubic meter, so realistic scale structures would be more power hungry if $k_1 > 100$ were required. But k_1 was chosen to be 100 arbitrarily. No doubt it can be much less than 100; indeed, likely less than 10. It depends on the initial value of ρ_o, which is almost always less than 10. That's a kilowatt per cubic meter, and a megawatt for a thousand cubic meters. And it's reactive, not dissipated, power. So, if all of the auspicious phasing can be engineered and Mach effects really exist, it seems that absurdly benign wormholes and warp drives lie in our future. And if they lie in our future and topology change is possible, then such things likely lie in our past as well.

AN ALTERNATE SCENARIO

You may be thinking to yourself, "Yeah, right." Turn your warp drive or wormhole generator on, ramp it up towards the threshold power, and then ZAP! You're on your way. Where's the bridge? The Brooklyn, not Einstein-Rosen type.

It's easy to believe that the just so conditions required to produce the effects that drive the singular behavior that enables gravitational decoupling may be impossible to generate,

Table 9.1 Density as a function of initial density and ωt

ωt (rad.)	Initial density (gm/cm^3)				
	$-.001$	-0.01	-0.1	-1.0	-10
0.002	-200	-20	-2.1	-1.2	-10
3.142	-663.4	-632.9	-632.6	-632.6	-632.7
6.284	-200.4	-26.3	-17.6	-17.5	-20.0
59.69	-664.5	-634.8	-634.6	-634.6	-634.6
62.83	-204.2	-56.2	-53.0	-53.0	-53.8
625.2	-675.2	-650.3	-650.1	-650.1	-650.2
628.3	-236.5	-151.8	-150.9	-150.9	-151.1
6280	-748.6	-735.7	-735.7	-735.7	-735.7
6283	-400.5	-375.9	-365.8	-375.8	-375.8

or impossible to make stable. So we ask: Is there another way to produce absurdly benign wormholes without using second-term Mach effects? Happily, the answer to this question is yes. The method is a bit more complicated than that already discussed. But it is still simple.

The method in question is a bootstrap method. It depends on the finite propagation time of gravinertial signals over finite distances owing to the relativistic light speed velocity limitation on such signals. We will assume that the singular behavior for the coefficients of the trigonometric terms in Eq. 9.16 when $\rho_o = 0$ is somehow suppressed. This may be due to some unspecified damping process, or the granularity of matter at the microscopic scale, or whatever. We do, however, assume that for this to happen in a system like that assumed in our discussion of singularity-driven exoticity exposure in the previous section. Since singularity behavior is assumed suppressed, our governing equation no longer has the second Mach effect term in it, as that is only important at or very near to the singularity. That is, in the present case Eq. 9.16 becomes:

$$\rho = \rho_o - \frac{k_1}{\rho_o} \sin \omega t \qquad (9.17)$$

Although we will not be taking advantage of singular behavior, we still arrange things so that $\sin \omega t$ and ρ_o go to zero and change sign together. That way, just beyond the suppressed singularity the trigonometric term in Eq. 9.17 will be negative, and because a very small ρ_o will occur in the denominator of its coefficient that negative term will be very large.

Since $\sin \omega t$ is both positive and negative, as the cycling of the AC power signal it describes continues, you might be inclined to think that ρ must always be an oscillating function that is both positive and negative. But you would be mistaken. Were k_1 and ρ_o constants, oscillation between positive and negative values would take place. But ρ_o is not a constant. In our numerical integration, ρ_o is just the computed value of ρ in the previous step of the calculation. We can show that the curious result $\rho(t) \leq 0$ for all t after a specified t by doing another numerical integration of the sort already invoked. We start at a milliradian of ωt beyond $\sin \omega t = 0$ and assume a wide range of initial densities. The results of these calculations, assuming $k_1 = 100$ for a set of typical multiple of π

Table 9.2 Density as a function of k_1 and ωt for initial density $= -0.1$ g/cm^3

ωt (rad.)	k_1		
	25	50	100
3.142	−316.3	−447.3	−632.6
6.284	−8.8	−12.4	−17.6
59.69	−317.3	−448.7	−634.6
62.83	−26.5	−37.5	−53.0
625.2	−325.1	−459.7	−650.1
628.3	−75.4	−106.7	−150.9
6280	−367.8	−520.2	−735.7
6283	−187.9	−265.7	−375.8

radians, are given in Table 9.1. The same sort of behavior is found when the value of k_1 is varied (for an initial density of −0.1 g/cm^3), as shown in Table 9.2.

Not much reflection on the density values in Tables 9.1 and 9.2 is required to discern that the negative going evolution is not a linear function of ωt. One might have to cycle such a system for some ridiculously long time in order to get to the point where the bare masses of nearby matter would be exposed. Does this mean that we have to go back to the singularity-driven model of a wormhole generator? Not necessarily. Remember the bootstrap business? We can augment the behavior captured in Tables 9.1 and 9.2 by, in effect, doubling down. That is, instead of relying only on the production of a Mach effect in our Fig. 9.2-type system, we either duplicate that system or we make another circuit element of the power system, an inductor that produces resonance in the power system say, produce the same sort of Mach effect behavior. We arrange these two Mach effect parts of the complete system so that the time delay between them puts them 180° out of phase. In this way, each of the two Mach effect subsystems see the state of the other at an earlier time when the other subsystem was in the state it is at any given moment. This way, when each subsystem becomes negative, it does so in the presence of a negative gravitational potential created by the other subsystem a half-cycle earlier, making its negativity – as determined by distant observers – more negative than it otherwise would be. In this way we should be able to finesse the negativity issue.

Much of the foregoing was already worked out long ago and published as "Twists of Fate: Can We Make Traversable Wormholes In Spacetime?" [*Foundations of Physics Letters* **10**, 153–181 (1997)], a paper marred by typographical errors in some of the equations. The bootstrap method just outlined was developed there in terms of the ADM electron model in a bit greater detail. That part of that paper is reproduced here as an addendum at the end of this chapter. If you had tried to build a wormhole generator on the basis of the material presented in that paper, it wouldn't have worked. The importance of the "bulk acceleration" condition in generating Mach effects was not then fully appreciated, so no provision for acceleration of the material to which the AC power signal is applied was noted. And in 1997 no experiments had been done to try to see the second Mach effect. That was not attempted until after Tom Mahood's Master's thesis defense where Ron Crowley and Stephen Goode pointed out that one should be able to see such effects in the lab. Now some experimental evidence suggests that the second Mach effect really does exist.

At least one issue remains to be addressed here: What about quantum gravity? What does it have to do with all this? Well, in terms of the sort of approach Ashtekar motivated with the ADM electron model many years ago – loop quantum gravity, as it is now known – nothing. The negative bare mass, spin-modified ADM model describes objects much larger than the Planck scale, and quantum gravity is supposed to be significant only at or near the Planck scale. Taking the magnetic self-energy of the electron into account yields a factor of $\sim 10^{11}$ for the square root of A, the coefficient of the electrical self-energy term in the ADM model we introduced to account for spin, and the radius of the spin modified negative bare mass ADM electron is larger than the classical ADM electron by this amount. That is, this model gives the electron a radius on the order of 10^{-23} cm, several orders of magnitude smaller than the present level of detectability but a decade of orders of magnitude larger than the Planck length. Electrons, with their humongous exotic bare masses, are *not* firmly locked in the embrace of quantum gravity, it would seem. So perhaps there is hope that they may be exposed to make stargates without first mastering quantum gravity.

ADDENDUM

From "Twists of Fate: Can We Make Traversable Wormholes in Spacetime?"

First we note that in relativistic gravity, the Newtonian gravitational potential propagates as light speed. So the changing instantaneous mass of each of the circuit elements is only detected at the other circuit element after a finite time has elapsed. We make use of this fact by adjusting the distance between the circuit elements, mindful of the signal propagation delay. The trick is to adjust the distance between the L and C components so that just as one component – say, C – is reaching its peak transient negative mass value, the delayed (that is, retarded) gravitational potential of the other component – L – seen by C is also just reaching its peak negative value at C. As far as *local* observers in proximity to the L and C components are concerned, this will appear to make no difference, for the *locally measured* value of the total gravitational potential, like the vacuum speed of light, is an invariant [Woodward, 1996a]. But distant observers will see something different, since neither of these quantities are global invariants.

To see what will happen from the point of view of distant observers we employ the ADM solution, Eq. [7.23]. Since this solution, as previously noted, is obtained for isotropic coordinates, we can use it unmodified for distant observers. We now remark that to get back the electron's mass (to within 10%) we must have:

$$\varphi_u + \varphi_b = 1, \tag{5.14}$$

yielding:

$$m \approx \frac{\sqrt{e^2/G}}{\frac{2c^2}{\varphi_u+\varphi_b}}, \tag{5.15}$$

as long as $\varphi_u + \varphi_b \ll c^2$. As mentioned above, $\varphi_b \approx -c^2$ because the dust bare mass is negative and concentrated at its gravitational radius. This does not change (for either local or distant observers). $\varphi_u \approx c^2$ doesn't change for local observers, either. But for distant observers φ_u does change because it, for them, is the sum of the potential due to cosmic matter, φ_c, and the potential due to the companion circuit element, φ_{ce}, that is, the potential produced by L at C in the case we are considering.

We next write:

$$\varphi_u + \varphi_b = \varphi_c + \varphi_{ce} + \varphi_b. \tag{5.16}$$

This expression, if $\varphi_{ce} = 0$, is just equal to one [as in Eq. (5.14)]. But if the mass of L seen at C is negative, then $\varphi_{ce} < 0$ and the expression is less than one. To see the effect of L at C on m we take $\varphi_c + \varphi_b = 1$ in Eq. (5.16), substitute in to Eq. (5.15) and do a little rearranging to get:

$$m \approx \frac{\sqrt{e^2/G}}{2c^2} (\varphi_{ce} + 1). \tag{5.17}$$

As φ_{ce} goes from zero to increasingly negative values, m first decreases to zero and then becomes increasingly negative, too. This effect of the gravitational potential produced by L at C affects all of the elementary particles that make up C. It follows that distant observers see the mass of C made more negative by the action of L than it would be due to the transient effect in C per se alone. Local observers in immediate proximity to either of the circuit elements, however, will be completely unaware of this effect.

But this is only part of the story. As the periodic mass fluctuations in the L and C components proceed, the mass of the L next becomes negative. The mass of L now is affected by the gravitational potential at L produced by C, which was affected by L in the previous cycle, and so on. *For distant observers* a bootstrap process appears to operate driving the mass of each of the components more and more negative as the device continues to cycle. If the amplitude of the effect driven in L and C is sufficiently large, some finite, reasonable number of cycles should be all that are required to attain the condition of Eq. (5.12) [bare mass exposure] – assuming, of course, that the forming TWIST [traversable wormhole in space-time] does not blow itself apart.

Reprinted from James F. Woodward, "TWISTS of Fate: Can We Make Traversable Wormholes in Spacetime?" *Foundations of Physics Letters*, vol. 10, pp. 153–181 (1997) with permission from Springer Verlag.

10

The Road Ahead

In the previous chapters we have seen that when inertia is understood as a gravitational phenomenon encompassed by general relativity theory, surprisingly large "Mach" effects that should occur at Newtonian order are predicted in systems containing accelerating objects undergoing internal energy changes. Although these effects are only transients, they can be engineered so as to produce thrusts without the ejection of material propellant, and to drive the formation of absurdly benign wormholes should the ADM electron model with negative bare mass and spin capture the essence of reality.

These are more than just theoretical speculations. Experimental evidence suggests that Mach effects are facts of reality. The experiments that indicate Mach effects are real, however, are at a level of development that does little more than show the presence of the effects in far from ideal systems. So, the question at this juncture seems to be, how should we proceed? In this chapter we look at how we might proceed, both in the short and longer terms. And we deal with some of the obvious speculations that arise when the prospect of real, absurdly benign wormholes is addressed.

THE SHORT TERM

The first Mach effect can be used, as discussed in Chaps. 3, 4, and 5, to produce thrust in simple systems without the ejection of propellant. The experimental work described in those chapters had as its goal the demonstration of the reality of Mach effects without much concern for practical applications. However, this effect, in itself, has potential practical applications. At even modest levels, it can be used for space propulsion, satellite station keeping and orbital transfer.

Unlike conventional systems used for these tasks, devices based on the first Mach effect never run out of propellant. With extended use, Mach effect devices will eventually degrade and have to be replaced. But a continually replenished supply of propellant is not required. We're not talking about a free lunch, of course. Energy must still be supplied to make these devices work. But for near Earth operations, arrays of solar panels can be used for power. And in the case of space propulsion, a power source can be carried with the craft.

J.F. Woodward, *Making Starships and Stargates: The Science of Interstellar Transport and Absurdly Benign Wormholes*, Springer Praxis Books, DOI 10.1007/978-1-4614-5623-0_10,
© James F. Woodward 2013

Since there is no need to lug along a stellar mass, or some-such, of propellant, a reasonable amount of energy will likely suffice for even long distance trips.

The steps that must be followed to transform laboratory scale devices intended to test theoretical expectations into practical devices are fairly straightforward. The key thing to keep in mind is that even the simplest experimental systems are actually quite complicated and often behave in curious and unexpected ways. In the case of the PZT stacks used in the most recent experimental work, design could profit from detailed knowledge of the materials used and modeling so that various configurations could be explored before real devices are built to be tested. More extensive instrumentation of the test systems is also desirable. Whereas the first stacks built in 1999 were equipped with one accelerometer embedded in the most active part of each stack, more recent stacks have been constructed with two additional accelerometers, one at each end of the stacks in anticipation of enhanced instrumentation.

Thermal information is also clearly desirable as the operation of the devices depends on the operating temperature. At the outset, such information was collected with a thermistor epoxied to the aluminum mounting bracket that supported the device. Although such information is helpful, owing to the location of the thermistor, a noticeable time delay occurs in its recording of temperature changes, making it of modest help for potential feedback systems. Recently, this situation has been improved by moving the thermistor to the aluminum cap on the stack and adding a second thermistor embedded in the brass reaction mass.

This helps markedly, but a more elegant approach would be to do thermal imaging. This would mean abandoning the Faraday cage used to enclose the devices. But once Mach effects are established as genuine, this step should not present a problem. Very expensive video systems for thermal imaging are available, but others are available at significantly less cost. The reach of the these into the IR spectrum is limited, though, so a cost/benefit analysis might be in order. Either a vacuum chamber of adequate dimensions is required to accommodate the imaging hardware, or an IR transparent window must be put into the vacuum chamber wall.

Another way to speed the building of Mach effect devices suitable for space propulsion is to develop modeling codes so that designs can be tested virtually before real hardware is actually fabricated. Such codes should make allowance for a variation of several factors, the most obvious being the physical dimensions of the elements and their mechanical and electrical properties. Thermal behavior should also be accommodated, and eventually provision for devices being part of arrays of devices should also be made. The modeled parameters also suggest that feedback and control schemes should be incorporated into the design and modeling of these devices. For example, active cooling to maintain a stable operating temperature is desirable. And since the mechanical resonance of these devices depends on the working preload of the stacks, designing devices that incorporate a way to actively adjust the preload via a feedback circuit to tune their operation makes sense.

An obvious question to be addressed is: Should one design Mach thrusters as single, large devices? Or should they be designed as small devices to be used in arrays? Both paths here can be pursued. But small devices used in arrays seem the more practical approach. Should one of the devices in an array fail, it need not be the end of the world. All one need do is eventually plug in a modular replacement device. Failure of a single, large device might be the end of the world for those depending on it. The use of arrays of small devices has the added advantage of making the operation of the system responsive to the phasing of the individual units. The enhanced maneuverability this would enable would be well worth the price of a more elaborate system than a single device. An array has the mundane advantage of being easier to cool as well. Given the temperature sensitivity of composite ferroelectric materials, this is not a negligible consideration.

From the theoretical point of view, several issues seem worthy of some attention. Perhaps the simplest is working through the mathematics of the full tensor version of the Newtonian order approximations made in Chaps. 2 and 3. Although there is no reason to expect any surprises in such an endeavor, its execution as a matter of formal completeness makes sense.[1] In another vein, looking at systems other than those involving ferroelectric actuators and capacitive energy storage elements likewise merits investigation. The simple and obvious is not always the best way to proceed. And the modeling activity mentioned above in conjunction with optimization of present experimental systems would be helpful.

More difficult than the forgoing theoretical activities is investigation of the way in which Mach effects are generated. That is, the detailed examination of how changes in the internal energies of materials take place, and how that relates to the production of Mach effects should be examined. Although it is clear that internal energy is stored in the interatomic bonds of the dielectric materials in the capacitors involved in the experiments described in Chaps. 4 and 5, it is not clear how that process produces the Mach effects predicted, or where exactly the mass fluctuations take place. A related issue is the visualization of Mach effects. The equations (in Chap. 3) do not have an obvious, easily visualized mechanical model – unlike the zero point fluctuation interpretation of the quantum vacuum. It's easy to imagine that the vacuum is surfeit with frothing virtual particles, and hope that we might be able somehow to get some purchase on them to extract energy and propulsion. There are two problems with this. First, the frothing virtual particle view of the vacuum is merely an "interpretation." As Peter Milonni and others showed decades ago, a simple reordering of commuting operators in quantum calculations leads to another interpretation where the vacuum is empty and real particles interact via an action-at-a-distance mode (as in Wheeler-Feynman absorber theory). That is, there are no freely propagating virtual electromagnetic fields in the vacuum. Almost everyone, however, prefers to regard the vacuum as the plenum of virtual stuff.

The only problem with this is the second problem: the vacuum is measured, as a matter of fact, to be empty – with a mass-energy density of $\sim 2 \times 10^{-29}$ g/cm^3. That is, the imagined sea of virtual particles isn't really there at all.[2] The business of models and the physics contained in the formalism for processes can be tricky and sometimes misleading.

As became clear in an investigation of the question of whether the locally measured speed of light is invariant in accelerating frames and gravity fields (when the measurement is made in a frame that is not in free fall), a model that helps one to understand Mach

---

[1] After the experience related in Chap. 3 regarding the second Mach effect, I would be the last person to blow this off as inconsequential.

[2] Part of my lack of concern about finding a good model for Mach effects stems from my experience with zero-pointers who can be utterly irrational in their belief that zero point fluctuations are real, when obviously, they aren't. Handing people a cute picture as a substitute for reality should only be done with extreme caution.

effects can be constructed. It is an analog of the circumstances that produce the Mössbauer effect. The Mössbauer effect is the radiation of a 14.4 keV gamma ray photon by the radioactive isotope of iron: Fe^{57}.

The remarkable feature of the Mössbauer effect is that the iron atoms are locked in their lattice in such a way that when the gamma photon is emitted, the lattice responds rigidly, so the recoiling mass, the entire block of iron containing the radiating atom, is effectively infinite. The result is the emission of photons that all have almost exactly the same energy (and frequency). This makes possible timing with exquisite accuracy – to a part in 10^{17}. This led to several tests of relativity theory in the early 1960s not even imagined possible before Mössbauer's discovery of his effect in 1957.

So what? Well, the thing that makes Mössbauer radiation so exact is the fact that the recoiling object is not just an atom that is free to move in its lattice location. The lattice is rigidly locked, so the entire block has to recoil together. Now think about Mach effects. When the material being accelerated moves rigidly, no internal energy is stored, and the transient Mach effects vanish. Using the Mössbauer effect as an analogy, when the acceleration takes place and the gravity of distant matter acts to resist the acceleration, the field acting on any part of the accelerating object does not at the same time act on the rest of the object because the lattice forces do not make the object respond rigidly. Only after the internal energy being stored "tops off" does the object begin to respond like a rigid body. So the total reaction summed over the object is not that of a rigid object, and transiently the mass of the object differs from that of a rigid object. Keep in mind, though, that this is just an analogous, easily envisioned, mechanical model. It is not the reality. The physics is in the mathematics, not in the model.

THE LONGER TERM

People were trying to figure out how to go really fast, and preferably break the "light speed barrier" long before Thorne, prodded by Sagan, worked through the wormhole physics that must be mastered if we are actually to get around the universe in reasonable time frames. Usually, these efforts started with a literature search targeted to reports of anomalous observations that might hint at a coupling between gravity and electromagnetism that might be exploited somehow to achieve rapid transport. Such searches often also included works of theory where something beyond the usual Einstein-Maxwell equations were investigated. Not much of this work made it into mainstream journals, for even when heavily disguised, the underlying motivation for the work was usually easy to detect. Some, however, did get published in respectable journals. Thorne's wormhole work, in addition to laying out the physics of wormholes, made publication of more of this work at least marginally respectable. Most of the work that was produced and published related fairly directly to the details of wormholes and whether they could be turned into time machines. Essentially none of the published work related to how one might actually go about building an absurdly benign wormhole generator.

Almost all of those who had been looking for anomalous observations and off-beat theories before Thorne continued to do much the same things after Thorne's benchmark wormhole work. No doubt, the reason why almost no one invested any effort in trying to figure out how to actually make traversable wormholes is that the physical requirements seem utterly preposterous on the face of it. And, aside from the Casimir effect, there appeared to be no way to acquire the negative rest mass matter required for their construction − at least not if we take the standard canon of physics seriously.

It seemed that if we are to make wormholes, some radically new theory of gravity and/or matter must first be developed. The high visibility failure of the physics community to develop a quantum theory of gravity (that everyone could agree was the correct theory) has made it easy to dismiss serious investigations of wormhole technology. The reinterpretation of a carefully selected subset of the contents of the canon however makes it possible to show, in terms of well-established theory, that wormhole technology may well lie within our grasp. Isolation of that subset was essentially a series of accidents accompanied by a large dose of dumb luck. But in a sense, much of our experience is accidental. So, perhaps that is not too surprising.

The issues addressed in Chap. 9 are but a very crude sketch of how to proceed. One thing that is clear, though, is that experiment and theory will have to proceed in tandem. The semi-classical ADM model of electrons (and quarks) is hardly a complete theory of matter. And the obverse-reverse relationship of gravity and inertia is not yet really appreciated. Those are but the first steps in the development of wormhole technology. And there are a host of related issues. Will the energy conditions that some think prohibit traversable wormholes apply to artificial constructions of the sort envisaged with Mach effects? What engineering problems will be encountered in trying to realize actual devices? What sort of radiation shielding will be needed? What sources of power will be required? Can they be made sufficiently compact to carry on a ship of reasonable size? And, of course, hyperspace navigation is uncharted territory.

In the longer term wormhole technology is a subject that must be carefully investigated. The motivating issues are now commonplace: the inevitability of an extinction level asteroid impact event, or the arrival of predatory aliens, not to mention our own actions that could make the planet unlivable. These are now in the mainstream media and taken seriously by at least some. But setting aside fantasies about deep black projects carried out by some conspiratorial cabal, or cabals, no one seems to be serious about investing the resources needed to try to make wormhole technology a reality.

Recently, NASA and DARPA jointly sponsored the "100 Year Starship" program, a short-term project aimed at creating a vehicle for private resources to fund a project that might take 100 years or more to complete. The government sponsors of the project deemed a program of this duration too long given the vicissitudes of the political process for any government support to be steadily sustained. A half-megabuck of the initial funding was bestowed on a consortium of private entities to organize the effort envisaged by NASA and DARPA. It's too early to tell if the recipients of the 100 Year Starship program funds will successfully get their act together and create the institutional support required. It is difficult to be optimistic.[3]

---

[3] You may be thinking that Arthur Clarke's law applies here: "When a distinguished but elderly scientist says something is possible, he is almost certainly right. When he says something is impossible, he is very probably wrong." But it doesn't.

TIME TRAVEL

Aside from some casual remarks, the business of time travel has been ignored so far. Perhaps a few comments are in order. The fact of the matter, as Thorne discovered at the outset of his investigations of traversable wormholes, is that wormholes enable time travel. If topology change is forbidden, then wormhole time machines are devices of our future at best (or worst, depending on how you look at these things). If topology change is possible, then wormhole time machines may well be features of our past as well.

Future only, or past and future, time machines are problematic at best. So problematic that they became the chief focus of wormhole physics in the years after Morris and Thorne's classic paper on traversable wormholes. Indeed, as an aging experimentalist, this author viewed all of that with the deepest skepticism for several years, until Ron Crowley insisted that I do the Mach effect calculation to sufficient accuracy to bring out the second effect term. The negativity of that term and the non-linearity of the equation provided an obvious path to wormhole technology. Nonetheless, even then I remained deeply skeptical.

In the early fall of 1993, walking off a mountain, I experienced a "gestalt shift."[4] It appears that my unconscious mind had been working on the "paradoxes" posed by time travel in "background" mode. And on the long trip off Mt. San Jacinto the results of all of that thought came bubbling to the surface, particularly the realization that were time travel possible – and knowing of the implications of the second term Mach effect I was prepared to believe that it was – the paradoxes usually raised to argue that it wasn't possible were obviously wrong. There will be here no long discussion of time travel paradoxes. You'll find many books that deal with them in excruciating detail. Far and away the best of these is *Time Travel and Warp Drives* by Allen Everett and Thomas Roman.[5] These authors do an outstanding, up-to-date job on the issues of time travel. So, here we'll mention two or three, those that happen to be Stephen Hawking's favorites.

The first is usually called the "grandfather" paradox. It's where you go back and kill your grandfather before he met and married your grandmother, and so on. Actually, this paradox was discussed at considerable length by Wheeler and Feynman in their second paper on absorber electrodynamics in 1949. They called it the "bilking" paradox. If you think there's anything to the grandfather paradox, you should read Wheeler's and Feynman's paper.

The second is the "cumulative audience" paradox. This alleges that were time travel to the past possible, hordes of temporal tourists would show up at historical events of significance: the nativity, the San Francisco earthquake, the detonation of the first atomic bomb — pick your significant event. And the presence of the hordes of temporal tourists would have been recorded for posterity. This, of course, presupposes that future folk are really stupid. But it does raise an interesting question: Do future folk discretely travel to

---

[4] Until then, I had regarded gestalt shifts as pseudo-psychological techno–babble. Skeptical experimentalists are inclined to regard anything they can't experience for themselves deeply askance. Especially when those things are associated with pop psychology.

[5] Published in late 2011 by the University of Chicago Press.

interesting events and times in the past to observe, taking care not to be observed themselves?

Several years ago I reviewed a book with a contribution by Hawking where he mentioned the cumulative audience paradox and went on to assert that he would not take bets on time travel because his opponent might be from the future and know in fact that it is possible. Having then recently read that Hawking's outdoors activities were often attended by groupies who would observe him from a distance, I couldn't resist remarking that Hawking's experience gave the lie to the paradox, for who knows *when* his groupie observers might be from? That got me some blowback, though not from Hawking himself. We don't move in the same circles.[6] It was from one of his grad students with a message from Hawking intended to tell me he appreciated the joke.

In the review where I tweaked Hawking on the cumulative audience paradox, I also mentioned that some students at MIT had then recently advertised a time travel conference they had organized. Part of the advertisement was an invitation for future folk to show up to discuss time travel with those from our era. None did, of course. The reason why is pretty obvious. If you were a denizen of the future and had the opportunity to travel to the past, would you choose to spend that time with a bunch of geeks at MIT? I wouldn't. Indeed, I mentioned in the review that given the chance, I would visit the time of the primes of Fernanda de Utrera and Juan Maya.[7]

When Hawking did a TV show on time travel not long ago, he chose the prime of Marilyn Monroe and watching Galileo looking through his telescope. Taste is an individual matter, I guess. Oh, and he used the invitation to a meeting ploy, in his case, a party, to try to discredit time travel, too. No one showed up for his event either. Evidently, either time travel to the past is impossible, or future folk are not stupid.

If you take time travel seriously, you may think that I'm treating the subject with inappropriate levity. Well, yes. Time travel really is a serious matter.[8] The fact of the matter is that I really do think that not only is time travel possible, so, too, is topology change and travel to the arbitrarily distant past. Partly, I hold this view because neither time travel nor topology change are convincingly prohibited by any law of physics that merits acceptance as a truly fundamental law of reality. And as a noted physicist once said, "Anything that is not forbidden is mandatory."

However, I am an experimentalist. I don't believe laws that pretend to the status of fundamentality unless there is compelling experimental or observational evidence that they deserve that status. So when I went through the gestalt shift thing, I started looking for the sorts of events that one might expect if future folk really were screwing around. I could tell you stories, but I won't. You wouldn't find them credible or convincing if you are certain that time travel is not done. They are, after all, just stories. Instead, I encourage you to consider the possibility that time travel is a part of our reality and look for yourself for

---

[6] Actually, I try to not move in circles. It doesn't get you anywhere.

[7] Fernanda de Utrera was the unique cantaora of flamenco of the past century, and Juan Maya easily the best flamenco guitarist of that period.

[8] But it's hard to pass up Groucho Marx's comment on time: Time flies like an arrow. Fruit flies like a banana.

the sorts of things that might happen if the future's past, our present, is being messed with. Just because we haven't mastered wormhole technology yet doesn't mean that it hasn't been done in the future. But if someone offers you a chance to buy stock in a time travel company, don't invest yet. Those making the offer may know how to make absurdly benign wormholes, but the odds are overwhelming that they don't and you're being conned.

NO JOHNS

We have come a long way. But there is much farther to go. Progress may be slow, at times glacially slow. For example, 15 years elapsed between the publication of MUSH and the writing of this book. The missing piece of the puzzle partly responsible for the delay was getting spin into the ADM model for the electron. You might think this was a problem of seeing how to modify the ADM equation, or finding Asim Barut's quantization scheme for the charged leptons. But you would be mistaken. All of that was known by the summer of 1995, and most was known before MUSH, was written in the summer and fall of 1994.

Indeed, I was sufficiently confident that the solution to that problem would be found quickly that instead of following my customary practice of only sending reprints to those who requested them, I sent a dozen or so, unsolicited, to people I wanted to read the paper. Among the recipients of those unsolicited reprints was Carl Sagan. After thinking things over carefully, I inscribed his reprint with "No message, no johns" to be sure of getting his attention. In his novel *Contact* he had made a point of noting that restroom facilities were not included in the design of the vehicle that transported his protagonists to and from the center of the galaxy through traversable wormholes. The "message" being instructions for the construction of the vehicle was, of course, the central plot device of the novel. He didn't respond to me. But the rumors from other quarters indicated that I had gotten his attention.

I would like to be able to tell you that the reason for the 15-year delay in nailing down the last really fundamental issue in making wormholes was due to the very subtle, exceedingly difficult nature of the problem. But that's not the case. The problem was that I got it into my mind that the spin issue should be solved in a particular way – and no matter how it was cast, that way always led to dead ends. This was so frustrating that I just decided to ignore the problem. Without convincing evidence that Mach effects were real, I didn't deem the problem important enough to merit the sort of irritation I had gone through in my first pass at trying to solve it. No one seemed to care whether it might be possible to make real wormholes in the near future. Why should I?

When I finally decided to have another go at the problem in the summer of 2010, I spent several weeks repeating the stupidities of yesteryear. After bloodying myself on that blank wall, I finally resolved to analyze what had to be done to solve the problem from scratch. There were two simple, key elements that had to be part of any successful solution. Once identified, resolution of the problem followed quickly.

Now, if a lot of people are working on a problem, the likelihood that 15-year delays will happen because someone can't see how to analyze a critical issue are very unlikely. But it would be a mistake to assume that the opposite is the case. Just because a lot of smart people are working on a problem doesn't mean that they will not collectively ignore the best way to deal with it. Keep in mind that it took 20 years for quantum mechanics to develop the renormalization program to deal with the infinities of relativistic quantum electrodynamics. And there were a lot of very smart people attuned to the problem. So the message of all this is, as Yogi Berra put it, "It's tough to make predictions, especially about the future." Nonetheless, I now think it fair to say:

No message. No johns. ...

Bibliography

Popular Books

Barrow JD, Tipler FJ (1986) The anthropic cosmological principle. Oxford University Press, Oxford

Bohm D (1983) Wholeness and the implicate order. Ark Paperbacks, New York

Crease RP, Mann CC (1986) The second creation: makers of the revolution in 20th century physics. Macmillan, New York

Davies P (1995) About time: Einstein's unfinished revolution. Simon and Schuster, New York

Davies P (1977) The physics of time asymmetry. University of California Press, Berkeley

Davies P (2001) How to build a time machine. Penguin Putnam, New York

Everett A, Roman T (2011) Time travel and warp drives. University of Chicago Press, Chicago

Greene B (2004) The fabric of the cosmos: space, time, and the texture of reality. Knopf, New York

Gribbin J (1995) Schrödinger's kittens and the search for reality: solving the quantum mysteries. Little Brown, New York

Hawking S (1996) The illustrated a brief history of time. Bantam Books, New York

Herbert N (1988) Faster than light: superluminal loopholes in physics. Plume, New York

Kennefick D (2007) Traveling at the speed of thought: Einstein and the quest for gravitational waves. Princeton University Press, Princeton

Kuhn T (1962) The structure of scientific revolutions. University of Chicago Press, Chicago

Pais A (1982) Subtle is the lord: the science and the life of Albert Einstein. Oxford University Press, Oxford

Popper K (1959) The logic of scientific discovery. Routledge, London

Thorne KS (1994) Black holes and time warps: Einstein's outrageous legacy. Norton, New York

Wilczek F (2008) The lightness of being: mass, ether, and the unification of forces. Basic Books, New York

Technical Books

Adler R, Bazin M, Schiffer M (1965) General relativity and gravitation. McGraw-Hill, New York

Barbour J, Pfister H (1994) Mach's principle: from Newton's bucket to quantum gravity, vol 6, Einstein studies. Birkhauser, Boston

Bernstein J (1995) An introduction to cosmology. Prentice Hall, Englewood Cliffs

Berry MV (1976) Principles of cosmology and gravitation. Cambridge University Press, Cambridge

Ciufolini I, Wheeler JA (1995) Gravitation and inertia. Princeton University Press, Princeton

J.F. Woodward, *Making Starships and Stargates: The Science of Interstellar Transport and Absurdly Benign Wormholes*, Springer Praxis Books, DOI 10.1007/978-1-4614-5623-0,
© James F. Woodward 2013

Einstein A (1955) The meaning of relativity, 5th edn. Princeton University Press, Princeton
Horowitz P, Hill W (1989) The art of electronics. Cambridge University Press, Cambridge
Hoyle F, Narlikar JV (1974) Action at a distance in physics and cosmology. Freeman, San Francisco
Millis MG, Davis EW (2009) Frontiers of propulsion science, vol 227, Progress in astronautics and aeronautics. AIAA, Reston
Milonni P (1994) The quantum vacuum: an introduction to quantum electrodynamics. Academic, San Diego
Misner C, Thorne K, Wheeler JA (1973) Gravitation. Freeman, San Francisco
Rindler W (1991) Introduction to special relativity, 2nd edn. Oxford University Press, Oxford
Taylor EF, Wheeler JA (1992) Spacetime physics. Freeman, New York
Visser M (1995) Lorentzian wormholes: from Einstein to Hawking. AIP Press, Woodbury

Technical Articles

Alcubierre M (1994) The warp drive: hyper-fast travel within general relativity. Class Quantum Gravity 11:L73–L77
Arnowitt R, Deser S, Misner CW (1960a) Gravitational-electromagnetic coupling and the classical self-energy problem. Phys Rev 120:313–320
Arnowitt R, Deser S, Misner CW (1960b) Interior schwarschild solutions and interpretation of source terms. Phys Rev 120:321–324
Ashtekar A (1989) Recent developments in quantum gravity. Ann N Y Acad Sci 571:16–26
Assis AKT (1989) On Mach's principle. Found Phys Lett 2:301–318
Barut AO (1979) Lepton mass formula. Phys Rev Lett 42(1251)
Bondi H, Samuel J (1997) The Lense-Thirring effect and Mach's principle. Phys Lett A 228:121–126
Brans CH (1962) Mach's principle and the locally measured gravitational constant in general relativity. Phys Rev 125:388–396
Champney DC, Isaak GR, Khan AM (1963) An 'Aether Drift' experiment based on the Mössbauer effect. Phys Lett 7:241–243
Cook RJ (1975) Is gravitation a result of Mach's inertial interaction? Il Nuovo Cimento 35B:25–33
Costa de Bequregard O (1961) Principe d'equivalence, masses negatives, repulsion gravitationnelle. Comptes Rendus de l'Académie des Sciences (séance du 20 Mars 1961), pp 1737–1739
Cramer J (1986) The transaction interpretation of quantum mechanics. Rev Mod Phys 58:647–687
Einstein A (1952) Does the inertia of a body depend upon its energy content? In: The principle of relativity: a collection of original memoires on the special and general theory of relativity. Dover, New York, pp 69–71
Feynman RP (1966) The development of the space-time view of quantum electrodynamics. Phys Today 19 (8):31–44
Forward RL (1989a) Space warps: a review of one form of propulsionless transport. J Br Interplanet Soc 42:533–542
Forward RL (1989b) Negative matter propulsion. J Propuls Power 6:28–37
Hawking SW (1992) Chronology protection conjecture. Phys Rev D 46:603–611
Luchak G (1953) A fundamental theory of the magnetism of massive rotating bodies. Can J Phys 29:470–479
Morris MS, Thorne KS (1988) Wormholes in spacetime and their use for interstellar travel: a tool for teaching general relativity. Am J Phys 56:395–412
Moyer M (2012) Is space digital? Scientific American, Feb 2012, pp 31–37
Nordtvedt K (1988) Existence of the gravitomagnetic interaction. Int J Theor Phys 27:1395–1404
Pasqual-Sánchez J-F (2000) The harmonic gauge condition in the gravitomagnetic equations. arXiv:gr-qc/0010075
Peebles PJ, Dicke RH (1962) Significance of spatial isotropy. Phys Rev 127:629–631

Price R (1993) Negative mass can be positively amusing. Am J Phys 61:216–217

Raine DJ (1975) Mach's principle in general relativity. Mon Not R Astron Soc 171:507–528

Raine DJ (1981) Mach's principle and space-time structure. Rep Prog Phys 44:1152–1195

Rindler W (1994) The Lense-Thirring effect exposed as anti-Machian. Phys Lett A 187:236–238

Sciama DW (1953) On the origin of inertia. Mon Not R Astron Soc 113:34–42

Sciama DW (1964) The physical structure of general relativity. Rev Mod Phys 36:463–469; erratum: 1103

Signore RL (1996) Nuovo Cimento B 111:1087; B 112:1593 (1997)

Sultana J, Kazanas D The problem of inertia in Friedmann Uinverses. arXiv:1104.1306v Int J Mod Phys D (submitted)

Weinberg S (1989) The cosmological constant problem. Rev Mod Phys 61:1–23

Wheeler JA, Feynman R (1945) Interaction with the absorber as the mechanism of radiation. Rev Mod Phys 17:157–181

Wheeler JA, Feynman R (1949) Classical electrodynamics in terms of direct interparticle action. Rev Mod Phys 21:425–433

Articles by the Author

(1995) Making the Universe safe for historians: time travel and the laws of physics. Found Phys Lett 6:1–39

(1996) Killing time. Found Phys Lett 9:1–23

(1997) Twists of fate: can we make traversable wormholes in spacetime? Found Phys Lett 10:153–181

(2003) Are the past and future really out there? Annales de la Fondation Louis de Broglie 28:549–568

(2004) Flux capacitors and the origin of inertia. Found Phys 34:1475–1514

(2011) Making stargates: the physics of traversable absurdly benign wormholes. Physics Procedia. In: Proceedings of the SPESIF 2011, University of Maryland, 1 Oct 2011

Author Index

A
Alcubierre, Miguel, 185, 235, 243, 246
Aristotle, 6, 7
Arnowitt, Richard, 217, 218, 244
Ashtekar, Abhay, 215–218, 223, 224, 244, 256
Assis, Andre K.T., 114, 236
Augustine, Norman, 188

B
Baker, Robert, 188
Bakos, Jim, 106, 107
Barbour, Julian, 31, 53, 236
Barut, Asim O, 232–234, 246, 266
Bell, John, 91, 92
Bertotti, Bruno, 236
Bigelow, Robert, 187
Bohr, Niels, 200, 209, 211, 225, 228, 231–233, 246
Bolden, Charles, 188
Brandenburg, John, 188, 246
Brans, Carl, 32, 37, 38, 41–43, 57–61, 82, 238, 239
Brito, Hector, 118–122
Buldrini, Nembo, 124, 132, 133, 243, 251

C
Champney, D.C., 72, 237, 240
Chiao, Ray, 105
Clauser, John, 91, 92
Cole, John, 185
Cook, Nick, 186, 187
Corum, James, 118, 119, 188
Cramer, John, 52, 75, 105, 127, 132, 147, 230
Crowley, Ron, 69, 75, 106, 255, 264
Cumming, Duncan, 129, 144, 146

D
Davis, Eric, 124, 147, 187, 189
Day, Jeff, 101
Deser, Stanley, 217, 218, 244
Dirac, Paul, 33, 47, 193, 211–213, 215, 226, 227
Dornheim, Michael, 106
Dyson, Freeman, 213

E
Eddington, Arthur, 229
Einstein, 4–7, 9, 10, 13, 15–18, 21–26, 29–33, 38–42, 44, 46, 53–55, 65, 66, 70, 73, 80, 82, 89, 92, 185, 192, 198–200, 203, 204, 207–209, 213–215, 217, 221, 225–227, 233, 235–241, 244
El Genk, Mohamed, 146, 189

F
Fearn, Heidi,
Feynman, Richard, 47–49, 52, 80, 212, 213, 236, 239, 261, 264
Forward, Robert, 184, 230
Fourier, Joseph, 203
Freeman, Morgan, 189

G
Galileo, 6–8, 21, 54, 236, 265
Ghosh, Amitabha, 114
Goldin, Daniel, 104, 185, 188, 244
Goode, Stephen, 69, 75, 106, 255
Goudsmit, Samuel, 210, 226
Griffin, Michael, 188, 189

J.F. Woodward, *Making Starships and Stargates: The Science of Interstellar Transport and Absurdly Benign Wormholes*, Springer Praxis Books, DOI 10.1007/978-1-4614-5623-0, © James F. Woodward 2013

H
Haisch, Bernard, 186, 196, 197
Hamilton, David, 125–128, 215
Hathaway, George, 143, 144
Hawking, Stephen, 20, 70, 192, 264, 265
Hudson, Gary, 133, 143, 178

I
Isaac, G.P., 72, 237, 240

K
Kennefick, Daniel, 44
Khan, M., 72, 237, 240
King, Don, 126, 127
Kuhn, Thomas, 90, 91

L
Lazar, Bob, 106
Luchak, George, 66–69

M
Mahood, Tom, 78, 105, 126, 134, 144, 170, 255
March, Paul, 111, 126, 127, 129–131, 133, 143, 144, 146
Mathes, David, 133, 146, 178
Maxwell, James Clerk, 193, 194
McKeever, John, 126
Mead, Franklin, 143, 187, 189
Meholic, Greg, 147, 189
Millis, Marc, 104, 124, 147, 186, 189
Milonni, Peter, 105, 197, 207, 213, 261
Misner, Charles, 18, 20, 22, 27, 28, 38, 71, 217, 218, 244
Murad, Paul, 189
Myrabo, Liek, 184

N
Noether, Emmy, 15

O
O'Keefe, Sean, 188
O'Neil, Graham, 111

P
Palfreyman, Andrew, 143, 146
Pauli, Wolfgang, 209, 210

Peoples, Jim, 75, 111, 133, 143, 146, 178, 186
Planck, Max, 6, 72, 187, 196, 198–202, 208, 209, 213, 214, 225, 226, 228, 244–246, 256
Popper, Karl, 90
Pound, R.V., 71, 72
Puthoff, Harold, 186, 188, 196, 226

R
Raine, Derek, 236
Rebka, G.A., 71, 72
Rindler, Wolfgang, 53, 67, 79, 236
Robertson, Glen, 30, 31, 189
Rueda, Alfonso, 186, 196, 197

S
Sagan, Carl, 262, 266
Sarfatti, Jack, 201, 202, 204, 207, 244
Schwinger, Julian, 213
Sciama, Dennis, 33–41, 44–46, 53, 55, 56, 80–83, 236, 238–240
Shawyer, 144, 145
Sheehan, Daniel, 147
Slepian, Joseph, 118–121
Sloan, Jim, 99, 126
Smith, Joshua, 106
Sobzak, Greg, 105
Stoney, G. Johnstone, 199

T
Tajmar, Martin, 122–124, 188
Thorne, Kip, 3, 20, 69, 185, 207
Tomonaga, Sin-itiro, 213
Tuttle, Bruce, 99, 126

U
Uhlenbeck, George, 210, 211, 226

V
Visser, Matt, 188, 230

W
Wanser, Keith, 77
Wesson, Paul, 215
Wheeler, John, 18, 24, 47, 200, 239
White, Sonny, 143, 144
Wilczek, Frank, 15–17, 52, 92, 236, 237
Winglee, Robert, 184, 186

Subject Index

A

Absolute space, 4, 8–11, 15, 80
Absolute time, 8
Absurdly benign wormhole, 20, 21, 53, 54,
    69, 187, 190–193, 201, 207, 246–248,
    259, 262
Acceleration, 3, 8, 9, 15, 16, 18, 23, 31, 32, 33,
    35, 38, 40–43, 45–47, 50–52, 70–74, 76,
    77, 80, 82, 83, 83, 90, 94, 95, 127, 132,
    148, 171, 174–177, 214, 219, 237, 238,
    242, 243, 249, 250, 252, 255, 262
Accelerometer, 116, 136, 140, 151, 152, 154,
    155, 163, 164, 167, 168, 170, 177, 260
Acoustic impedance, 107, 108
Acoustic waves, 107
Action-at-a-distance, 48, 50–52, 78, 80, 197,
    239, 261
ADM model of electrons, 147, 263
AIAA. *See* American Institute of Aeronautics
    and Astronautics (AIAA)
Air Force, 133, 143, 178
American Institute of Aeronautics
    and Astronautics (AIAA), 124, 147, 189
Analog Devices, 97, 140, 142
Angular momentum, 209, 211, 226–229, 232, 246
Anomalous experimental results, 92
Area, 34, 90, 106, 185, 186
Austrian Research Center (now Austrian
    Institute of Technology), 122, 133

B

Background independence, 40, 217
Bare mass, 25, 212, 213, 216–222, 226, 227, 230,
    231, 233, 244–248, 252, 255–257, 259
Black hole, 20, 24, 25, 32, 33, 45, 71, 72, 230,
    245, 248

Boeing, 186, 187
Bohr quantization condition, 228, 231, 233
Bose-Einstein condensate, 202
Bounded from below, 193, 194, 233
Breakthrough Propulsion Physics, 188
Bulk acceleration, 102, 132, 174, 243, 255

C

Calibration coils, 136, 138, 157, 161
California State University
    Fullerton (CSUF), 69, 75, 101, 105, 106,
    124–126, 129, 133, 143, 243
Canetics, 95, 96, 142, 143
Casimir effect, 194, 195, 197, 263
Cavity resonator, 71, 72, 144
Center frequency, 151, 162, 165–167,
    169, 170
Chronology protection conjecture, 70
Classical wave equation, 48, 67, 68, 70, 81,
    86, 241
Closed timelike curve (CTC), 20, 21, 192
Conservation of momentum, 119
Coordinates, 11–14, 19, 22, 23, 82, 84, 143,
    218, 229, 230, 245, 256
Coordinate speed of light, 23, 229, 230
Coriolis force, 22, 26, 237
Cosmological constant, 30, 31, 80, 194,
    196, 198
Coupling coefficients, 175, 201, 202
CSUF. *See* California State University
    Fullerton (CSUF)
CTC. *See* Closed timelike
    curve (CTC)
Cumulative audience paradox, 264, 265
Curie temperature, 107
Cutoff, 196, 213

J.F. Woodward, *Making Starships and Stargates: The Science of Interstellar Transport
and Absurdly Benign Wormholes*, Springer Praxis Books, DOI 10.1007/978-1-4614-5623-0,
© James F. Woodward 2013

D

d'Alembertian operator, 67, 70, 241
Damper, 109, 110, 136, 139, 140, 144, 165, 168
DARPA, 133, 178, 263
Dean drive effect, 153, 164–170
Degrees of freedom, 50, 197
Demonstrator, 133, 150, 178
Department of Energy, 125
Dummy capacitor, 101, 157, 161–163

E

EDO Ceramics, 99
Einstein Equivalence Principle (EEP), 22, 23,
    41–43, 55, 89, 93, 221, 237, 239
Einstein-Rosen bridge, 200, 215
Electromechanical, 76, 94, 109, 110, 115, 140,
    151, 161, 162, 164, 175, 252
Electron, 23, 47, 48, 145, 146, 193, 199, 200,
    208–218, 220–223, 225–234, 236,
    244–247, 255, 256, 259, 266
Electronic data acquisition, 95
Electron self-energy, 212
Electron spin, 210–212, 226, 230, 246
Electrostriction, 102, 111, 115, 150, 176
Elementary particles, 25, 38, 42, 70, 78, 89, 92,
    199, 202, 206, 207, 214–217, 222, 228,
    233, 244, 245, 247, 257
Embedding diagram, 19, 20
Energy, kinetic, 74, 232
Energy, potential, 22, 32, 37–39, 41–43, 55, 81,
    85, 89, 193, 194, 221, 238–240
Equivalence Principle, 5, 7, 8, 18, 22, 25, 31, 38,
    41, 82, 89, 123, 196, 219–221, 237, 239, 245
Euclidean, 12–14
European Space Agency, 186
Event horizon, 20, 24, 25, 32, 33, 201
Exotic matter, 3, 20, 30, 55, 75, 192, 197–199, 201,
    207–224, 230, 235, 244, 246–248, 250

F

Faraday cage, 100, 120, 124, 135, 137, 145, 149,
    153, 159–162, 164, 165, 168, 171, 243, 260
Fictitious forces, 22, 25, 26
Fifth dimension, 66
Fine structure constant, 199, 209, 228, 231
Flexural bearings, 134, 154, 168, 243
Force, 4, 30, 65, 89, 133, 184, 208, 229, 237, 264
Forward/reversed protocol, 168
Four-divergence, 68, 80, 84, 85
Frame of reference, 5, 7, 10, 11, 13, 18, 22, 23, 34,
    42, 71, 78, 89, 196, 227, 237

G

Galinstan, 135, 136, 243
Gamma rays, 71, 72, 262
Gauss' divergence theorem, 68, 241
General covariance (principle), 25, 29, 237
General relativity theory (GRT), 5, 17–19, 21,
    23–26, 30–33, 36, 37–46, 53, 65, 66, 79,
    80, 82, 83, 89–91, 93, 123, 190, 192,
    198, 200, 203, 207, 215–217, 221, 230,
    235, 237, 259
Glenn Research Center, 104
GNACHOS. *See* Gravitationally negative
    astronomical compact halo objects
    (GNACHOS)
Grandfather paradox, 264
Gravitational collapse, 20
Gravitational field, 16, 21, 22, 24, 30, 35, 38,
    39, 41–45, 47, 65, 68, 70, 71, 77, 80,
    82–85, 89, 237, 241
Gravitationally negative astronomical
    compact halo objects (GNACHOS), 230
Gravitational mass, 22, 198, 222, 237, 245
Gravitational potential energy, 22, 32, 38, 41,
    43, 55, 81, 85, 89, 193, 221, 238
Gravitational radius, 24, 25, 220, 222, 230, 245,
    246, 257
Gravitational redshift, 71
Groom Lake Interceptors, 105, 106
Group velocity,

H

Hall effect, 96
Higgs particle, 91, 92
Hockey pucks, 126, 127
Hyperspace, 19, 20, 200, 263
Hypothetico-deductive method, 90

I

Indian Institute of Technology, 114
Inertia, 3–29, 65, 89, 168, 186, 214, 235, 261
Inertial effects, 51, 95, 243
Inertial reaction force, 8, 9, 15, 22, 26, 35–37,
    40–47, 50, 52, 53, 55, 65, 68, 70, 72, 77,
    78, 80–84, 89, 90, 221, 238, 240, 241
Instantaneous rest frame, 74, 84
Institute for Advanced Studies at Austin, 187
Interfering waves, 107
Internal energy, 70, 74, 90, 94, 95, 132, 175,
    250, 259, 261, 262
Ion engine, 183, 184
Isotropic coordinates, 218, 229, 245

K

KD Components, 101
Kharagpur, 114
Kinetic energy, 74, 232

L

Large Hadron Collider (LHC), 91, 92, 200
Lead-manganes-niobate (PMN), 127
Lead-zirconium-titanate (PZT), 76–78, 99,
    101–104, 107–115, 119, 126, 127, 141,
    148–154, 158, 170–172, 174–177, 243,
    252, 260
LHC. *See* Large Hadron Collider (LHC)
Lightcone, 14
Lightspeed, 75, 111, 113, 114, 119, 126, 129,
    186, 214, 239, 246, 252, 254, 256, 262
Liquid metal contacts, 99, 135, 136, 243
Localization of gravitational energy, 22, 27
Lockheed-Martin, 75, 111, 113, 133, 153,
    186–188
Lorentz factor, 67, 84
Lorentz force, 118, 119

M

Mach effect, 4, 10, 65–86, 90, 92–96,
    100–103, 105, 106, 109, 111, 112, 114,
    115, 119, 121, 123, 124, 127, 129, 132,
    133, 142, 143, 145–147, 149, 152, 153,
    156, 164, 166, 173–175, 177–179, 214,
    242, 243, 249, 250, 252–255, 259–262,
    264, 266
Mach-Einstein-Sciama (laws of inertia), 55, 65,
    93, 240
Mach-Lorentz-thrusters (MLTs), 118–125,
    143–146
Mach's guitar, 129
Mach's principle, 4–5, 10, 16, 25, 26, 29–64,
    68, 69, 75, 79–82, 85, 90, 92, 93, 100,
    101, 114, 115, 118, 198, 214, 218, 221,
    231, 235–241, 244
Magnetic moment, 210, 211, 226–229,
    245, 246
Measurement problem, 105
Metamaterials, 202–204, 207, 244
Millennium Projects, 75, 111, 113, 153,
    186, 187
MLTs. *See* Mach-Lorentz-thrusters (MLTs)
Momentum, 9, 10, 15–18, 22, 24, 30, 43, 50,
    67, 68, 77, 78, 119, 199, 206, 209, 211,
    212, 225–229, 232, 241, 243, 246
Mössbauer effect, 71, 72, 237, 262

N

NASA Administrator, 104, 185, 188, 244
National Institute for Advanced Concepts
    (NIAC), 186, 188, 189
National Institute for Discovery Science
    (NIDS), 187
Negative energy, 193–197, 202–204, 212, 244, 245
Negative mass, 20, 45, 85, 114, 198, 207, 213,
    214, 219, 220, 230, 244–247, 256
Neutrinos, 91, 218
Neutron, 17, 197
Newton, action-at-a-distance, 50
Newton, bucket, 9, 31, 53
Newtonian gravity, 21, 26, 32, 46, 66–68, 82,
    237, 240
Newton, laws of mechanics, 15, 43
Newton's second law, 15, 18, 21, 22, 26, 77, 127
NIAC. *See* National Institute for Advanced
    Concepts (NIAC)
NIDS. *See* National Institute for Discovery
    Science (NIDS)
Normal science, 91
Nuclear rockets, 183
Nuclei, 17, 200, 209, 225

O

Oak Ridge National Laboratory (ORNL),
    125–127
Optical gradient density filter, 110
Optical position sensor, 127, 129, 136
Origin of inertia, 3–28
Origin of mass, 81, 92, 214, 221
Orion project, 183, 185
ORNL. *See* Oak Ridge National Laboratory
    (ORNL)

P

Paradigm, 55, 90–92
Parametric amplification, 100
Phase velocity, 176, 204
Photon, 18, 20, 22, 183, 194–196, 198, 209,
    212–214, 225, 262
Photovoltaic cell, 110
Piezoelectric, 102, 103, 175–177, 252
Planck units, 72, 196, 200
PMN. *See* Lead-manganes-niobate (PMN)
Poincaré stresses, 208, 220, 221
Positive energy theorem, 193, 194, 205, 218, 244
Preload, 150, 177, 260
Proper measurements, 13
Proper source density, 73, 94, 242

Propulsion, advanced, 104, 105, 111, 118, 146, 147, 183–205, 214, 215, 243
Propulsion, exotic, 147
Propulsion, non-chemical, 185, 243
Proton, 17, 197
PZT. *See* Lead-zirconium-titanate (PZT)
PZT stacks, 99, 101, 103, 109–115, 119, 141, 148–153, 170–172, 174–177, 260

Q

Quadratic formula, 216, 220
Quantum gravity, 31, 53, 91, 192, 199–201, 215, 224, 226, 244–246, 256
Quantum mechanics, 52, 91, 92, 105, 145, 147, 192–194, 199, 200, 208, 211–213, 217, 225, 226, 235, 244, 245, 267
Quantum vacuum, 146, 193, 196–199, 201, 204, 244, 261
Quarks, 17, 200, 263
Quick Basic, 96

R

Rare earth magnets, 112
Relativistic Heavy Ion Collider (RHIC), 200
Relativistic quantum field theory, 91, 199, 240, 244
Relativity of inertia, 5, 29, 30, 236–238
Renormalization, 37, 212, 213, 217, 222, 244, 245, 267
Repolarization, 107
Residual air, 150, 158–160
Restmass, 16, 17, 23, 24, 29, 34, 67, 90–92, 192, 214, 237, 247, 263
RHIC. *See* Relativistic Heavy Ion Collider (RHIC)
Rubber pads, 111, 112, 126

S

Sandia National Laboratory, 125, 126
Scalar, 10, 15, 24, 32–36, 38, 40, 41, 46, 50, 66, 68, 70, 73, 74, 80, 85, 89, 193, 201, 238, 239, 241
Schuster-Blackett conjecture, 66
Semiconductor strain gauges, 96
Signal averaging, 122, 156, 157
Singularity, 20, 24, 51, 192, 198, 200, 207, 227, 249, 253–255
Softening spacetime, 201–204
Space Technology Applications International Forum (STAIF), 111, 112, 114, 133, 134, 143, 146, 189

Spacetime foam, 198–202, 244
Spatial flatness, 23, 31, 54, 55, 89
Special relativity theory (SRT), 3, 5, 10–18, 23, 29, 32, 47, 67, 77, 83–85, 89, 91, 92, 196, 211, 237
Speed of light, 5, 10, 12–18, 23–25, 30, 33–36, 46, 47, 52, 55, 66, 71, 72, 80, 83, 89, 91, 196, 204, 208, 218, 221, 226, 229, 230, 236–238, 240–242, 245–247, 249, 256, 261
Spin, 25, 45, 51, 147, 210–212, 217, 226–233, 245, 246, 256, 259, 266
SRT. *See* Special relativity theory (SRT)
STAIF. *See* Space Technology Applications International Forum (STAIF)
Standard Model, 91, 92, 207, 213–215, 218, 227, 236, 244
Stargate, 3, 25, 26, 29, 54–55, 69, 75, 78, 147, 198, 202, 206, 217, 221, 222, 226, 227, 233, 235–257
Starship, 26, 54–55, 69, 75, 147, 178, 179, 236, 240, 246–251, 263
Stationary force, 76, 77, 101, 107, 171, 175, 242, 243
Steiner-Martins, 149–151, 162, 167, 173, 175, 177
Stern-Gerlach experiment, 210
Superstring theory, 192, 215

T

Tachyonic particles, 91
Tensor, 10, 11, 24, 30, 32, 35, 44, 65, 66, 201, 261
Thermistor, 111, 136, 141, 153, 173, 260
Thrust balance, 124, 125, 133–137, 140, 143, 149, 153, 164, 165, 243
Time, direction, 14
Time travel, 69, 70, 183, 226, 264–266
Topology change, 192, 199, 201, 253, 264, 265
Transactional interpretation (of quantum mechanics), 52, 147
Traversable, 3, 39, 54, 69, 193, 198, 201–203, 222, 246, 247, 255–257, 263, 264, 266
Traversable wormhole in space-time (TWIST), 255–257
Trout turbines, 187
TWIST. *See* Traversable wormhole in space-time

U

U–80, 96, 144
Unexpected results, 93
Unimeasure Corporation, 96

V

Vacuum chamber, 107, 108, 110, 114, 115, 123, 129, 131, 150, 158–160, 165, 260
VASIMIR, 184
Vdm/dt, 77, 78, 127
Vector, 10–15, 17, 21, 33, 35–39, 45, 46, 54, 66, 82, 83, 238, 239
Vector field, 24, 32, 36, 50, 68
Vector formalism, 32, 238

W

Warp drive, 3, 4, 18, 30, 75, 185, 190, 194, 204, 235, 243, 244, 246, 250, 253, 264

Wilkinson Microwave Anisotropy Probe (WMAP), 37, 42, 47, 53, 54, 239
Wormhole, 3, 30, 69, 89, 133, 183–207, 230, 235, 261

Y

100 Year Starship, 178, 179, 263

Z

Zero point fields (ZPF), 186, 187, 195, 196

5556554R00166

Made in the USA
San Bernardino, CA
11 November 2013